Geodetic and Geophysical Effects Associated with Seismic and Volcanic Hazards

Edited by
José Fernández

2004

Birkhäuser Verlag
Basel · Boston · Berlin

Reprint from Pure and Applied Geophysics
(PAGEOPH), Volume 161 (2004), No. 7

Editor:

José Fernández
Instituto de Astronomía y Geodesia
Facultad de Ciencias Matemáticas
Ciudad Universitaria
Pza. de Ciencias, 3
28040 Madrid
Spain

e-mail: jose_fernandez@mat.ucm.es

A CIP catalogue record for this book is available from the Library of Congress,
Washington D.C., USA

Bibliographic information published by Die Deutsche Bibliothek:
Die Deutsche Bibliothek lists this publication in the Deutsche Nationalbibliographie; deatailed
bibliographic data is available in the internet at <http://dnb.ddb.de>

ISBN 3-7643-7044-0 Birkhäuser Verlag, Basel – Boston – Berlin

© 2004 Birkhäuser Verlag, P.O.Box 133, CH-4010 Basel, Switzerland
Part of Springer Science+Business Media

Printed on acid-free paper produced from chlorine-free pulp
Printed in Germany
ISBN 3-7643-7044-0

9 8 7 6 5 4 3 2 1   www.birkhauser.ch

Contents

1301 Introduction
     J. Fernández

1305 Stress Changes Modelled for the Sequence of Strong Earthquakes in the South
     Iceland Seismic Zone since 1706
     F. Roth

1329 3-D Modelling of Campi Flegrei Ground Deformations: Role of Caldera
     Boundary Discontinuities
     F. Beauducel, G. De Natale, F. Obrizzo, F. Pingue

1345 Comparison of Integrated Geodetic Data Models and Satellite Radar
     Interferograms to Infer Magma Storage During the 1991–1993 Mt. Etna
     Eruption
     A. Bonaccorso, E. Sansosti, P. Berardino

1359 GPS Monitoring in the N-W Part of the Volcanic Island of Tenerife, Canaries,
     Spain: Strategy and Results
     J. Fernández, F. J. González-Matesanz, J. F Prieto, G. Rodríguez-Velasco,
     A. Staller, A. Alonso-Medina, M. Charco

1379 Far-field Gravity and Tilt Signals by Large Earthquakes: Real or Instrumental
     Effects?
     G. Berrino, U. Riccardi

1399 Study of Volcanic Sources at Long Valley Caldera, California, Using Gravity
     Data and a Genetic Algorithm Inversion Technique
     M. Charco, J. Fernández, K. Tiampo, M. Battaglia, L. Kellogg, J. McClain,
     J. B. Rundle

1415 Gravity Changes and Internal Processes: Some Results Obtained from
     Observations at Three Volcanoes
     G. Jentzsch, A. Weise, C. Rey, C. Gerstenecker

1433 New Results at Mayon, Philippines, from a Joint Inversion of Gravity and
     Deformation Measurements
     K. F. Tiampo, J. Fernández, G. Jentzsch, M. Charco, J. B. Rundle

1453 The Interpretation of Gravity Changes and Crustal Deformation in Active
     Volcanic Areas
     M. Battaglia, P. Segall

1469   Intrusive Mechanisms at Mt. Etna Forerunning the July-August 2001
       Eruption from Seismic and Ground Deformation Data
       A. Bonaccorso, S. D'Amico, M. Mattia, D. Patanè

1489   Methods for Evaluation of Geodetic Data and Seismicity Developed
       with Numerical Simulations: Review and Applications
       K. F. Tiampo, J. B. Rundle, J. S. Sá Martins, W. Klein, S. McGinnis

1509   A Free Boundary Problem Related to the Location of Volcanic Gas Sources
       J. I. Díaz, G. Talenti

1519   High CO_2 Levels in Boreholes at El Teide Volcano Complex (Tenerife, Canary
       Islands): Implications for Volcanic Activity Monitoring
       V. Soler, J. A. Castro-Almazán, R. T. Viñas, A. Eff-Darwich,
       S. Sánchez-Moral, C. Hillaire-Marcel, I. Farrujia, J. Coello, J. de la Nuez,
       M. C. Martín, M. L. Quesada, E. Santana

1533   Simulation of the Seismic Response of Sedimentary Basins with Vertical
       Constant-Gradient Velocity for Incident *SH* Waves
       F. Luzón, L. Ramírez, F. J. Sánchez-Sesma, A. Posadas

1549   The Use of Ambient Seismic Noise Measurements for the Estimation of
       Surface Soil Effects: The Motril City Case (Southern Spain)
       Z. Al Yuncha, F. Luzón, A. Posadas, J. Martín, G. Alguacil, J. Almendros,
       S. Sánchez

1561   Results of Analysis of the Data of Microseismic Survey at Lanzarote Island,
       Canary, Spain
       A. V. Gorbatikov, A. V. Kalinina, V. A. Volkov, J. Arnoso, R. Vieira, E. Velez

1579   Microtremor Analyses at Teide Volcano (Canary Islands, Spain): Assessment
       of Natural Frequencies of Vibration Using Time-dependent Horizontal-to-
       vertical Spectral Ratios
       J. Almendros, F. Luzón, A. Posadas

1597   Tilt Observations in the Normal Mode Frequency Band at the Geodynamic
       Observatory Cueva de los Verdes, Lanzarote
       A. V. Kalinina, V. A. Volkov, A. V. Gorbatikov, J. Arnoso, R. Vieira,
       M. Benavent

Pure appl. geophys. 161 (2004) 1301–1303
0033–4553/04/071301–3
DOI 10.1007/s00024-004-2505-6

❙ **Pure and Applied Geophysics**

Geodetic and Geophysical Effects Associated with Seismic and Volcanic Hazards

Introduction

At present, one of the priorities of research should be to study all the aspects correlated with natural hazards, in particular those of geological origin, in an endeavour to reduce the vulnerability associated with them and therefore improve the quality of life, especially if our society at least purports to be a welfare society. Two of the natural hazards of geological origin that cause the greatest impact and pose the biggest risk to society are volcanic and seismic activity (e.g., ISDR, 2002; EM-DAT, 2003). Consequently, any advances in ascertaining the different physical processes linked to the several stages associated to these phenomena are highly important and clearly apply in day-to-day monitoring techniques. The final implication of any method capable of detecting and even predicting natural hazard precursory phenomena is that it could help to prevent the damage to people and property that such events might produce. Achieving this goal, which obviously has clear economic advantages, involves ascertaining every possible aspect of precursory phenomena (zone dependency, estimated size, etc.). The experimental and theoretical aspects of this task are highly complex, and must be combined if the best possible research results are to be attained. Volcanic and seismic hazards are very hard to predict, and though in recent years significant progress has been made with current monitoring systems, much remains to be done before such phenomena can be detected accurately (e.g., RUNDLE et al., 2000; SIGURDSSON et al., 2000; STEIN et al., 2000; MATSU'URA et al., 2002a,b; ONU; 2003; SPARKS, 2003; USGS, 2003). Both kind of phenomena produce effects before, during, and after the activity, and even between events. On the basis of this fact and the high levels of precision attainable, many geophysical and geodetic techniques have proven to be necessary and powerful tools in the monitoring of volcanic and seismic activity. Applying such techniques to routine monitoring of active zones inevitably involves data processing and subsequent final interpretation of observed records. The advent of new techniques, such as the space-based geodetic techniques SAR Interferometry (InSAR) and continuous GPS, or the use of continuous gases measurements, has provided very powerful sources of information for gaining in-depth knowledge of these phenomena, and at a reasonable cost. However, more sophisticated and realistic mathematical models, as well as modern techniques for solving the inverse problem, are required to understand the new and more complex records.

It was in this framework that the International Complutense Seminar (Seminario Internacional Complutense) "Geodetic and geophysical effects associated with seismic and volcanic hazards, Theory and observation." was organized and held at the School of Mathematics of University Complutense of Madrid from 8 to 11, October 2001. This Special Issue contains eighteen papers, most of which were presented at the International Seminar, and addresses different topics: geodetic, geophysical and geochemical effects caused by seismic and volcanic activity; monitoring of volcanic and seismic processes using space and terrestrial techniques; complementarity of these techniques; theoretical modelling of volcanic and seismic processes; inverse problem; interpretation of observations; hazards; seismicity patterns and application. Other articles presented in the Seminar were published in the Complutense University journal *Física de la Tierra* (FERNÁNDEZ and LUZÓN, 2002).

The following reviewers are acknowledged for their assistance: F. Amelung, J.-P. Avouac, P. Baldi, G.W. Bawden, F. Beauducel, R. Bermejo, M.J. Blanco, M. Bonafede, S. Bonvalot, M. Bouchon, A.G. Camacho, B. Capaccioni, V. Cayol, F. Cornet, A. Correig, P.M. Davis, M. Diament, A. Donnellan, M. Dravinski, A. Folch, J. Fonseca, G.R: Foulger, A. Gudmundsson, S. Gurrieri, R. Hanssen, T. Jahr, P. Keary, J. Langbein, P. Lundgren, R. Madariaga, I. Main, K. Makra, J. Martí, C.L. Moldoveneanu, F. Mulargia, M. Navarro, D, Pyle, L. Rivera, R. Scandone, P. Segall, N. Segovia, O. Sotolongo-Costa, R. Stein, K.F. Tiampo, P. Vincent, G. Wadge and W. Zürn. In particular, I would like to thank F. Cornet for his assistance and support in editing this Special Issue. His useful advice and suggestions are sincerely appreciated.

I would like to take this opportunity to thank the Vicerrectorate for International Relations of the Complutense University of Madrid for its support in organizing the International Seminar, and its entire staff and personnel for their support before and during the seminar. I am also grateful for the support given by the Institute of Astronomy and Geodesy, a Spanish Council for Scientific Research-University Complutense of Madrid Joint Research Centre. Finally, the editor wishes to thank all the authors of this Special Issue for their contributions.

REFERENCES

EM-DAT (2003), The OFDA/CRED International Disaster Database – www.cred.be/emdat - Université Catholique de Louvain, Brussels, Belgium.

FERNÁNDEZ, J., and LUZÓN, F. (Eds.) (2002), *Geodetic and Geophysical Techniques, Models and Applications*, Física de la Tierra *14*.

ISDR (2002), *Living with Risk: A Global Review of Disaster Reduction Initiatives.* http://www.unisdr.org/unisdr/Globalreport.htm

MATSU'URA, M., MORA, P., DONNELLAN, A., and YIN, X. C. (Eds.) (2002a), *Earthquake Processes: Physical Modelling, Numerical Simulation and Data Analysis. Part I*, Pure Appl. Geophys. *159*(9), 1905–2168.

MATSU'URA, M., MORA, P., DONNELLAN, A., and YIN, X. C. (Eds.) (2002b), *Earthquake Processes: Physical Modelling, Numerical Simulation and Data Analysis. Part II*, Pure Appl. Geophys. *159*(9), 2169–2536.

ONU (2003), http://www.unisdr.org

RUNDLE, J. B., TURCOTTE, D. L., and KLEIN, W. (Eds.) (2000), *GeoComplexity and the Physics of Earthquakes*, Geophysical Monograph Series, AGU, Washington, 284 pp.

SIGURDSSON, H., HOUGHTON, B., MCNUTT, S. R., RYMER, H., and STIX, J. (Eds.) (2000), Encyclopedia of Volcanoes, Academic Press, San Diego.

SPARKS, R. S. J. (2003), *Forecasting Volcanic Eruptions*, Earth Planet. Sci. Lett. *210*, 1–15. DOI:10.1016/S0012-821X(03)00124-9.

STEIN, S., HAMBURGER, M., DIXON, T., and OWEN, S. (2000), *UNAVCO Conference Explores Advances in Volcanic Geodesy*, EOS Transactions, AGU *81*, 121, 126.

USGS (2003), http://www.usgs.gov

Editor
José Fernández
Instituto de Astronomia y Geodesia (CSIC-UCM)
Facultad de Ciencias Matematicas
Ciudad Universitaria
Pza. de Ciencias, 3
28040-Madrid
Spain
E-mail: jose_fernandez@mat.ucm.es

Pure appl. geophys. 161 (2004) 1305–1327
0033–4553/04/071305–23
DOI 10.1007/s00024-004-2506-5

© Birkhäuser Verlag, Basel, 2004

❘ Pure and Applied Geophysics

Stress Changes Modelled for the Sequence of Strong Earthquakes in the South Iceland Seismic Zone Since 1706

FRANK ROTH[1]

Abstract—The South Iceland seismic zone is, roughly speaking, situated between two sections of the mid-Atlantic ridge, i.e., the Reykjanes Ridge southwest of Iceland and the Eastern Volcanic Zone on the island. It is a transform zone, where earthquakes are expected to occur on E-W-trending left-lateral shear faults, equivalent to conjugate, N-S-oriented right-lateral, rupture planes. In fact, earthquakes take place on en-échelon N-S-oriented faults, which is indicated by the distribution of main shock intensities, aftershocks as well as by surface fault traces. The stress field continuously generated in the fault zone by opening of the adjacent ridges is computed and superimposed on the stress field changes induced by a series of 13 earthquakes (M > 6) between 1706 and 2000. The level of the pre-seismic stress field is analysed as well as the size of the area under high stress. Finally, the post-seismic stress field of June 2000 is analysed, to see where high stresses might have accumulated. The modelling indicates that the rupture planes located on separated parallel N-S-striking zones are dense enough to lead to an area-wide stress release by the series of events. The obtained pre-seismic stress level for most events is high and stable with the exception of situations when several strong shocks occur over a time span of several days, i.e., display typical main shock-aftershock patterns. The size of areas under high stress aside from of the rupture plane, i.e., where no event occurs at the specific time, is of medium to small size.

Key words: Earthquake series, rifting, shear stress, modelling, dislocations.

1. Introduction

Seismic events often occur after a period of quiescence; a pattern that is repeated several times at a fault zone. This can be explained by stresses accumulated by plate motion and released in earthquakes, aseismic slip or inelastic creep. In the following, a model is described that accounts for plate motions and earthquakes on Iceland. In the framework of the European Community funded project 'Earthquake Prediction Research in a Natural Laboratory', a model study was performed to obtain models of the stress field and stress changes in space and time for the South Iceland seismic zone (SISZ). This comprised the changes in crustal stress due to rifting, earthquakes, and aseismic movement in the interacting fault system. Usually, for earthquake hazard estimation, the location, the magnitude and the statistically estimated

---

[1] Section: Natural Disasters. Department: Physics of the Earth. GeoForschungsZentrum Potsdam, Telegrafenberg, D-14473 Potsdam, Germany. E-mail: roth@gfz-potsdam.de

recurrence period of former events is used. To improve this, here the rupture length and width as well as the tectonic setting and the crustal deformation rates are considered while calculating the space time development of the stress field. The targets were to achieve a better understanding of the distribution of seismicity, its clustering and migration, and finally to improve the forecasting of future events in this populated and economically important region of Iceland, i.e., to indicate at least areas of stress concentrations if the specification of a time window for their occurrence is not possible. We tried to answer the following questions: Do these events, placed on parallel faults, release all the energy stored in the 3-D volume of the SISZ? Do the earthquakes always take place in areas of high stress? What is the critical stress level? How large is its variability? Where are the highest stresses at present?

2. The Tectonic Setting

The South Iceland seismic zone (SISZ) is situated between two sections of the mid-Atlantic ridge, the Reykjanes ridge (RR), especially its transition along the Reykjanes peninsula into the Western Volcanic Zone, and the Eastern Volcanic Zone (EVZ; cf. Fig. 1a). Even though there is no clear expression of an E-W trending fault and the angle between the SISZ and the neighbouring ridges is far from 90°, it is considered as a transform zone (cf. BERGERAT; and ANGELIER, 2000; BJARNASON et al., 1993; GUÐMUNDSSON, 1995, 2000; GUÐMUNDSSON and HOMBERG, 1999; SIGMUNDSSON et al., 1995). Following this hypothesis, left-lateral shear stress is expected along the E-W striking zone. This is equivalent to right-lateral shear stress on N-S oriented rupture planes. In fact, earthquakes seem to occur on N-S trending en-échelon faults as can be seen in Figure 1b (cf., BELARDINELLI et al., 2000; EINARSSON et al., 1981; HACKMAN et al., 1990, and further references there). They are located between the Hengill-Ölfus triple junction, where the RR meets the Western Volcanic Zone (WVZ), and Hekla volcano, in the EVZ (cf., EINARSSON et al., 1981). As we further know from the analysis of earthquake fault plane solutions (cf. ANGELIER et al., 1996; BERGERAT

▶

Figure 1
(a) Map of Iceland and surrounding area. Thick lines indicate mid-Atlantic ridge segments, as used in the models. The smaller box shows the region of the model on the South Iceland seismic zone. The SISZ extends approximately between (21.4°W, 64°N) to (18.8°W, 64°N). – RR: Reykjanes ridge, RP: Reykjanes peninsula, KR: Kolbeinsey ridge, WVZ/EVZ: Western/Eastern Volcanic Zone, SISZ: South Iceland seismic zone, TFZ: Tjörnes fracture zone. (b) The South Iceland seismic zone showing mapped surface breaks and (shaded) regions in which over half of the buildings were destroyed in historic seismic events (redrawn after EINARSSON et al., 1981). The north-south dashed line near Vatnafjöll (19.8°W, 63.9°N) indicates the estimated location of the fault on which the May 25, 1987, earthquake occurred (after BJARNASON and EINARSSON, 1991). The structural features and the coastline are after EINARSSON and SÆMUNDSSON 1987.

a

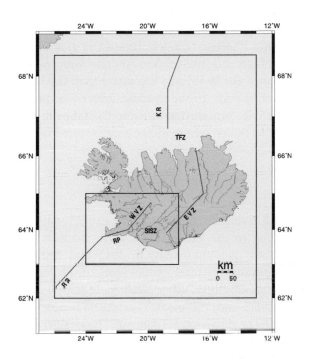

b

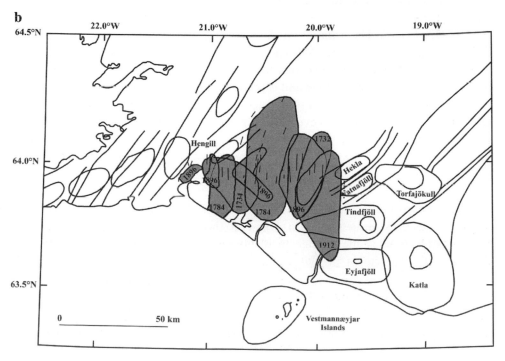

and ANGELIER, 2000; BERGERAT *et al.*, 1999; EINARSSON, 1991; RÖGNVALDSSON and SLUNGA, 1994; STEFÁNSSON *et al.*, 1993) the larger horizontal principal stress is NE-SW, i.e, it fits to an active N-S or E-W trending fault. This stress orientation seems to have been constant since Pliocene time (cf., BERGERAT *et al.*, 1999). It indicates that the SISZ is not a weak fault zone like the San Andreas fault (cf., ZOBACK *et al.*, 1987). Concerning the rift zones, the WVZ was less active than the EVZ in historical times (cf., EINARSSON *et al.*, 1981). From geodetic observations from 1986 to 1992, SIGMUNDSSON *et al.* (1995) conclude that 85 ± 15 % of the relative plate motion takes place in the EVZ.

3. The Earthquake Data

All events with M ≥ 6 since 1706 are used, see Table 1. All ruptures were set to be oriented N-S, according to the isolines of damage intensity and surface ruptures shown in Figure 1b. As only the events in 1912 and 2000 were instrumentally recorded, the source parameters (average coseismic displacement, fault length and width) were usually determined using scaling laws relating observed seismic intensities to magnitudes and magnitudes to seismic moment. This is not very accurate—a problem to be discussed further on. The position of the epicentres and the rupture planes used in the models is given in Figure 2. The rupture width is set to 14 km except for those events very close to the rift segments, where 7 km is chosen. Doing so, we follow the depth distribution of SISZ seismicity shown in STEFÁNSSON *et al.* (1993, Fig. 9). The thinning of the brittle, seismogenic crust obviously reflects the higher heat flow near the rifts. The asymmetry can be explained by the fact that the crust in the SISZ was still created in the WVZ before the rifting activity jumped to the EVZ. The seismicity at Hengill is presently high but events since 1706 were not stronger than M = 6; at the eastern end, the seismicity is generally substantially lower and there have been no events with M ≥ 6 since 1706 east of 20°W.

4. The Method

The forward modelling of stress fields is done by applying a static dislocation theory to geodetic data and data obtained through the seismic moments from seismograms. Besides the change in displacement during the event, the changes caused by the movement of plates is included (for further details see e.g., ROTH, 1989).

Stresses were computed for a homogeneous half-space, as a starting model. Although surface stress changes are calculated, these should be representative for crustal stresses using values for the moduli (Lamé's constants $\lambda = \mu = 39$ GPa), that are typical for oceanic crust (see DZIEWONSKI *et al.*, 1975) and not for sedimentary

Table 1

Earthquakes $M \geq 6$ since 1706 in the South Iceland seismic zone following HALLDÓRSSON *et al. (1984; after* HACKMAN *et al., 1990),* STEFÁNSSON *and* HALLDÓRSSON, *(1988) and* STEFÁNSSON *et al. (1993). The catalogue assumed to be complete from 1706 for these earthquakes.*

Date[1]	Magnitude[1]	Epicenter[1]		South end of rupture[2]		Coseismic slip[3]	Rupture length[4]
		Lat.°N	Long.°W	x [km]	y [km]	U_0 [m]	L [km]
1706	6.0	64.0	21.2	131	−5	0.30	10
1732	6.7	64.0	20.1	184	−11	0.77	22
1734	6.8	63.9	20.8	152	−23	0.96	25
14.08.1784	7.1	64.0	20.5	168	−18	1.9	35
16.08.1784	6.7	63.9	20.9	145	−22	0.77	22
26.08.1896	6.9	64.0	20.2	180	−14	1.2	28
27.08.1896	6.7	64.0	20.1	184	−11	0.77	22
05.09.1896	6.0	63.9	21.0	140	−16	0.30	10
05.09.1896	6.5	64.0	20.6	161	−9	0.48	18
06.09.1896	6.0	63.9	21.2	131	−16	0.30	10
06.05.1912	7.0	63.9	20.0	189	−27	1.5	32
17.06.2000	6.5	64.0	20.4	169	−12	0.9	16
21.06.2000	6.4	64.0	20.7	154	−13	1.1	18

Remarks:
[1] Data taken from STEFÁNSSON *et al.* (1993); for the events in 1706, 1732, and 1734 no exact date is known. For the 1912 event, BJARNASON, *et al.* (1993) give $U_0 = 2$ m and a rupture length of ≥ 20 km. Data on the events of June 2000 with magnitudes as moment magnitudes are from STEFÁNSSON, GUÐMUNDSSON, and HALLDÓRSSON (priv. comm., see also PEDERSEN, *et al.* (2001) and BERGERAT *et al.* (2001) for details on surface faulting).
[2] Position in the model coordinate system with origin at 64 °N, 23.8 °W.
[3] Calculated via the magnitude moment relationship $\log M_0$ [dyne cm] $= 1.5 \, M_S + (11.8-\log(\sigma_a/\mu))$ with the apparent stress $\sigma_a = 1.5$ MPa and the shear modulus $\mu = 39$ GPa (after KANAMORI and ANDERSON, 1975, eq. 26), followed by using the values of μ above, the rupture length as given in the table as well as a vertical fault width of 14 km east of 21°W and 7 km between 21°W and 21.2°W. Finally, the values for all events before those in 2000 were reduced by a factor of 2, following the discussion of HACKMAN *et al.* (1990). All this does not apply to the June 17 and 21, 2000 earthquakes, for which sound instrumental data are available, and e.g.,a maximum rupture depth of 9 and 7 km were given, respectively. This is in good agreement with PEDERSEN *et al.* (2001), who modelled InSAR observations. They found as best model parameters for the June 17 event: a maximum slip of 2.40 m, a fault length 16 km, a fault width 10 km and for the June 21 event: a maximum slip of 2.15 m, a fault length 15 km, a fault width 9 km, while the slip on single segments varied from 25 to 125 cm and between 45 and 105 cm, respectively.
[4] Besides for the June 2000 events, calculated using $\log L$ [km] $= 0.5 \, M - 2$ (after QIAN, 1986) which results in slightly lower values compared to e.g., (SCHICK, 1968).

layers at the surface. We restricted the model to a homogeneous elastic half-space for three reasons: (i) We are not looking for absolute stresses but only for the spatial distribution of stress changes. Therefore the absolute value of Lamé's constants is not essential and only mentioned for reference. (ii) We expect symmetric stress fields around the vertical strike-slip sources, and (iii) we concentrated on the strongest stress changes, i.e., the elastic interaction of the stress sources neglecting post-seismic inelastic stress transfer. As the faults are introduced vertically into the unlayered environment, the stresses vary little with depth, except for the lower end of the fault.

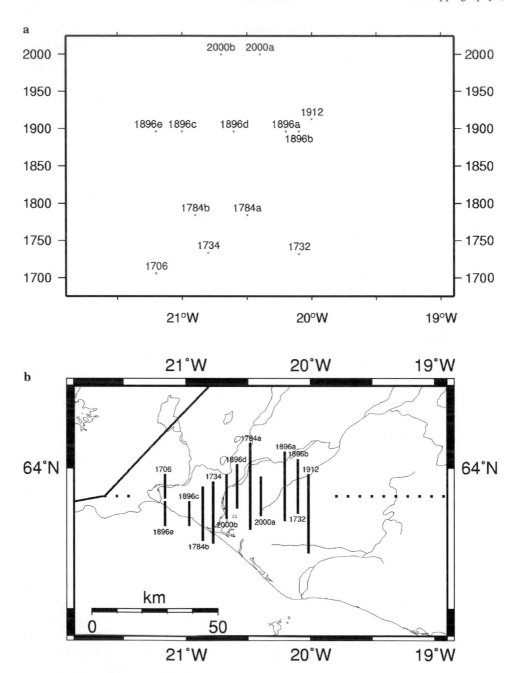

Figure 2
(a) Location of the earthquakes in space and time, cf. Table 1. As the events have been located on N-S trending faults in an E-W trending fault zone, their location is very accurately displayed in this graph. Events occurring in one year are numerated by letters a, ..., e. (b) The position of the rupture planes of the earthquake sequence as used in the models, cf. Table 1 and Figure 3.

Results were calculated for 280×220 test-points covering 280 km in E-W direction and 220 km in N-S direction. The area investigated extends from 18 to 24°W and from 63 to 65°N. The origin is set to 23.8°W, 64°N (cf., Fig. 1) and it includes the SISZ $\pm$ 1° north and south of 64°N, the SW edge of the EVZ, and the northeastern part of the RR. The reference frame is aligned along the SISZ, i.e., E-W, with the x-axis pointing east and the y-axis pointing north. This means that all stresses calculated are stresses with respect to this coordinate system.

4.1 The Initial Stress Field

The initial stress field is determined as follows: A tensional stress acting N103°E (nearly parallel to the SISZ; cf., DeMets et al., 1990) is assumed, due to the opening of the mid-Atlantic ridge in the region adjacent to the transform fault. While this rifting induces mainly shear stresses in the SISZ with a small opening component, the rift segments (RR and EVZ) are modelled with tensional stress and a small shear stress contribution. Tensional stresses at both ridges are modelled as constantly being released to ultimately result in zero values at the rifts. This induces stress in the transform zone. The stress magnitude, which is unknown, is set to a value that produces left-lateral shear stresses in E-W direction as large as the stress drop determined for the largest event (M = 7.1) in the studied earthquake sequence.

West of the SISZ, between the Hengill-Olfus triple junction and the Reykjanes ridge, no earthquakes strong enough to affect appreciably the stress field inside the SISZ occurred during the period modelled here. Therefore, these segments were modelled as an aseismic rift with oblique slip (mainly normal faulting with a smaller component of left-lateral strike-slip), to better fit the ridge slowly bending from NE-SW to E-W between the SW tip of the Reykjanes peninsula to the triple junction (Fig. 1a). Inside the SISZ, at the eastern and western tips, two areas with steady stress release were introduced (see dashed lines in Fig. 3). This could occur by a high rate of small events and, maybe, by creep. This is likely the case as these areas did not show strong events (M $\geq$ 6) since 1706, as stated above.

4.2 The Iterative Stress Release And Build-Up

On this initial field, the stress changes due to earthquakes are iteratively superposed as well as the stress changes due to further spreading at the ridge segments. From global geodetic measurements an opening of 1.94 cm/year is found, e.g., in DeMets et al. (1990), Sigmundsson et al. (1995, updated to 1.86 cm/year in DeMets et al., 1994). We used 2 cm/year as a zeroth order approach but reduced it to only 1 cm/year, as discussed later. Further, as the simplest assumption, lacking other data, the spreading rate is taken to be constant during the modelled time period, even though this can be questioned as for instance the debate on the stress increase in the New Madrid Seismic Zone demonstrates (cf., Newman and Stein, 1999; Schweig and Gomberg, 1999). The stress field

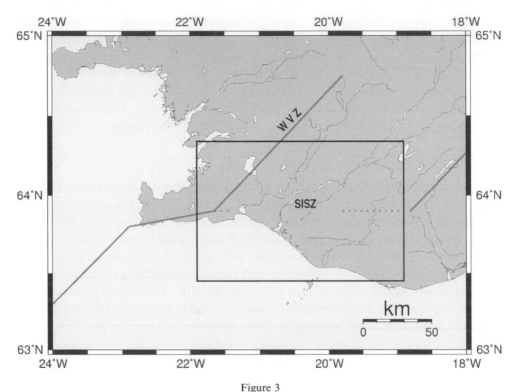

Figure 3

The box indicates the area in SW-Iceland used in the modelling as indicated with the small box in Figure 1a. The small box here indicates the area where the earthquakes struck and for which blow-ups of the stress distributions are displayed in the following so that details are clearly visible.

before every event is thus the superposition of the initial field, the stress changes by all preceding events, and the plate tectonic stress build-up since the starting time of the model.

5. The Results

The stress fields at 20 dates were calculated: the pre- and post-seismic situation for all 13 events. The time before 6 events was too short to accumulate appreciable plate tectonic stresses since the preceding event. In these cases, the post-seismic

▶

Figure 4

(a) Shear-stress field in the South Iceland seismic zone and its surroundings as assumed in 1706, the starting field for the model calculations. The shear stress is between 1.4 and 2.0 MPa in the SISZ. – Here, only the central region of the modelled area is displayed, so that the details inside the SISZ ((124 km,−5 km)– (199 km, 5 km)) are clearly visible, cf. Figure 3. (b) Shear-stress field in the South Iceland seismic zone and its surroundings as assumed in 1706 The same as part (a) with coloured areas for stresses that are given in MPa. The white dashed lines delineate the are a of steady stress release, cf. Figure 3.

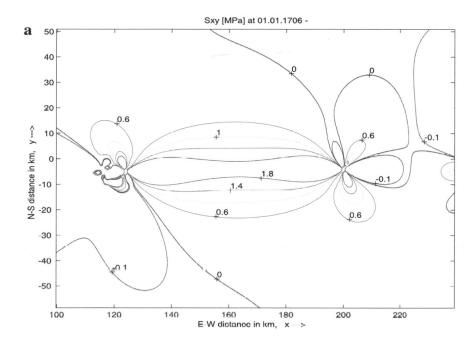

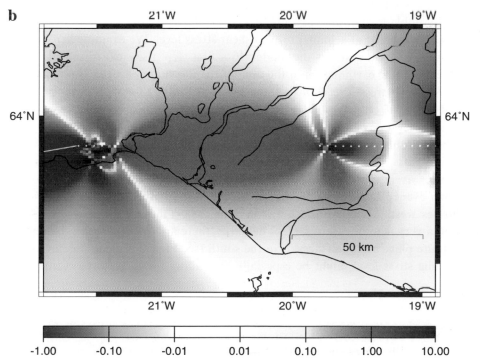

stress field of the preceding event was treated as the pre-seismic stress field of these earthquakes. Figure 4a gives the central part of the initial stress field in 1706. The rift segments, lower left and upper right (outside the figure), with about zero tensile stresses, also show low shear stresses. The area of the SISZ exhibits increased left-lateral E-W shear stresses. The SISZ as a strike-slip fault zone with strong events extends here approximately from 21.2°W, 63.9°N ((124 km, -5 km) in model coordinates) to 19.7°W, 63.9°N (199 km, 5 km), cf. also Figure 1. Figure 4b gives the initial stress field again with the dark red areas subject to the highest shear stress.

A selection of the results obtained by this model is given in Figures 5 to 8. Figure 5a presents the stress field after the first three events (at 21.2, 20.1, and 20.8°W) to give a good impression of what changes are caused by the earthquake stress release in the model. Moreover, it displays the high stresses around 20.5°W before the strongest event of the series occurred there. Figure 5b displays the situation after this and the last event in the 18th century, in which five earthquakes ruptured almost everywhere in the SISZ. Not earlier than 112 years later (1896) the next damaging events occurred. In 1912 the following event occurred. The situation before and after the earthquake is given in Figure 6.

Originally, an extrapolation was done from the 1912 earthquake to spring 1999. After the two earthquakes took place June 2000, at the end of the PRENLAB project, it was updated to June 17, 2000 (Fig. 7) and the effect of both events were determined, see Figure 8.

As a simple assumption one might expect that earthquakes in a certain fault zone normally occur at a similar critical shear stress level. We examine here whether such an expectation matches the model. In a first step, the mean shear stress level before each of the earthquakes in the area of the impending event is summarised in Figure 9a. The stress level is near the average (1.8 MPa) for many events. This is true for most, if only the first events in a year are considered. The second 1784 earthquake (two days after the first, 0.4 in magnitude smaller, 23 km away) might have been an aftershock; at least, as the rupture starts far south (cf., Figs. 1a and 2b), large parts of its rupture are situated in an area of very inhomogeneous stress with lower average stress (1.6 MPa). A similar argument applies to the second of the 1896 events: It occurred in the stress remnants of the first shock, just a few kilometres west. Also, the first event in June 2000—though no aftershock—took place in a very inhomogeneous stress field, i.e., there are low stresses in the north of the rupture plane and high stresses in the south (see Fig. 7). Calculating an average pre-seismic stress level might be especially misleading for these three events, as all

▶

Figure 5

(a) The stress field before the Aug. 14, 1784, M = 7.1, earthquake occurred at (20.5°W, 64.0°N), i.e., (x = 187 km, y = −18 km to y = +17 km). (b) The stress field after the Aug. 16, 1784, M = 6.7, earthquake occurred at (20.9°W, 63.9°N), i.e., (x = 145, y = −22 to y = 0).

a

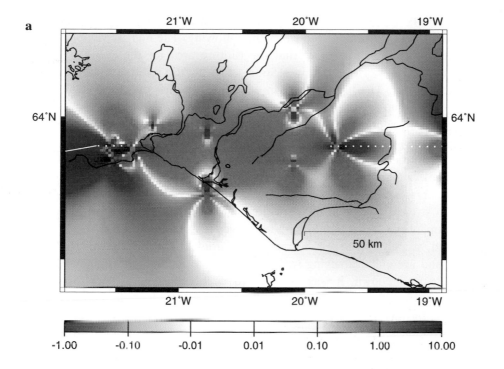

b

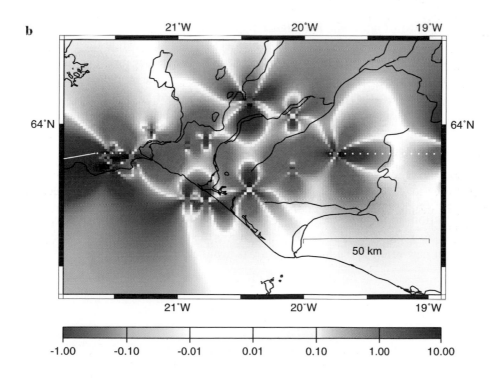

the test-points around the rupture plane are considered equally (no assumptions about asperities). A further problem is the following: The highest value (for 1896e) is mainly influenced by the fact that the rupture area for the event in 1706 was located completely north of that in 1896 (cf., Fig. 2b). So it did not fall into the area of reduced stress. The fourth event in 1896 and the first shock in 2000 are both influenced by the largest event in the series, i.e., the 1st one in 1784. Thus, the accuracy of the source parameter of the events, especially of the 1784 earthquake strongly influence the whole model.

In a second step to check the performance of the model in a quantitative way, we further examined whether each earthquake hit a high stress area and how large the high stress areas with no event were at the same time (the range in longitude with high stresses was summed when the north-south extension of the area was at least 5 km; from this, the longitude range for the event that occurred was subtracted, usually 0.1 to 0.2 degrees in longitude). The results can be found in Table 2 and are quite satisfying with respect to the named question. As the initial stress field is unknown, it was assumed to be homogeneous. Therefore the first series of events that ruptures once all across the fault zone is very strongly influenced by this assumption. After this tuning phase in the 18th century, most events hit high stress areas and the size of high stress areas with no event was rather small.

As stated earlier, the plate velocity used in the model was set to 1 cm per year, only half of that which is measured. The reason can be seen in Figure 9a in comparison with Figure 9b. The higher plate velocity leads to a steady and strong increase in the pre-seismic stress level, which is very unlikely, while the reduced rate entails an almost constant level. One reason for the discrepancy might be that 30% of the spreading takes place in the WVZ, thus not increasing the modelled shear stresses in the SISZ., cf. SIGMUNDSSON et al. (1995) who stated that $85 \pm 15\%$ of the relative plate motion takes place in the EVZ. Moreover, it might be that only a fraction of the stress build-up by plate tectonics is released seismically. For the rest, stable sliding or aseismic creep could be responsible.

In view of seismic hazard estimation, the present stress field is very interesting. This is very similar to that after the second event in June 2000, when stresses are concentrated around 19.9, 20.3, 20.6, and at 20.8–21.2°W (cf., Fig. 8b). The stress concentrations in the east and west of the SISZ were expected as there haven't been earthquakes since 1912 and 1896, respectively, this was reproduced by the model. The main new result is the high stress areas around 20.3 and 20.6°W.

▶

Figure 6
The stress field before (a) and after (b) the May 6, 1912, M = 7.0, earthquake occurred at (20.0°W, 63.9°N), i.e., (x = 187, y = −27 to y = +5).

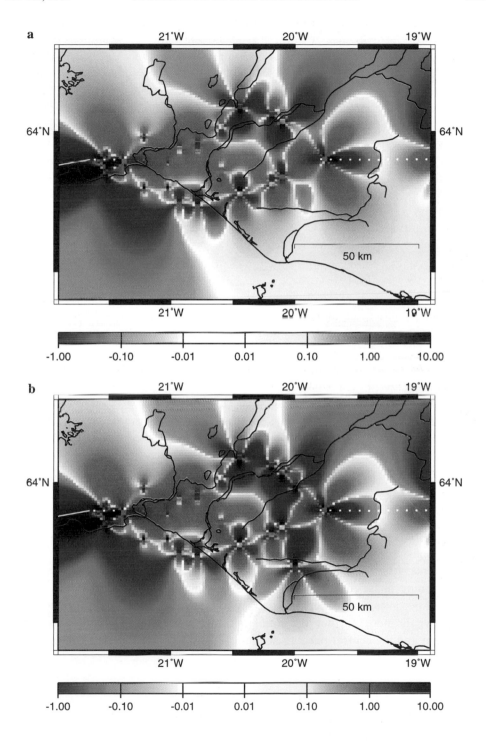

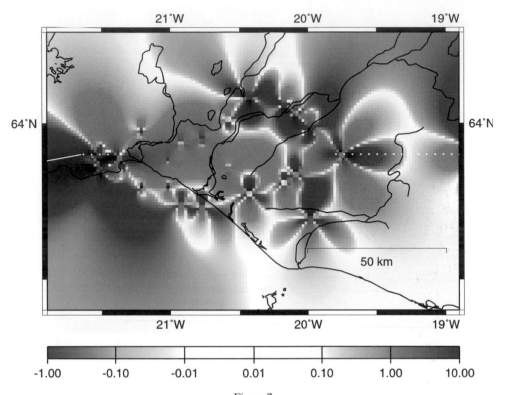

Figure 7
The stress field before the June 17, 2000, M = 6.5, earthquake occurred at (20.4°W, 64.0°N), i.e. (x = 169, y = −12 −y = +4).

6. Discussion

6.1 The Intial Stress Field

The initial stress field is unknown and was determined with very simple assumptions, i.e., rather homogeneous besides the tip effects of the transitions to the rift segments. It was necessary to consider a long series of events that provided us with a 'tuning phase' until the initial oversimplifications are washed-out and the stress field becomes increasingly more likely to represent a real, inhomogeneous one. To improve the model further, the initial unknown stress field of 1706 was reduced in the eastern and the central parts to account for the fact that the first events did not

▶

Figure 8
(a) The stress field after the June 17, 2000, M = 6.5, earthquake occurred at (20.4°W, 64.0°N), i.e., (x = 169, y = −12 −y = +4). (b) The stress field after the June 21, 2000, M = 6.4, earthquake occurred at (20.7°W, 64.0°N), i.e., (x = 154, y = −13 to y = +5).

a

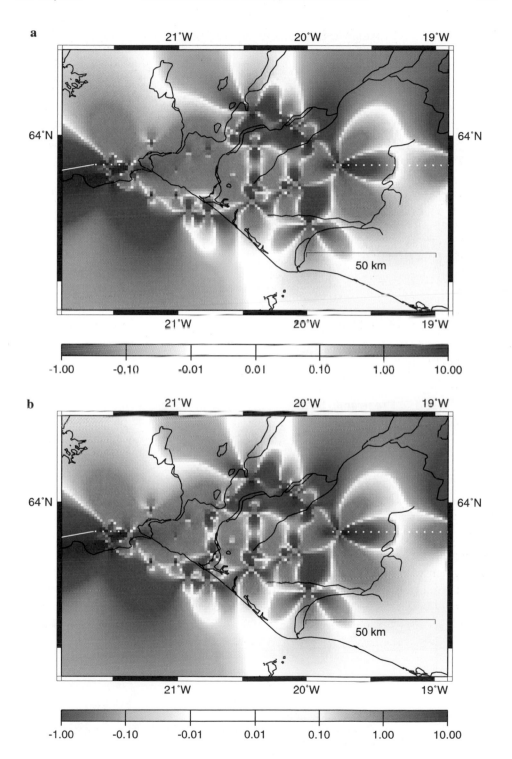

a

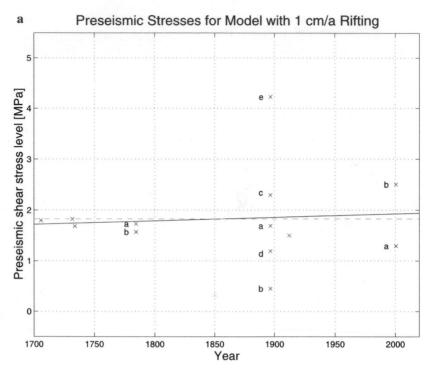

b

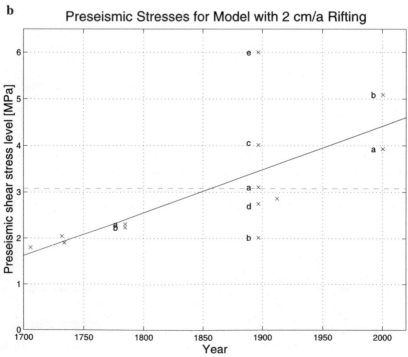

Table 2

Qualitative evaluation of the model performance

Event Time	Pre-seismic stress high at °W	Earth-quake at °W	Hit high stress area?	Size of other areas with high stress	Remarks
1706	19.8–21.2	21.2	yes	large	tuning phase[1]
1732	19.8–21.1	20.1	yes	large	tuning phase
1734	19.8–20.0, 20.2–21.1	20.8	yes	large	tuning phase
1784a	19.8–20.0, 20.2–20.7, 20.8–21.1	20.5	yes	medium	tuning phase
1784b	19.8–20.0, 20.9–21.1	20.9	yes	small	tuning phase
1896a	19.8–20.3, 20.8–21.2	20.2	yes	medium	other high stress area hit 10 days later
1896b	19.8–20.0, ± 20.6, 20.8–21.2	20.1	no	medium	other high stress area hit 9 days later
1896c	19.8–20.0, ± 20.6, 20.8–21.2	21.0	yes	medium	other high stress area hit next day
1896d	19.8–20.0, ± 20.6, 20.8–21.2	20.6	yes	medium	other high stress area hit on the same day
1896e	19.8–20.0, 20.8–21.2	21.2	yes	small	
1912	19.8–20.0, 20.8–21.2	20.0	yes	small	
2000a	± 19.9, 20.5–21.2	20.4	no/ yes	small	hit short (N-S) high stress area[2]
2000b	± 19.9, 20.6–21.2	20.7	yes	small	

Comments:

[1] As the initial stress field is unknown, it was assumed to be homogeneous. Thus the first series of events with ruptures through out the fault zone is very strongly influenced by this assumption.

[2] The stress field before the 1st earthquake in June 2000 is very inhomogeneous, i.e., there are low stresses in the north of the rupture plane and high stresses ones in the south (see Fig. 7). This situation is still influenced by the rupture position and magnitude of the first event in 1784 (cf., Fig. 2b). Note that the historical data have some uncertainty (cf., Fig. 1b).

occur before 1732 and 1734, respectively. This decreased the stress field in general and therefore also the pre-seismic stress level to result in 1.65 MPa. Although the inhomogeneity along the impending rupture of the events in 1784 and 1896 is slightly decreased, the overall variation remained essentially unchanged (± 0.88 MPa).

◄

Figure 9

Cross plot of the pre-seismic shear stress level at the site of the impending earthquakes vs. occurrence time. The stress values at 2 to 14 test-points at 0.5 km distance to the surface trace of the rupture plane were averaged. (a) Instead of a plate velocity of 2 cm/year only 1 cm/year was used (see text). Letters 'a' through 'e' denote the events in one year in temporal sequence. The dashed line provides the average stress (1.83 ± .9 MPa), the solid line represents the trend found by a linear least-squares fit. (b) Same as part (a), but a plate velocity of 2 cm/year was used.

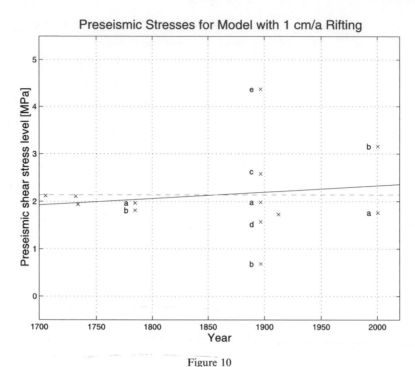

Figure 10
Cross plot of the pre-seismic shear stress level at the site of the impending earthquakes vs. occurrence time.
Here, a maximum rupture depth of 10 km was used. For further explanations see Figure 9.

6.2 Stress Concentrations in the East and West of the SISZ

In general, the origin of the stress concentrations at the end of the SISZ is (i) the finite length of the transform zone leading to increased stresses at the ends and (ii) the fact that the ridges do not extend to infinite depth, but are assumed to reach only 7 km depth and enter then an inelastic, hot region not capable of supporting stresses for periods of years. Deeper penetration of the brittle layer (elastic behaviour of the visco-elastic material) there would homogenise the stress field between rift tips. This unlikely for the times of several hundred years considered here, only for the stress release periods of tenths of seconds during the earthquakes. However, a similar effect would be produced if 'rifting' as the source for stress transfer is replaced by drag at the base of the adjacent plates. We checked this in various models, however due to reason (i) such a redistribution of stress would not lead to considerably lower stresses at the ends of the SISZ.

As stated at the beginning, strong earthquakes have not been observed in these very areas of stress concentration. Therefore and as we do not model thousands of medium and small size events, we introduced in the model that stresses are steadily released at the areas (marked with dashed lines) where the SISZ approaches the ridge segments. Their lengths are chosen mainly from the known location of strong (M ≥ 6) events, i.e. from the rift tips to the region of the closest strong event.

Possibly, there might have been stronger events before 1706 in the period in which the historical catalog is less accurate than later, cf. HALLDÓRSSON (1991). Looking into the increased precise, instrumentally recorded seismicity, there are several indications that the stress release indeed mainly takes place in small and medium events at the ends of the SISZ: In 1987 there was a medium-sized earthquake (M_S = 5.8) at 63.91°N 19.78°W (198,-9) near Vatnafjöll (see Fig. 1b) at the east end of the SISZ, and in 1998, there were two similar events at the Hengill-Ölfus triple junction at the western end of the SISZ: June 4 (M = 5.1) and November 13 (M = 5) both accompanied by numerous smaller events. Therefore, we believe that the stress concentrations given in section 'results' represent those for future strong earthquakes and not the regions east of 19.9 or west of 21.2°W up to the rifts.

6.3 Inhomogeneous Stress Fields

Some of the rupture planes (1734, 1784a,b, 1896a, and 1912) taken from the catalogue extend rather far to the north and south of the SISZ, i.e., into areas of low stresses due to plate tectonics. This leads to inhomogeneous stresses before the respective events, entailing low pre-seismic stresses which reduce the average pre-seismic stress level and increase its scattering. To check this variability, we used two models different from the one mentioned above. Firstly, a model was calculated that uses the same seismic moment of the events, but cuts the fault length to 50% while doubling the coseismic displacement. It leads to stronger changes in the stresses as the moment release of all events is concentrated in smaller areas. It follows an increase of the pre-seismic stress level to 2.49 ± 2.26 MPa, as the events now fall fully into the high stress region, and a general increase of the stress level in areas previously not hit by an earthquake. Unfortunately, the scattering in pre-seismic stresses nevertheless escalates mainly due to those earthquakes occurring near previous ones (cf. 1896b or 2000a).

Secondly, a model confining the rupture width to 10 km was examined. The processing of the June 2000 earthquakes led to maximum rupture depth of less than ten kilometres. Even though there are smaller events located to 13 km deep (cf., STEFÁNSSON et al., 1993), the assumption that all ruptures extend to no more than a depth of 10 km seems to be reasonable. Such a model was calculated too, replacing the maximum depth of 14 km for most events (cf. Table 1) by 10 km. In this case too, the stress release by the events is higher, as it is concentrated in an area closer to the rupture plane. Moreover, the interaction of the events is lower due to this concentration in space. However, the stress level before the main events remains in a similar range as before (average now 2.1 MPa instead of 1.8 MPa), and the variation in the pre-seismic stress differs little (the standard deviation is 0.87 MPa instead of 0.88 MPa) from the model with deeper ruptures, cf. Figure 10.

6.4 Source Locations

From the results discussed and Figure 2b it becomes clear that the main uncertainty in the model is the source parameters of the events, especially the longitude and N-S extension of the rupture planes: The events in 1706 and 1896e have the same longitude but are non-overlapping in the N-S direction, 1896a and b are very close together, etc. Due to the strong gradients in the stress fields (see for example Fig. 8b) a slight shift of some events towards east or west will strongly alter the consecutive stress fields.

The advantage based in the long series of strong seismic events indicates problems here. The damage areas from historical records are not gathered by scientists and are usually biased by uneven population density. Consequently the magnitudes and locations are not very accurate, as stated earlier. As we mentioned in the footnotes of Table 1, there are doubts concerning the correct rupture size when compared to global relations between magnitude and rupture length. Here, we adhere to the data found in the catalogue; a good basis until the historical events might be re-evaluated.

6.5 General aspects

Pursuing the predictive power of the modelling approach, strong limitations are visible: The main problem is, why some events did not occur earlier (at lower stress), just when the stress in their rupture passed the 'threshold in pre-seismic stress' (i.e., here: the average pre-seismic stress). This critical value is either unstable or the model accuracy is still too low to confine it sufficiently. This becomes very clear when the variability in stress is recalculated to give the uncertainty in the occurrence time. The stress build-up by plate motion is very low in these models as the spreading rate is not very high (compare the stress field in 2000 in Fig. 7 with that of 1912 in Fig. 6b). Therefore, an uncertainty in stress of 0.9 MPa corresponds to a time which ranges several years, depending on the position with respect to the SISZ tips.

7. Conclusion

The modelling shows that the rupture planes located on separated parallel N-S-striking zones are dense enough to lead to an area-wide stress release by the series of events. A tendency with time towards higher values of pre-seismic stress was found. It is an indication that the stress increase due to rifting might have been assumed too high, i.e., not all of the stress increase in South Iceland due to the spreading rate of 2 cm/year (assumed to be constant between 1706 to 2000) was released by earthquakes. The assumption that only half of the stress is accumulated inside the SISZ led to a rather constant pre-seismic stress level.

In all models, the pre-seismic stress level for most main shocks is high and fairly stable. This is also true for the events at the end of the sequence, namely those of 1912, of June 20, 2000, and—with minor reservations—of June 17, 2000. It indicates that the rather simple model can already explain the main features of the behaviour of the SISZ. This is especially astonishing, when the fact is kept in mind, that most (all but three) events employed are not instrumentally recorded. Before the June 2000 events, the SISZ seems to have been prepared for rupturing at the specific locations. However, the method cannot determine the occurrence time of future events with enough accuracy.

Several variations in model parameters did not lead to essentially different results, i.e., the model is rather stable. In general, the models go beyond the standard earthquake moment release and seismic hazard analysis as they include the spatial location and extension of the events, as they quantify the amplitudes of stress changes by earthquakes and by plate motion on the faults, as well as providing an extrapolation to the present stress situation. Considering the stress transfer between earthquakes means that a time-dependent change in the probability for future earthquakes is included.

Acknowledgements

I like to extend thanks to the Icelandic Meteorological Office, especially P. Halldórsson and Ragnar Stefánsson, for exceptional support regarding the seismicity of Iceland. The paper profited considerably from the detailed reading of two anonymous reviewers and their suggestions. I am also indebted to F. Lorenzo Martín and Fernando J. Lorén Blasco for assisting in the graphics and reviewing the text. The project was funded by the European Commission (DG XII/D2) as project ENV4-CT97-0536, and by GeoForschungsZentrum Potsdam. The preparation of this publication was also supported by the Deutsche Forschungsgemeinschaft under grant SFB 526 (Collaborative Research Centre 'Rheology of the Earth – from the Upper Crust to the Subduction Zone', subproject B2).

REFERENCES

ANGELIER, J., RÖGNVALDSSON, S. T., BERGERAT, F., GUÐMUNDSSON, Á., JAKOBSDÓTTIR, S., and STEFÁNSSON, R. (1996), *Earthquake focal mechanisms and recent faulting: A seismotectonic analysis in the Vördufell area, south Iceland seismic zone*. In B. Thorkelsson and European Seismological Commission and Icelandic Meteorological Office and Ministry for the Environment and University of Iceland (eds.), Seismology in Europe. Papers presented at the XXV General Assembly of ESC, September 9–14, 1996 in Reykjavík, Iceland (pp. 199–204).

BELARDINELLI, M. E., BONAFEDE, M., and GUÐMUNDSSON, Á. (2000), *Secondary Earthquake Fractures Generated by A Strike-Slip Fault in the South Iceland Seismic Zone*, J. Geophys. Res. 105(B6), 13,613–613,629.

BERGERAT, F. and ANGELIER, J. (2000), *The South Iceland Seismic Zone: Tectonic and Seismotectonic Analyses Revealing the Evolution from Rifting to Transform Motion*, J. Geodyn. 29(3–5), 211–231.

BERGERAT, F. and ANGELIER, J. (2001), *Mécanismes des failles des séismes des 17 et 21 Juin 2000 dans la Zone sismique sud-islandaise, d'après les traces de surface des failles d'Árnes et de l'Hestfjall*, Comptes Rendus Acad. Sci. Paris, Sciences del la Terre et des planèts *333*, 35–44.

BERGERAT, F., GUðMUNDSSON, Á., ANGELIER, J., and RÖGNVALDSSON, S. T. (1999), *Seismotectonics of the Central Part of the South Iceland Seismic Zone*, Tectonophys. *298* (4), 319–335.

BJARNASON, I. and EINARSSON, P. (1991). *Source Mechanism of the 1987 Vatnafjöll Earthquake in South Iceland*, J. Geophys. Res. *96* (B3), 4313–4324.

BJARNASON, I. T., COWIE, P., ANDERS, M. H., SEEBER, L., and SCHOLZ, C. H. (1993), *The 1912 Iceland Earthquake Rupture: Growth and Development of a Nascent Transform System*. Bull. Seismol Soc. Am. *83*, 416–435.

DEMETS, C., GORDON, R. G., ARGUS, D. F., and STEIN, S. (1990), *Current Plate Motions*, Geophys. J. Int. *101*, 425–478.

DZIEWONSKI, A. M., HALES, A. L., and LAPWOOD, E. R. (1975), *Parametrically Simple Earth Models Consistent with Geophysical Data*, Phys. Earth Plan. Int. *10*, 12–48.

EINARSSON, P. (1991), *Earthquakes and Present-day Tectonism in Iceland*, Tectonophys. *189*, 261–279.

EINARSSON, P., BJÖRNSSON, S., FOULGER, G., STEFÁNSSON, R. and SKAFTADÓTTIR, T. "*Seismicity pattern in the South Iceland Seismic Zone*", in *Earthquate Prediction*, ed. D. W. Simpson and P. J. Richards, Maurice Ewing Ser., vol. 4, pp. 141–151, AGU, Washington, D.C., 1981.

EINARSSON, P. and SÆMUNDSSON, K. (1987) *Earthquake epicenters 1982–1985 and Volcanic Systems in Iceland*. Map accompanying the Festschrift "Í hlutarins eðli", scale 1:750000, Menningarsjóður Reykjavík.

GUðMUNDSSON, Á. (1995), *Infrastructure and Mechanics of Volcanic Systems in Iceland*. J. Volc. Geoth. Res. *64*, 1–22.

GUðMUNDSSON, Á. (2000). *Dynamics of Volcanic Systems in Iceland: Example of Tectonism and Volcanism at Juxtaposed Hot Spot and Mid-ocean Ridge Systems*, Ann. Rev. Earth Planet. Sci. *28*, 107–140.

GUðMUNDSSON, Á. and HOMBERG, C. (1999), *Evolution of Stress Fields and Faulting in Seismic Zones*, Pure Apple. Geophys. *154*, 257–280.

HACKMAN, M. C., KING, G. C. P., and BILHAM, R. (1990), *The Mechanics of the South Iceland Seismic Zone*. J. Geophys. Res. *95* (B11), 17,339–17,351.

HALLDÓRSSON, P. (1991), *Historical earthquakes in Iceland until 1700*. In (K. Kodzak, ed.), Proc. 3rd Internatl. Symp. on Historical Earthquakes in Europe, Prague, pp. 115–125

HALLDÓRSSON, P., STEFÁNSSON, R., EINARSSON, P. and BJÖRNSSON, S. *Mat á jardskjálftahættu: Dysnes, Geldinganes, Helguvík, Vatnsleysuvík, Vogastapi og Thorlákshöfn*, Stadarvalsnefnd um idnrekstur, Idnadarráduneytíd, Reykjavík, 1984. (In Icelandic.)

KANAMORI, H. and ANDERSON, D. L. (1975), *Theoretical Basis of Some Empirical Relations in Seismology*, Bull. Seismol. Soc. Am. *65*, 1073–1096.

NEWMAN, A. V. and STEIN, S. (1999), *Reply: New Results Justify Open Discussion of Alternative Models*, EOS, Trans. Am. Geophys. Un. *80* (17), 197, 199.

PEDERSEN, R., SIGMUNDSSON, F., FEIGL, K. L., and ÁRNADÓTTIR, T. (2001), *Coseismic Interferograms of two $M_s = 6.6$ Earthquakes in the South Iceland Seismic Zone*, June 2000, Geophys. Res. Lett. *28*(17), 3341–3344.

QIAN, H. (1986), *Recent displacements along Xianshuihe Fault Belt and its Relation with Seismic Activities*, J. Seismol. Res. *9*, 601–614.

RÖGNVALDSSON, S. T. and SLUNGA, R. (1994), *Single and Joint Fault Plane Solutions for Microearthquakes in South Iceland*, Tectonophys. *237*, 73–86.

ROTH, F. (1989), *A Model for the Present Stress Field along the Xian-shui-he Fault Belt*, NW Sichuan, China, Tectonophys. *167*, 103–115.

SCHICK, R. (1968), *A Method for Determining Source Parameters of Small Magnitude Earthquakes*, Zeitschr. für Geophys. *36*, 205–224.

SCHWEIG, E. S. and GOMBERG, J. S. (1999), *Comment: Caution Urged in Revising Earthquake Hazard Estimates in New Madrid Seismic Zone*. EOS, Trans. Am. Geophys. Un. *80* (17), 197.

SIGMUNDSSON, F., EINARSSON, P., BILHAM, R., and STURKELL, E. (1995), *Rift-transform Kinematics in South Iceland: Deformation from Global Positioning System Measurements*, 1986 to 1992, J. Geophys. Res. *100*, 6235–6248.

STEFÁNSSON, R., BÖÐVARSSON, R., SLUNGA, R., EINARSSON, P., JAKOBSDÓTTIR, S., BUNGUM, H., GREGERSEN, S., HAVSKOV, J., HJELME, J., and KORHONEN, H. (1993), *Earthquake Prediction Research in the South Iceland Seismic Zone and the SIL Project*, Bull. Seismol. Soc. Am. *83*, 696–716.

STEFÁNSSON, R. and HALLDÓRSSON, P. (1988), *Strain Release and Strain Build-up in the South Iceland Seismic Zone*, Tectonophys. *152*, 267–276.

ZOBACK M. D., ZOBACK, M. L., MOUNT, V. S., SUPPE, J., EATON, J., HEALY, J. H., OPPENHEIMER, D., REASENBERG, P., JONES, L., RALEIGH, C. B., WONG, I. G., SCOTTI, O. and WENTWORTH, C. (1987), *"New evidence on the state of stress on the San Andreas Fault System"*, Science, *238*, pp. 1105–1111.

(Received March 25, 2002, revised January 27, 2003, accepted February 10, 2003)

 To access this journal online:
http://www.birkhauser.ch

Pure appl. geophys. 161 (2004) 1329–1344
0033–4553/04/071329–16
DOI 10.1007/s00024-004-2507-4

| Pure and Applied Geophysics

3-D Modelling of Campi Flegrei Ground Deformations: Role of Caldera Boundary Discontinuities

FRANÇOIS BEAUDUCEL[1], GIUSEPPE DE NATALE[2], FRANCO OBRIZZO[2], and FOLCO PINGUE[2]

Abstract—Campi Flegrei is a caldera complex located west of Naples, Italy. The last eruption occurred in 1538, although the volcano has produced unrest episodes since then, involving rapid and large ground movements (up to 2 m vertical in two years), accompanied by intense seismic activity. Surface ground displacements detected by various techniques (mainly InSAR and levelling) for the 1970 to 1996 period can be modelled by a shallow point source in an elastic half-space, however the source depth is not compatible with seismic and drill hole observations, which suggest a magma chamber just below 4 km depth. This apparent paradox has been explained by the presence of boundary fractures marking the caldera collapse. We present here the first full 3-D modelling for the unrest of 1982–1985 including the effect of caldera bordering fractures and the topography. To model the presence of topography and of the complex caldera rim discontinuities, we used a mixed boundary elements method. The *a priori* caldera geometry is determined initially from gravimetric modelling results and refined by inversion. The presence of the caldera discontinuities allows a fit to the 1982–1985 levelling data as good as, or better than, in the continuous half-space case, with quite a different source depth which fits the actual magma chamber position as seen from seismic waves. These results show the importance of volcanic structures, and mainly of caldera collapses, in ground deformation episodes.

Key words: Campi Flegrei, deformations, caldera, 3-D, boundary elements, levelling.

Introduction

Campi Flegrei is a trachytic caldera located close to Naples (Southern Italy), formed by several episodes of caldera collapses. The first series of collapses, forming the outer caldera rim, probably started about 35000 years B.P. and are called Grey Tuff eruptions from the typically erupted products. The most recent collapse, dated 12000 years B.P., produced the Yellow Tuff, whose deposits cover entirely the caldera and the province of Naples, constituting the main ancient building material for the towns. Ground movements at Campi Flegrei caldera had been recognised since Roman times. The time evolution of slow ground movements (called "bradisisma"

---

[1] Observatoire Volcanologique de la Soufrière, IPGP, Le Houëlmont, 97113 Gourbeyre, Guadeloupe (FWI). E-mail: beauducel@ipgp.jussieu.fr
[2] Osservatorio Vesuviano, INGV, Via Diocleziano 328, Naples, Italy. E-mail: pino@or.ingr.it

from a Greek term for "slow seism") has been reconstructed from traces of marine deposits on the ancient monuments. The most recent and complete reconstruction was made by DVORAK and MASTROLORENZO (1991), and is summarised in Figure 1. The secular trend of ground movement is subsidence of the caldera, at a rate of about 1 to 1.4 cm/year. Superimposed on this long-term trend, some fast and intense episodes of ground uplift occurred, culminating in eruption in the case of the 1538 episode (Monte Nuovo). The most recent unrest episodes started in 1970 with a fast uplift of 1.5 m and, after a slight decrease from 1972 to 1982, produced a maximum uplift of about 1.8 m in two years (see detail in Fig. 1). From late 1984 onwards, the caldera underwent a slower subsidence, with an average rate of about 5 cm/year. The intense ground uplift episode 1982–1984 also involved considerable seismicity (more than 15,000 recorded earthquakes, located in the first 3 km of the crust), with maximum magnitude slightly larger than 4 (DE NATALE and ZOLLO, 1986; DE NATALE *et al.*, 1995). Modern research on Campi Flegrei ground movements began after the episode of 1982–1984. The long-term deformation processes have been interpreted within the framework of regional structures and geodynamics (LUONGO *et al.*, 1991; CUBELLIS *et al.*, 1995).

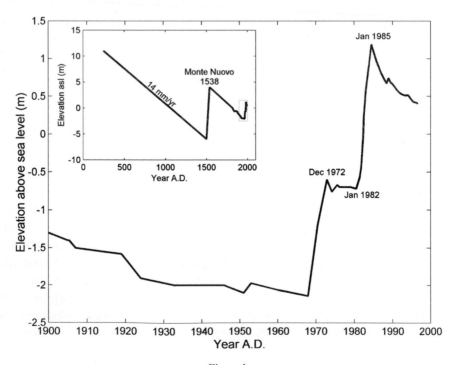

Figure 1

Elevation of the ground at Pozzuoli over the last 2,000 years, observed at the Serapeo Roman market. Vertical movements vary over 13 m and reach velocities up to 2.5 mm/day (modified after DVORAK and MASTROLORENZO, 1991).

The difficulties of applying simple mechanical models to explain Campi Flegrei deformation, similar to those hypothesised for basaltic volcanoes such as Kilauea (Hawaii), have been recognised since the mid 1970s (OLIVERI and QUAGLIARIELLO, 1969), and were summarised more recently by DE NATALE et al. (1991). They are mainly concerned with the low depth required for the source of overpressure to produce the observed, very localised ground deformation pattern. Furthermore, very large overpressures (on the order of several hundreds of MPa) are required to justify 1.8 m of uplift in two years. The very shallow depth required for the source (less than 2.5 km) is not compatible with realistic depths for the magma chamber. In fact, FERRUCCI et al. (1992) found evidence for a solid-plastic interface at about 4 km depth, interpreted as the top of the magma chamber (Figure 2b).

DE NATALE and PINGUE (1993), DE NATALE et al. (1997) and TROISE et al. (1997), demonstrated that, considering the effect of the caldera boundaries, modelled as ring faults almost free to shear under the effect of a deep source of overpressure, both the shape of ground deformation and the seismicity during unrest episodes could be interpreted in terms of a source of overpressure below 4-km depth. Such a model seems to explain most of the observations of Campi Flegrei unrests.

This paper represents the first attempt at detailed, three-dimensional modelling of the ground deformation pattern observed during uplift episodes at Campi Flegrei, using these caldera-bounding discontinuities. The boundary element method is used to simulate the ground deformation field. The vertical displacement data of the 1982–1984 unrest, measured by levelling (DVORAK and BERRINO, 1991), have been inverted for location and overpressure of the magma chamber, and for detailed geometry of collapse structures (ring fault dip, fault depth and width).

Forward Problem: Observations and Model Used

For modelling we use the Mixed Boundary Elements Method (MBEM) (CAYOL and CORNET, 1997), which allows the solving of 3-D problems taking into account topography, free surface and medium discontinuities (fractures) structures, without the problems of complex meshing of the finite-element methods. The MBEM approach optimises time computation, combining two different boundary element methods: (1) the direct method (RIZZO, 1967; LACHAT and WATSON, 1976), based on Betti's reciprocal theorem and the solution of Kevin's problem of a point force in an infinite body, and well suited for modelling topography and cavities; (2) the Displacements Discontinuity method (CROUCH, 1976), based on the analytical solution of a single displacement discontinuity in an infinite space, and well suited for modelling fractures.

The mesh used to describe the surface topography, made by Delaunay triangles, is shown in Figure 4. The elastic parameters of the area, taken by DE NATALE and PINGUE (1990) and DE NATALE et al. (1991), have been assumed to be constant

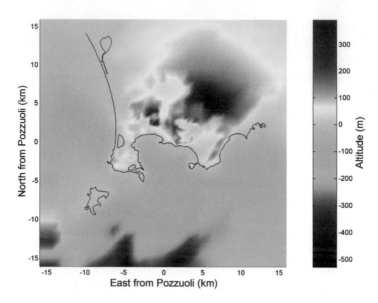

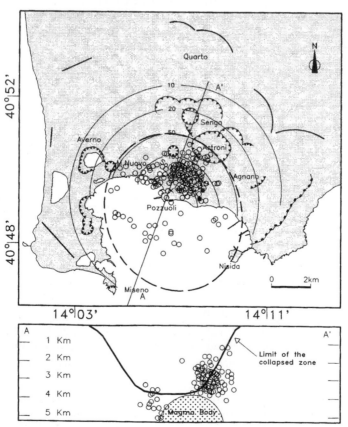

spatially, namely rigidity $\mu = 5$ GPa and Poisson's coefficient $v = 0.3$. The depth of the magma chamber has been constrained to the range 3.5–4.0 km on the basis of the results of FERRUCCI et al. (1992), who found the depth of a reverse rigidity contrasts from the observations of P- to -S wave conversions on the seismograms of regional earthquakes and local man-made shots. In this study, the magma source is represented as a sphere undergoing isotropic pressure (volume) increase. Based on previous studies of ground deformation modelling, we fixed the magma source depth and size, assumed spherical, of radius $r = 1$ km, with the centre at 4.5 km of depth. The geometry of caldera discontinuities has been taken from the gravity model as follows: the surface trace of the high hydrothermalised zone has been digitised and extended in depth to form a 3–D ring fracture. Due to the high uncertainty of shape and location of this ring, parameters such as size, depth of top and bottom limits, and dip angle have been set to a priori values but will be free to vary into the inversion process. Boundary conditions on these discontinuities are free to shear (no friction), forbidden to interpenetrate. These simple conditions have been used in this study to simplify the modelling and focus the inversion on the geometry of discontinuities and location of magma source. We are aware that they are not realistic and it would be illogical, for instance, to look at the stress field in this model.

Inverse Problem. Monte Carlo Serial Sampling

All the previous data on magma chamber, surface topography and collapse structures have been incorporated into a highly flexible inversion scheme, able to invert various parameters within given limiting ranges, describing the intrinsic uncertainty of each parameter. The method used is a Monte Carlo serial sampling (MOSEGAARD and TARANTOLA, 1995), using random parameters chosen, at each iteration, near the previous best solution. The method aims to minimise, iteratively, the least-squares misfit function of theoretical versus observed vertical displacement data. The iteration stops when a significantly small value for the misfit function is reached, in this case, the equivalent misfit as obtained with the best Mogi source model (MOGI, 1958) of 60 mm, corresponding to a point source located at 2.8 km of depth just under the maximum displacement point (Pozzuoli Porto).

The horizontal position and pressure change at the source are considered unknown, and inverted for. Other parameters considered unknown are the dip angle,

◄

Figure 2

a) Digital elevation model of the Pozzuoli Bay compiled from topographic and bathymetric data. Elevation varies from −550 to +400 m, showing that topographic effects on deformation may be significant. b) Map with earthquake locations and vertical displacements (black curves); in the depth section A-A' the collapsed zone and earthquakes are shown (from DE NATALE et al., 1995).

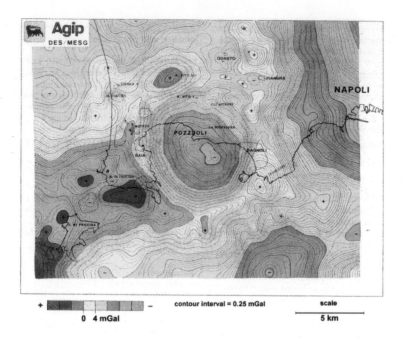

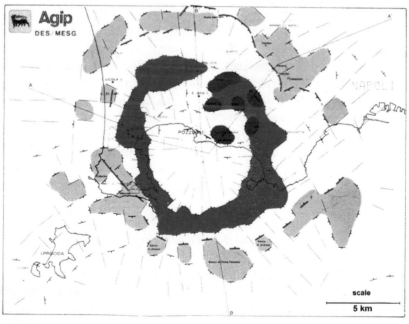

maximum depth and total height of the collapse discontinuities. Also the size of the ring discontinuities has been allowed to vary, within ±1/3 of the *a priori* size taken from gravity modelling (preserving the shape but allowing the scale to vary), taking into account the uncertainty in the gravity modelling. The total number of unknown model parameters is then 7, including source and collapse structures, while the observations (vertical displacement values on the levelling network) have about 80 independent values.

The iterative inversion of vertical displacement data of the 1982–1984 period required approximately 4000 iterations, i.e., about one month of computing on the Sparc Ultra 20 workstation. Figure 5 shows the evolution of the RMS residual during the iterations. Figure 6 shows the fit of the obtained 3-D model compared to the observed displacements, and the best fitting Mogi model. The final RMS residual obtained was 54 mm, with parameters indicated in Table 1. Overpressure is found equal to 35 MPa, a value three times smaller than those required by a homogeneous Mogi model. The mesh used to describe the collapse structures and magma chamber as resulting from the final model is shown in 3-D in Figure 7 and they are also shown in plane view in Figure 4. The location of the center of the spherical magma chamber is some kilometers south of Pozzuoli, beneath the sea. The dip of the ring discontinuities is steep, about 70°.

On the basis of the obtained model, it is possible to reconstruct a 3-D field of model vertical displacements (Fig. 8). This figure is remarkable, because it shows a displacement pattern in the most deformed area rather far from the circular geometry previously hypothezised (CORRADO *et al.*, 1976; BERRINO *et al.*, 1984; BONAFEDE *et al.*, 1986; BIANCHI *et al.*, 1987). The maximum deformation appears shifted into the Gulf of Pozzuoli, where most of the deformed area is contained. The vertical deformation pattern on land appears to be markedly elliptical, with the major axis oriented about N50°W. Such a trend is very similar to the elliptical trend of earthquake epicentres (see Fig. 2b). We stress here that the asymmetry of the vertical displacement pattern obtained is due to the dominant effect of ring discontinuities marking the caldera collapse structure.

Discussion

The 3-D displacement model obtained in this study opens new interesting perspectives for the study of Campi Flegrei unrest, as well as for other similar calderas. The first new insight concerns the non-symmetric pattern of expected

◄

Figure 3

Gravity observation and modelling (from *AGIP*, 1987). a) Bouguer anomaly map. b) 3-D structure modelling constrained by geological observations and geothermal drill data. The high hydrothermalized zone is used in this work to define the global geometry of caldera discontinuities.

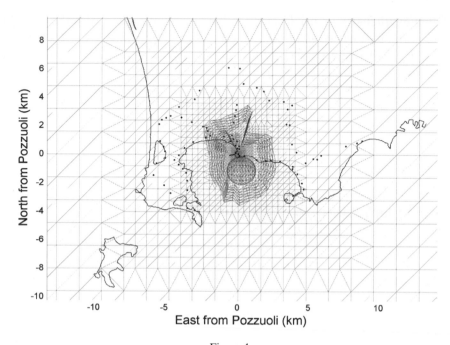

Figure 4
Mesh of free surface topography, caldera rim discontinuity and magmatic source structure, as in the best fit
model (Delaunay triangles). Black dots stand for levelling benchmarks.

deformation. The maximum displacement and most of the deformed area is
expected to be in the Gulf of Pozzuoli. Also, the elliptical orientation of the
displacement pattern on land is very similar to the elliptical geometry of epicentres
of local earthquakes that occurred in the same period (1982–1984). In view of
current models for local seismicity (TROISE *et al.*, 1997, 2003) this correspondence is
expected, due to the earthquakes which occur on the collapse discontinuities. It is
remarkable here, that this correspondence derives naturally from the inversion of
solely static displacement data, without any input information regarding seismicity.
Moreover, the elliptical symmetry on-land and the maximum deformation at sea
both occur from the inversion of sparse levelling data only collected on-land. The
sparse nature of data from levelling lines obviously disallows interpolation of the
original data to define a slightly elliptical pattern, nor is it possible to give any
information about displacement offshore. The essential element dominating the
character of the model displacement pattern is the presence of ring discontinuities
marking the inner caldera.

   Two different lines of discussion, however, can be taken to demonstrate the
effectiveness of the obtained solution, and, finally, of the hypothesis of dominant
control by the collapse discontinuities. The first one, already stated, is that half-space
solutions like the Mogi model provide results for the source depth that are

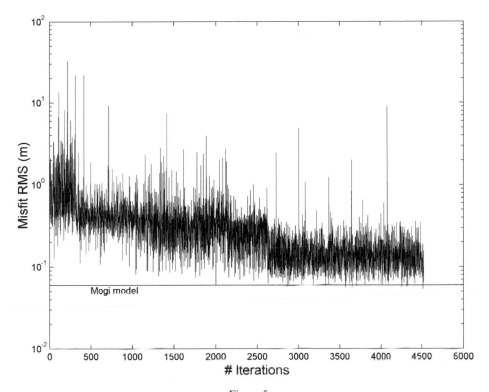

Figure 5

Misfit evolution versus number of iterations for the model inversion using a Monte Carlo serial sampling method, which slowly converged to more probable solutions. Horizontal line stands for Mogi model misfit (60 mm).

incompatible with physical constraints. This argument was made first by CASERTANO et al. (1976). Secondly, we discuss the effectiveness of modelled displacement patterns in the light of the actual data set. The best comparison for the modelled pattern would be with a dense displacement data set, making it possible to reconstruct detailed 2-D displacement maps. InSAR data (MASSONNET and FEIGL, 1998) would meet this need, especially for a better constraint on ground displacements next to the caldera discontinuities themselves, however, unfortunately they were not available during the 1982–1984 unrest. The observed subsidence trend at Campi Flegrei follows the same pattern as the uplift phase and, since 1992, it has been recorded by InSAR images. AVALLONE et al. (1999) computed the first model of subsidence based on InSAR data from the 1993–1996 period (Fig. 9), and this can be compared to the modelled pattern obtained in this study. A reconstruction of ground displacements over the period 1992–2000 from InSAR data has been obtained by LUNDGREN et al. (2001). Apart from the amount and sign of displacement, the patterns on land are very similar. They both show an elliptical trend oriented N50°W. Also the InSAR maximum displacement zones at Pozzuoli do not close on land. The maxima are

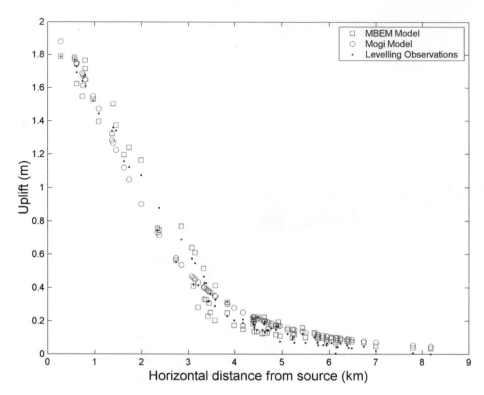

Figure 6

Comparison of levelling observations (black dots), Mogi model (circles) and MBEM model (squares) versus horizontal distance from the source. Minimum misfit is given by the 3-D model, i.e., 54 mm.

Table 1

Final parameters of best model of caldera and source

Parameters	Values
Inner caldera upper border depth[1]	1.16 km
Inner caldera lower border depth[1]	3.1 km
Inner caldera border dip angle	75°
Spherical source radius (fixed)	1 km
Spherical source depth (fixed)	4.5 km
Spherical source position West from Pozzuoli[2]	0.19 km
Spherical source position South from Pozzuoli[2]	1.33 km
Source pressure variation	+35 MPa
Best misfit	54 mm RMS

[1] Depth relative to mean sea level.
[2] Pozzuoli city coordinates = 40°49'15" N, 14°07'27" E.

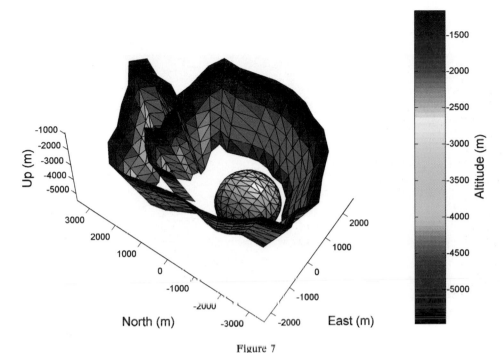

Figure 7

Three-dimensional view of caldera rim and magmatic spherical source mesh used in the modelling. Colour scale corresponds to depth below surface (all units in metre).

located at sea. The InSAR uplift data reinforce the ring discontinuity model proposed here.

The dominant role of these structural features on static deformations, was previously recognised starting from DE NATALE and PINGUE (1993), thereafter in the papers by DE NATALE et al. (1997), PETRAZZUOLI et al. (1999), and TROISE et al. (2004). This is the first time, however, that a detailed quantitative 3-D model taking accurately into account the caldera structure as modelled by gravity data has been created. The agreement with all the available data strongly supports the important role of the collapse structures in the model.

The dip of the system of inward discontinuities around the caldera is steep (around 70°) in agreement with the dip of corresponding fault planes of local earthquakes (DE NATALE et al., 1995; TROISE et al., 2003). The shape of the Bouguer anomaly, with the most negative area in the Gulf of Pozzuoli, may result from a rather asymmetric pattern of cumulative vertical displacements, with the area of greatest subsidence in the Gulf of Pozzuoli. The modelled location of the magma chamber is about 1 km offshore, south of Pozzuoli town. In the future, sea-floor measurements of displacement in the Gulf could improve our understanding of the deformed area and consequent models for unrest.

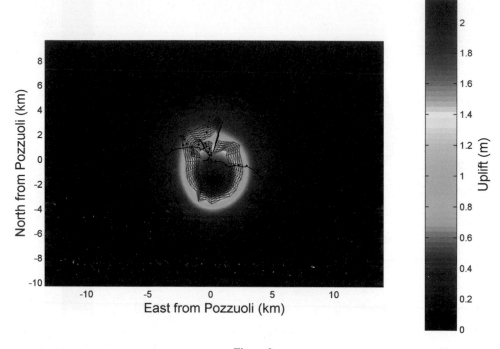

Figure 8

Vertical displacement field produced by the best fit model, caldera discontinuities (black mesh), and levelling benchmarks (black dots). Spatial displacement pattern is clearly influenced by the caldera geometry. Maximum uplift occurs about 2 km south of Pozzuoli City.

The magma chamber is only schematically described in our model with a simple spherical shape. Actually, the shape and volume of the chamber are not constrained by geophysical data. Only the depth of the top part has been well constrained by seismic evidence of *P*- to -*S* wave conversions by FERRUCCI *et al.* (1992) and DE NATALE *et al.* (2001b). The extension of the chamber, however, is a matter of speculation. Petrological estimates of the volume of the residual chamber (after the caldera formation) are highly variable, from a few to some hundreds of cubic kilometres of magma (ARMIENTI *et al.*, 1983; ORSI *et al.*, 1996). Modelling and observations of about 30 years of recent unrest suggests that at least the shallowest part of the magma chamber is spatially limited, with an area less than that of the inner caldera. It would be difficult to explain the localised nature of both ground deformation and seismicity otherwise. If the source of the huge caldera-forming ignimbrites (ROSI and Sbrana, 1987) was a magma chamber of large extent, recent observations would imply that the only active part is a shallower pluton a few cubic kilometres in volume.

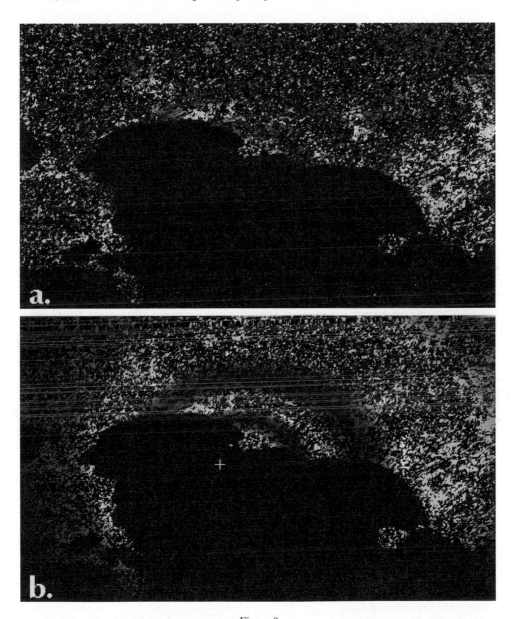

Figure 9
ERS InSAR data from the 1993–1996 period (AVALLONE *et al.*, 1999). (a) About 10 cm of subsidence are observed (3.5 fringes), and (b) Mogi model with 2.7-km source depth and horizontal location near Pozzuoli town (cross) which fits the observations satisfactorily. The spatial pattern is very similar to previous uplift and subsidence episodes.

Conclusions

Some of the results obtained in this work can be generalised to other similar areas. They imply that, when interpreting ground deformations at calderas, the structural features of the area must be carefully considered. In particular, the collapse structures play a fundamental role during unrest episodes, affecting both static deformation and seismicity. The presence of faults and fractures at volcanoes may strongly affect the ground deformation field, and must be taken into account when modelling unrest episodes. The use of extremely simple deformation models in homogeneous elastic media may be misleading, even if the apparent fit to the data is very good.

Regarding the Campi Flegrei area, most of the observations during unrest episodes can be modelled in terms of elastic deformation resulting from magma chamber overpressure in the presence of boundary caldera collapse structures. The deformation mechanism and temporal evolution, however, are likely to involve the further contributions of pressure/temperature changes in the geothermal system, as pointed out by several authors (DE NATALE *et al.*, 1991, 2001a; BONAFEDE, 1991; GAETA *et al.*, 1998).

Acknowledgements

The authors thank reviewers Geoff Wadge and Paul Segall for their useful comments and suggestions. Many thanks to Giuseppe Mastrolorenzo for providing some of the data, to Valérie Cayol for her support for the MBEM program and to François-Henri Cornet for his workstation. Work partially supported by GNV-INGV and EVG1-CT-2001-00047 'Volcalert' contracts, while FB was visiting researcher. IPGP contribution #1874.

REFERENCES

AGIP (1987), *Geologia e geofisica del sistema geotermico dei Campi Flegrei*, Int. Report, 17 pp.
ARMIENTI, P., BARBERI, F., BIROUNARD, H., CLOUCCHIATTI, R., INNOCENTI, F., METRICH, N., and ROSI, M. (1983), *The Phlegrean Fields: Magma Evolution within a Shallow Chamber*, J. Volcanol. Geotherm. Res. *17*, 289–312.
AVALLONE, A., BRIOLE, P., DELACOURT, C., ZOLLO, A., and BEAUDUCEL, F. (1999), *Subsidence at Campi Flegrei (Italy) Detected by SAR Interferometry*, Geophys. Res. Lett. *26*(15), 2,303–2,306.
BERRINO, G., CORRADO, G., LUONGO, G., and TORO, B. (1984), *Ground Deformation and Gravity Changes Accompanying the 1982 Pozzuoli Uplift*, Bull. Volcanol. *47*(2), 187–200.
BIANCHI, R., CORADINI, A., FEDERICO, C., GIBERTI, G., LANCIANO, P., POZZI, J. P., SARTORIS, G., and SCANDONE, R. (1987), *Modelling of Surface Ground Deformation in Volcanic Areas: The 1970–1972 and 1982–1984 Crises of Campi Flegrei, Italy*, J. Geophys. Res. *92*(B13), 14,139–14,150.
BONAFEDE, M. (1991), *Hot Fluid Migration: An Efficient Source of Ground Deformation: Application to the 1982–1985 Crisis at Campi Flegrei — Italy*, J. Volcanol. Geotherm. Res. *48*, 187–198.

BONAFEDE, M., DRAGONI, M., and QUARENI, F. (1986), *Displacement and Stress Fields Produced by a Centre of Dilatation and by a Pressure Source in a Viscoelastic Half-space: Application to the Study of Ground Deformation and Seismic Activity at Campi Flegrei, Italy*, Geophys. J. R. Astr. Soc. *87*, 455–485.

CASERTANO, L., OLIVERI DEL CASTILLO, A., and QUAGLIARIELLO, M. T. (1976), *Hydrodynamics and Geodynamics in the Phlegraean Fields Area of Italy*, Nature *264*, 161–154.

CAYOL, V. and CORNET, F. H. (1997), *3D Mixed Boundary Elements for Elastostatic Deformation Field Analysis*, Int. J. Rock Mech. Min. Sci. *34*(2), 275–287.

CORRADO, G., GUERRA, I., LO BASCIO, A., LUONGO, G., and RAMPOLDI, R. (1976/1977), *Inflation and Microearthquake Activity of Phlegraean Fields, Italy*, Bull. Volcanol. *40*(3), 169–188.

CROUCH, S. L. (1976), *Solution of Plane Elasticity Problems by the Displacement Discontinuity Method*, Int. J. Num. Meth. Eng. *10*, 301–343.

CUBELLIS, E., FERRIM M., and LUONGO, G. (1995), *Internal Structures of the Campi Flegrei Caldera by Gravimetric Data*, J. Volcanol. Geotherm. Res. *65*, 147–156.

DE NATALE, G. and ZOLLO, A. (1986), *Statistical Analysis and Clustering Features of the Phlegraean Fields Earthquake Sequence (May 1983 – May 1984)*, Bull. Seismol. Soc. Am. *76*(3), 801–814.

DE NATALE, G., PINGUE, F., ALLARD, P., and ZOLLO, A. (1991), *Geophysical and Geochemical Modelling of the 1982–1984 Unrest Phenomena at Campi Flegrei Caldera (Southern Italy)*, J. Volcanol. Geotherm. Res. *48*, 199–222.

DE NATALE, G., TROISE, C., and PINGUE, F. (2001a), *A Mechanical Fluid-dynamical Madel for Ground Movements at Campi Flegrei Caldera*, J. Geodynam. *32*, 487–517.

DE NATALE, G., TROISE, C., PINGUE, F., DE GORI, P., and CHIARABBA, C. (2001b), *Structure and dynamics of the Somma-Vesuvius volcanic complex*, Mineralogy and Petrology, *73*, 5–22.

DE NATALE, G., ZOLLO, A., FERRARO, A., and VIRIEUX, J. (1995), *Accurate Fault Mechanism Determinations for a 1984 Earthquake Swarm at Campi Flegrei Caldera (Italy) during an Unrest Episode: Implications for Volcanological Research*, J. Geophys. Res. *100*(B12), 24,167–24,185.

DE NATALE, G. and PINGUE, F., *Seismological and geodetic data at Campi Flegrei (Southern Italy): Constraints on volcanological models*. In *Volcanic Seismology* (eds. Gasparini ,P., Scarpa, R., and Aki, K.) (Springer-Verlag, 1990).

DE NATALE, G. and PINGUE, F. (1993), *Ground Deformations in Collapsed Caldera Structures*, J. Volcanol. Geotherm. Res. *57*, 19–38.

DE NATALE, G., PETRAZZUOLI, S. M., and PINGUE, F. (1997), *The Effect of Collapse Structures on Ground Deformations in Calderas*, Geophys. Res. Lett. *24*(13), 1,555–1,558.

DVORAK, J. J. and BERRINO, G. (1991), *Recent Ground Movement and Seismic Activity in Campi Flegrei, Southern Italy: Episodic Growth of a Resurgent Dome*, J. Geophys. Res. *96*, 2,309–2,323.

DVORAK, J. J. and MASTROLORENZO, G. (1991), *The Mechanism of Recent Vertical Crustal Movements in Campi Flegrei Caldera, Southern Italy*, Geol. Soc. Am., Special Paper, 263.

FERRUCCI, F., HIRN, A., VIRIEUX, J., DE NATALE, G., and MIRABILE, L. (1992), *P-SV Conversions at a Shallow Boundary beneath Campi Flegrei Caldera (Naples, Italy): Evidence for the Magma Chamber*, J. Geophys. Res. *97*(B11), 15,351–15,359.

GAETA, F. S., DE NATALE, G., PELUSO, F., MASTROLORENZO, G., CASTAGNOLO, D., TROISE, C., PINGUE, F., MITA, D. G., and ROSSANO, S. (1998), *Genesis and Evolution of Unrest Episodes at Campi Flegrei Caldera: The Role of Thermal-fluid-dynamical Processes in the Geothermal System*, J. Geophys. Res. *103*(B9), 20,921–20,933.

LACHAT, J. C. and WATSON, J. O. (1976), *Effective Numerical Treatment of Boundary Integral Equations: A Formulation for Three-dimensional Elastostatics*, Int. J. Numer. Meth. Eng. *24*, 991–1,005.

LUNDGREN, P., USAI, S., SANSOSTI, E., LANARI, R., TESAURO, M., FORNANO, G., and BERNARDINO, P. (2001), *Modeling Surface Deformation Observed with Synthetic Aperture Radar Interferometry at Campi Flegrei Caldera*, J. Geophys. Res. *106*(B9), 19,355–19,366.

LUONGO, G., CUBELLIS, E., OBRIZZO, F., and PETRAZZUOLI, S. M. (1991), *The Mechanics of the Campi Flegrei Resurgent Caldera — A Model*, J. Volcanol. Geotherm. Res. *45*, 161–172.

MASSONNET, D. and FEIGL K. L. (1998), *Radar interferometry and its Application to Changes in the Earth's Surface*, Rev. Geophys. *36*, 441–500.

MOGI, K. (1958), *Relations between the Eruptions of Various Volcanoes and the Deformations of the Ground Surfaces around them*, Bull. Earthquake Res. Inst. Univ. Tokyo *36*, 99–134.

MOSEGAARD K. and TARANTOLA, A. (1995), *Monte Carlo Sampling of Solutions to Inverse Problems*, J. Geophys. Res. *100*(B7), 12,431–12,447.

OLIVERI DEL CASTILLO, A. and QUAGLIARIELLO, M. T. (1969), *Sulla genesi del bradisismo flegreo*, Atti Associazione Geofisica Italiana, 18th Congress, Napoli, 557–594.

ORSI, G., DE VITA, S., and DI VITO, M. (1996), *The Restless, Resurgent Campi Flegrei Nested Caldera (Italy): Constraints on its Evolution and Configuration*, J. Volcanol. Geotherm. Res. *74*, 179–214.

PETRAZZUOLI, S. M., TROISE, C., PINGUE, F., and DE NATALE, G. (1999), *The Mechanics of Campi Flegrei Unrests as Related to Plastic Behaviour of the Caldera Borders*, Annali di Geofisica *42*(3), 529–544.

PINGUE, F., TROISE, C., DE LUCA, G., GRASSI, V., and SCARPA, R. (1998), *Geodetic Monitoring of the Mt. Vesuvius Volcano, Italy, Based on EDM and GPS Surveys*, J. Volcanol. Geotherm. Res. *82*, 151–160.

RIZZO, F. J. (1967), *An Integrated Equation Approach to Boundary Value Problems of Classical Elastostatics*, Q. Appl. Math. *25*, 83.

ROSI, M., and SBRANA, A (1987), *Phlegrean Fields*, Quaderni della Ricerca Scientifica, CNR Rome *9*(114), 175 pp.

SCANDONE, R., BELLUCCI, F., LIRER, L., and ROLANDI, G. (1991), *The Structure of the Campanian Plain and the Activity of the Neapolitan Volcanoes (Italy)*, J. Volcanol. Geotherm. Res. *48*, 1–31.

TROISE, C., PINGUE, F., and DE NATALE, G. (2003), *Coulomb Stress Changes at Calderas: Modeling the Seismicity of Campi Flegrei (Southern Italy)*, J. Geophys. Res., *108* (B6), 2292, doi: 10.1029/2002JB2006

TROISE, C., DE NATALE, G., PINGUE, F., and ZOLLO, A. (1997), *A Model for Earthquake Generation during Unrest Crises at Campi Flegrei and Rabaul Calderas*, Geophys. Res. Lett. *24*(13), 1,575–1,578.

TROISE C., DE NATALE G., and PINGUE F. (2004), *Non linear effects in ground deformation at calderas due to the presence of structural discontinuities*, Annales of Geophysics, in press

(Received March 26, 2002, revised November 26, 2002, accepted December 16, 2002)

 To access this journal online:
http://www.birkhauser.ch

Pure appl. geophys. 161 (2004) 1345–1357
0033–4553/04/071345–13
DOI 10.1007/s00024-004-2508-3

Pure and Applied Geophysics

Comparison of Integrated Geodetic Data Models and Satellite Radar Interferograms to Infer Magma Storage During the 1991–1993 Mt. Etna Eruption

ALESSANDRO BONACCORSO[1], EUGENIO SANSOSTI[2], and PAOLO BERARDINO[2]

Abstract—We compare the results obtained from the modelling of EDM, GPS, levelling and tilt data measured in the first part of the 1991–1993 eruption at Etna to the InSAR data acquired during the second part. The geodetic changes are very marked in the first half of the eruption and constrain a deflation source located at a few kilometers of depth ($\approx$ 3 km b.s.l.), in agreement with other independent geophysical evidence. SAR data, available during the second part of the eruption, were analysed for different time intervals in the second part of the eruption. The interpretation of SAR interferograms reveals a large-scale but less marked deflation of the volcano that could be caused by a deeper source. This second source is in accord with a second deeper anomaly revealed by recent seismic investigations. The combination of geodetic data modelling and SAR images suggests a complex plumbing system composed at least of two possible storage regions located at different depths.

Key words: Geodesy, SAR Interferometry, ground deformation, Mt. Etna volcano.

1. Introduction

During the 1991–93 Mt. Etna eruption, beginning December 14, 1991 and ending March 31, 1993, a clear and marked deflation of the edifice was recorded through different monitoring techniques (Fig. 1), namely EDM, GPS, levelling and continuous tilt (BONACCORSO, 1996). The eruption provoked a very significant deflation during the first months (comparisons made before eruption and in summer 1992), as evidenced by the EDM and GPS arcal contractions and by levelling subsidence. The vertical maximum lowering of about 11 cm was recorded on the western flank by high precision levelling and the most evident decreasing tilts were recorded on this flank. During the second part of the eruption (summer 1992–summer 1993 comparison) the volcano deflation decreased as evidenced by smaller EDM and

---

[1] Istituto Nazionale di Geofisica e Vulcanologia – Sezione di Catania. Piazza Roma, 2 – 95123 Catania, Italy. E-mail: bonaccorso@ct.ingv.it
[2] Istituto per il Rilevamento Elettromagnetico dell'Ambiente (IREA), National Research Council (CNR), Via Diocleziano, 328 - 80124 Napoli, Italy. E-mail: {sansosti.e, berardino.p}@irea.cnr.it

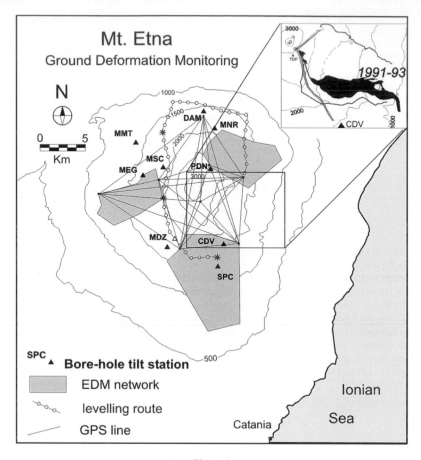

Figure 1
Mt. Etna topography map with the ground deformation monitoring networks operating during the 1991–1993 eruption.

GPS areal contractions and by levelling which did not show appreciable vertical variations. Throughout this second part of the eruption ERS-1 satellite synthetic aperture radar (SAR) data began to be collected, thus allowing the use of differential SAR interferometry (DIFSAR) for measuring ground deformation (e.g., GABRIEL *et al.*, 1989; ZEBKER *et al.*, 1994). This technique exploits the phase difference (interferogram) of SAR image data pairs and allows us to generate spatially dense deformation maps (the displacement field being measured in the radar line of sight) with a centimeter, in some cases millimeter, accuracy. The ground displacement pattern derived from DIFSAR interferograms showed a less marked, with respect to the first part of the eruption, deflation of the volcano edifice. Hereafter, we analyse the results of the previous modelling obtained from terrestrial and satellite geodetic data relative to the first half of the eruption to displacement patterns obtained from

SAR data recorded during the second half. A comparison between those data suggests the action of a complex plumbing system which could be composed of two magma storage regions located at different depths.

2. Previous Deformation Models at Mt. Etna for 1991–1993 Deflation

Terrestrial and spatial geodetic data (EDM, GPS, levelling and tilt) recorded in the first part of the eruption, i.e., before the eruption and in the summer of 1992, were analytically inverted by BONACCORSO (1996), who considered a double source composed of a shallow tensile crack, taking account of the intrusion accompanying the eruption onset, and by a deeper deflating source for the edifice deflation. The final solution gave an excellent fit to all the different measurements and defined the shallow tensile crack, which represents the intrusion inside the volcano edifice, and a 3-D ellipsoidal source (after DAVIS, 1986) centred at about 5 km depth (about 3 km b.s.l., considering a mean reference surface at 2 km a.s.l.) and undergoing deflation (Fig. 2). This depressurising source was interpreted as magma storage level and the negative pressure as due to mass loss through the eruption. The centre of the source was beneath the crater area but its asymmetry caused an expected displacement pattern with maximum effects further west from the crater area (Fig. 3) in agreement with the recorded EDM contractions, vertical lowering and tilt deflation which are maximum in this volcano sector.

In the second period of the eruption, i.e., between the summer 1992 and after the eruption end, small horizontal contractions (between −1 and −2 cm) were observed in the EDM and GPS networks (BONACCORSO et al., 1994; NUNNARI and PUGLISI, 1994). An attenuated variation of the tilt signals (BONACCORSO, 1996), and no appreciable vertical variation in the levelling route, which is circular and winds around the volcano at roughly the same height (MURRAY and MOSS, 1994). In this second period the modest geodetic variations do not permit a robust modelling of the data, since the small geodetic changes are of the same order of magnitude as the measurements error. Therefore the results of the inversion trials are not unequivocal and well constrained. However, the indications of the inversion trials show a clear tendency to set the depressurising source deeper. Thus the action of a deeper source (for instance, a source with a ca. 10 km depth) is called for to explain the observed deformation pattern, even if not well modelled with the small variations recorded.

Other source modelling was deduced from SAR interferograms covering the last months of the eruption (between September 1992 and the period after the eruption end). The first result obtained by Massonet et al. (1995) showed four near-concentric interferometric fringes. The authors used a simple point source, i.e. the Mogi model, and estimated, through trial and error, a depth of 16 ± 3 km for this source. A resemblance of the form of the apparent deformation derived by the interferograms to the topography is well evident and has been cited as evidence of a possible systematic error. DELACOURT *et al.* (1998) pointed out that SAR interferograms may be affected by tropospheric artefacts, whose correction could reduce the number of fringes associated with the deformation. In their study they claim that, for the Etna 1991–1993 eruption case, the tropospheric correction can reach two fringes for some interferograms. BEAUDUCEL *et al.* (2000) estimated a tropospheric correction in the

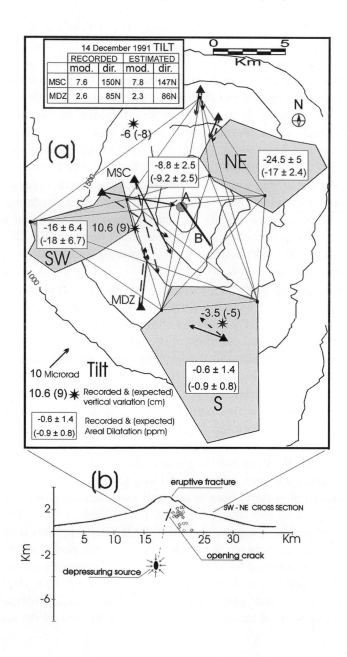

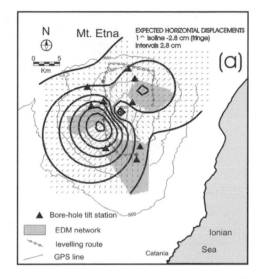
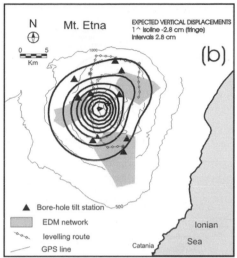

Figure 3

Deformation pattern expected from the depressuring ellipsoid obtained from the inversion of the geodetic
data (BONACCORSO, 1996) : (a) horizontal displacements, (b) vertical changes.

range of −2.7 to +3.0 (± 1.2) fringes. After this correction and by again using a
simple Mogi source, they obviously find a deflation source at a shallower depth with
respect to the one proposed by MASSONET et al. (1995) LANARI et al. (1998), by
applying a grid search method of inversion and without performing any tropospheric
correction, found that some interferograms of the considered period are fit well by a
point source located at a 9 km depth. It has also been noted (CAYOL and CORNETTE,
1998; WILLIAMS and WADGE, 1998, 2000) that topography could play a role in the
computation of the surface displacement for a given source and, therefore, should be
accounted for in the deformation source modelling process; this was not considered
in the previously cited works. However, this effect should be more evident in the
summit part of the edifice, while it should not be significant at greater distances, i.e.,
in the flatter lateral flanks (WILLIAMS and WADGE, 1998).

◄

Figure 2

(a) Map with comparison between recorded data (EDM, GPS, leveling and tilt) and estimates from
BONACCORSO (1996) model. For synthesising the horizontal pattern deformation, the areal dilatations,
three for EDM networks and one for GPS, are reported inside the little frames (the central frame is for
GPS). The co-eruptive tilt values of the variations recorded on 14 December, 1991 during the fracture
opening and the estimated ones are reported in the upper left inset (mod. = tilt vector amplitude in
microradians; dir. = azimuth of the tilt vector direction). The co-eruptive expected values are calculated
from the eruptive fracture single source. The surface projection of the ellipsoid centre (A) and of the
opening crack line (B) are also reported (redrawn after BONACCORSO, 1996). (b) Cross section showing the
location of the two deformation sources inferred from the Bonaccorso (1996) model: the deeper
depressuring 3-D ellipsoid and the shallowest opening crack. The location of the seismic events occurring
during the onset of the eruption on December 14, 1991, is also reported.

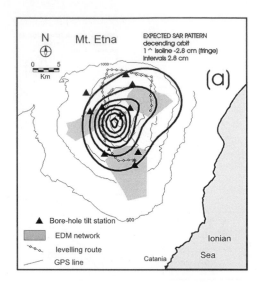
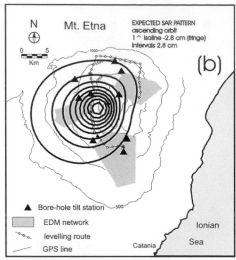

Figure 4

Deformation pattern expected from the depressuring ellipsoid obtained from the inversion of the geodetic data (BONACCORSO, 1996) : (a) SAR deformation as seen in the line of sight from the descending orbit, (b) SAR deformation as seen in the line of sight from the ascending orbit. The small cross represents the projection of the central crater. The source location is nearly under the crater area but provokes a displacement pattern with maximum effects further west, in accord with EDM contractions, vertical lowering and tilt deflation recorded in this volcano sector. The first isoline is 28 mm, the isoline step is one interferometric fringe corresponding to 28 mm.

It seems clear that in the first period of the eruption (first 6–7 months), marked and peaked deformations were recorded, indicative of the predominant effect of an intermediate source at about 3 km b.s.l. Instead in the second period of the eruption (last 6 months) smaller deformations were recorded. In this case DIFSAR measurements show a wide and regular fringe pattern whose modelled point source, found using trail and error methods, indicates the action of a deeper source, although not clearly fixed in depth.

3. Deformation Patterns Comparison

The biggest geodetic changes were recorded during the period of summer 1991–summer 1992, namely including the first six months of the eruption, while smaller geodetic changes were recorded in the second part of the eruption. The first period is not covered by SAR images, which instead are available from May 1992. This technique measures the projection of the deformation field along a direction from the observed site toward the satellite in its ascending or descending orbit. However, geodetic data furnish slightly different measurements, such as a limited number of horizontal displacements (GPS lines and EDM), vertical displacements (levelling points) and tilt. For this reason a direct comparison between DIFSAR results and geodetic measurements is no trivial matter; in order to obtain a homogeneous comparison between these measures we have projected the deformation pattern as expected from the deflation source of BONACCORSO (1996) along the direction of sight of the satellite. This result is shown in Figure 4.

In order to gain a deeper insight on the dynamic of the eruption and its deformation sources, we analysed a set of interferograms generated by using data acquired by the European ERS-1 satellite covering the period from May 1992 (first available ERS-1 acquisition) to August 1993. Since systematic topographic errors can be greatly reduced by limiting the (perpendicular) baseline, we performed a first interferogram selection on this basis. Among all possible interferograms we selected only those data with a baseline smaller than 110 m; indeed, for such interferograms, an error in the knowledge of topography on the order of 5–10 meters (which is the case of the DEM we used) would cause an error in the measure of the deformation of less than $1/8^{th}$ of fringe, i.e., less than 4 millimeters (LANARI et. al., 1998).

Using this selection criteria, results in the collection of five interferograms from ascending orbits and ten from descending ones are reported in Table 1. We have processed and analysed all the ascending orbit interferograms and most of the

Table 1

List of all the ERS-1 interferograms available on Mt. Etna site with perpendicular baseline smaller than 110 m and relative to the time period from May 1992 to August 1993. First and second columns report the acquisition dates, the third indicates the perpendicular baseline, the fourth the time difference (in days) between the two acquisitions. The last column reports the interferograms processed for this study. **In addition, the three interferograms shown in Figure 5 are highlighted.**

Date	Date	B perp.	Delta days	Processed
Track: 129 - Frame: 747 — Ascending Orbit				
23/08/1992	10/01/1993	−78	140	yes
23/08/1992	30/05/1993	47	280	yes
27/09/1992	17/10/1993	−33	385	yes
01/11/1992	**04/07/1993**	**−78**	**245**	**yes**
10/01/1993	08/08/1993	−85	210	yes
Track: 222 - Frame: 2853 — Descending Orbit				
21/06/1992	06/06/1993	−89	350	no
26/07/1992	30/08/1992	−85	35	yes
26/07/1992	13/12/1992	−76	140	yes
26/07/1992	28/03/1992	16	245	yes
30/08/1992	**13/12/1992**	**9**	**105**	**yes**
30/08/1992	28/03/1993	101	210	yes
04/10/1992	21/02/1993	−98	140	no
04/10/1992	02/05/1993	9	210	no
08/11/1992	21/02/1993	60	105	no
13/12/1992	**28/03/1993**	**92**	**105**	**yes**

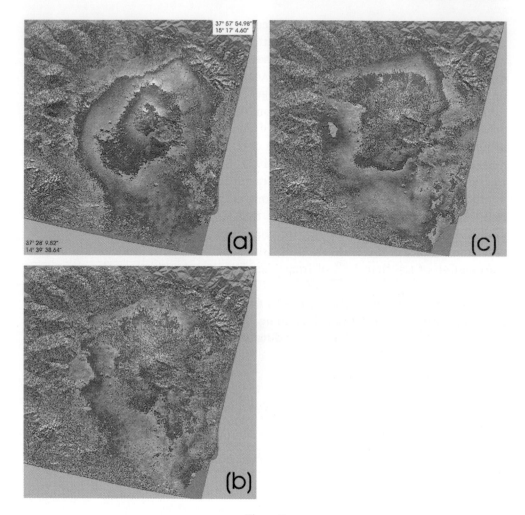

Figure 5
Selected differential interferograms of Mt. Etna in UTM projection with a shaded relief as a background image. Each fringe (a complete colour cycle) represents a change in phase corresponding to about 28 mm of deformation in the line of sight. Acquisition dates: (a) November 1, 1992 – July 4, 1993, ascending orbit, perpendicular baseline (B⊥) 78 m; (b) August 30, 1992–March 28, 1993, descending orbit, B⊥ = 101 m; (c) December 13, 1992–March 28, 1993, descending orbit, B⊥ = 92 m.

descending orbit ones, as shown in the last column of Table I. Four descending interferograms were not generated because the data were not available; however, the two interferograms involving the acquisition of February 21, 1993 are likely to be decorrelated because of the snow coverage usually present atop the volcano at that time of the year, thus reducing the possibly useful missing interferograms to two. For brevity, only a selection of three (most representative) interferograms is presented in Figure 5; these are also highlighted in Table 1. In particular, Figure 5a shows an interferogram acquired from ascending orbits and spanning the period from November 1, 1992 to July 4, 1993: about three concentric deformation fringes are clearly visible, corresponding to about 8 centimeters of deformation in the satellite line of sight. The first two fringes cover most of the volcano edifice, while the third fringe is limited to the summit craters area. Interferograms obtained by descending orbits are presented in Figures 5b and 5c and exhibit higher decorrelation noise,

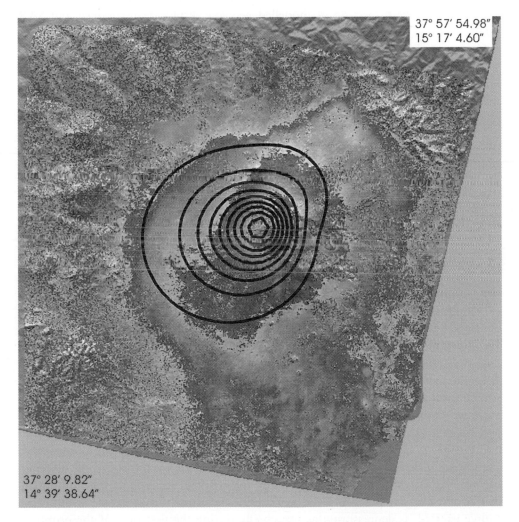

37° 57′ 54.98″
15° 17′ 4.60″

37° 28′ 9.82″
14° 39′ 38.64″

Figure 6

(a) modelled geodetic and recorded SAR ground deformation patterns. The isolines represent deformation pattern of the model obtained from geodetic measurements, recorded during the first half of the eruption, as expected along the direction of sight of the satellite. The first isoline is 28 mm, the isoline step is one interferometric fringe corresponding to 28 mm. The data of the model were recorded in a time interval comprising the first half of the eruption. The ascending SAR image (interval November 1, 1992–July 4, 1993) covers the second half of the eruption.

mostly located on the summit area. However, almost three fringes are still distinguishable in the interferogram in Figure 5b (from August 30, 1992 to March 28, 1993), thus confirming the average deformation rate during the considered eruptive period. In addition, the interferogram shown in Figure 5c presents less fringes, in accordance with its limited temporal span (from December 13, 1992 to March 28, 1993). These patterns, which show small deformation, are in concordance with the modest geodetic changes recorded during the second part of the eruption. In Figure 6 both the projection of modelled geodetic displacement pattern and SAR interferogram for the interval from November 1, 1992 to July 4, 1993 (ascending orbit) are shown. It is well evident that these two patterns, related to the first and second parts of the eruption respectively, are different and that the SAR interferogram constrains a deeper source. The latter aspect is also supported by the geodetic measurements pattern, showing small deformation on the entire volcano edifice which is compatible with the action of a deeper source.

4. Discussion

In the first half of the eruption SAR recording did not yet exist, while in the second half the small geodetic changes cannot be modelled in a robust inversion. Figure 6 depicts the two deformation patterns, respectively associated with the first half of the eruption (values expected from the modelling of the geodetic data) and with the second half (SAR interferogram). The two patterns appear to be caused by the action of two sources at different depths. The action of a shallower source (≈ 3 km b.s.l.) is sustained, in addition to the integrated modelling of all the geodetic changes in the first half of the eruption, also with other different evidence. In fact, for the interpretation of the GPS data a simple Mogi model with a depth of 1.5–3 km b.s.l. was proposed as a preliminary interpretation (NUNNARI and PUGLISI, 1994). Also the gravity changes accumulated before the eruption 1987–1989 suggested a source located in this depth range (BUDETTA and CARBONE, 1998). The geophysical evidence is also supported by petrological data which suggest that the depth of 3–5 km b.s.l. should represent the level of neutral buoyancy of magma (CORSARO and POMPILIO, 1998). Moreover, recent seismic analysis reveals presence of a possible chamber at a shallow depth (i.e., in the first km below sea level) through the mapping of the *b-values* (MURRU *et al.*, 1999) and through the analysis of the non-double couple mechanisms preceding the 1991–1993 eruption (SARAÒ *et al.*, 1999). This source exerted its depressuring action chiefly during the first part of the eruption. Therefore, the shallow chamber would represent one intermediate storage whose deflation caused a marked ground deformation. However, the recent mapping of the *b-values* (MURRU *et al.*, 1999) also highlighted a possible deeper chamber this supporting the hypothesis that at Mt. Etna a more complicated plumbing system could be present. The possible storage zones cannot have large dimensions since the recent seismic tomography studies (wavelengths of the resolved features ≈ 3 km) do not show large crustal melted magma chambers in the upper 18 km (CHIARABBA et al., 2000). A deeper source (top > 10 km) causes effects in a wider area however with more attenuated deformation (i.e., RUSSO et al., 1997). For example, the removal of 100×10^6 m^3 from a 10-km deep reservoir would produce a relative change of only 20 mm between the summit and points located about 4 km from the summit (DVORACK and DZURISIN, 1997). Subsequently for the second part of the eruption the DIFSAR technique reveals an attenuated but wider area of deformation which constrains a deeper source. Considering the pattern of the fringes, this source would be in agreement with a deeper anomaly characterised, beyond the shallower anomaly, from the mapping of the b-values (MURRU et al., 1999) and interpreted as a possible deeper storage zone.

5. Conclusion

In this paper we considered the shallow storage source (3–5 km b.s.l.) which acted strongly during the first half of the 1991 1993 Mt. Etna eruption and caused a marked deflation. This source's parameters are results constrained by the geodetic data modelling and are in accord with other geophysical observations. The shallower chamber would have induced maximum strain and stress in the upper western flank and this would coincide with the shallower seismic b-values anomaly and non-double couple mechanisms. However, the plumbing system of Mt. Etna may be more complex. SAR images were recorded during the second half of the eruption, allowing us to generate several interferograms of different time intervals. The analysis of these data highlights a weak but wider deflation signal which calls for a possible deeper source. This could be in accord with the seismic interpretations, which, through the deeper anomaly of the b value, reveal a possible zone of accumulation centered around 10 km b.s.l.. This source should be superimposed by the prevalent effects of the shallower intermediate storage during the first part of the eruption and would have acted as a deeper depressuring zone mainly in the second part. This means that the deeper source may have been active in the first part of the eruption but was masked by the larger effect of the shallower storage. This scenario would imply that a complex plumbing system exists at Etna, with different storage levels and giving rise to different deformation effects.

Acknowledgements

We are grateful to G. Puglisi for the SAR data which have been provided by the ESA project ERS AO3. 359 "Development of SAR Techniques aimed at managing Natural Disasters in geodynamically active areas," to E. Privitera and M. Murru for discussions on seismological investigations. We are indebted to R. Lanari for helpful suggestions during the review process. We thank P. Lundgren and an anonymous referee for their constructive reviews.

REFERENCES

BEAUDUCEL, F., BRIOLE, P., and FROGER, J. L. (2000), *Volcano-wide Fringes in ERS Synthetic Aperture Radar Interferograms of Etna (1992–1998): Deformation or Tropospheric Effect*, J. Geophys. Res. *105*, 16,391–16,402.

BONACCORSO, A. (1996), *A Dynamic Inversion for Modelling Volcanic Sources Through Ground Deformation Data (Etna 1991–1992)*, Geophys. Res. Lett. *23*, 5, 451–454.

BONACCORSO, A., CAMPISI, O., GAMBINO , S., FALZONE, G., LAUDANI, G., PUGLISI, B., ROSSI, M., VELARDITA, R., and VILLARI, L. (1994), *Ground Deformation Modelling of Geodynamic Activity Associated with the 1991-1993 Etna Eruption*, Acta Vulcanologia, *4*, 87–96.

BUDETTA, G. and CARBONE, D. (1998), *Temporal Variations in Gravity at Mt. Etna (Italy) Associated with 1989 and 1991 Eruptions*, Bull. Volcanol. *59*, 311–326.

CAYOL, V. and CORNETTE, F. H. (1998), *Effects of Topography on the Interpretation of the Deformation Field of Prominent Volcanoes — Application to Etna*, Geophys. Res. Lett. *23*, 1979–1982.

CHIARABBA, C., AMATO A., BOSCHI, E., and BARBERI, F. (2000), *Recent Seismicity and Tomographic Modeling of the Mount Etna Plumbing System*, J. Geophys. Res. 105/B5, 10923–10938.

CORSARO, R. and POMPILIO, M. (1998), *Il contrasto di densità tra magmi etnei e rocce incassanti: implicazioni sui meccanismi di risalita. In Progetto Etna: il vulcanismo dell' Etna in rapporto al suo contesto regionale*, Open File Report, IIV-CNR, Catania, 4/96, 27–32.

DAVIS, P. M. (1986), *Surface Deformation due to Inflation of an Arbitrarily Oriented Triaxial Ellipsoidal Cavity in an Elastic Half-space, with Reference to Kilauea Volcano*, J. Geophys. Res. *91*, 7429–7438.

DELACOURT, C., BRIOLE, P., and ACHACHE, J. (1998), *Tropospheric Corrections of SAR Interferograms with Strong topography. Application to Etna*, Geophys. Res. Lett. *15*, 2849–2852.

DVORACK, J. and DZURISIN, D. (1997), *Volcano Geodesy: The Search for Magma Reservoirs and the Formation of Eruptive Vents*, Rev. Geophysics 35, 3, 343–384.

Gabriel, A. K., Goldstein, R. M., and Zebker, H. A. (1989), *Mapping Small Elevation Changes over Large Areas: Differential Interferometry*, J. Geophys. Res. 94, 9183–9191.

LANARI, R., LUNDGREN, P., and SANSOSTI, E. (1998), *Dynamic Deformation of Etna Volcano Observed by Satellite Radar Interferometry*, Geophys. Res. Lett. *25*, 10, 1541–1544.

MASSONET, D., BRIOLE, P., and ARNAUD, A. (1995), *Deflation of Mount Etna Monitored by Spaceborne Radar Interferometry*, Nature *375*, 567–570.

MURRAY, J. B. and MOSS, J. (1994), *Results of Levelling at Mount Etna: Nov. 21 – Dec. 8, 1993*, 6 pp., Report for Osservatorio Vesuviano, Naples.

MURRU, M., MONTUORI, C., WYSS, M., and PRIVITERA, E. (1999), *The Locations of Magma Chambers at Mt. Etna, Italy, Mapped by b-values*, Geophys. Res. Lett. *26*, 16, 2553–2556.

NUNNARI, G. and PUGLISI, G. (1994), *The Global Positioning System as Useful Technique for Measuring Ground Deformations in Volcanic Areas*, J. Volcanol. Geoth. Res. *61*, 267–280.

RUSSO, G., GIBERTI, G., and SARTORIS, G. (1997), *Numerical Modeling of Surface Deformation and Mechanical Stability of Vesuvius Volcano, Italy*, J. Geophys. Res. *102*, 24,785–24,800.

SARAÒ, A., PANZA, G. F., PRIVITERA, E., and COCINA, O. (2001), *Non-double Mechanisms in the Seismicity Preceding the 1991–1993 Etna Volcano Eruption*, Geophys. J. Int. *145*, 1–25.

WILLIAMS, C. A. and WADGE, G. (1998), *The Effects of Topography on Magma Chamber Deformation Models: Application to Mt. Etna and Radar Interferometry*, Geophys. Res. Lett. *25*, 1549–1552.

WILLIAMS, C. A. and WADGE, G. (2000), *An Accurate and Efficient Method for Including the Effects of Topography in Three-dimensional and Elastic Models of Ground Deformation with Applications to Radar Interferometry*, J. Geophys. Res. *105*, 8103–8120.

ZEBKER, H. A., ROSEN, P. A., GOLDSTEIN, R. M., GABRIEL, A., and WERNER, C. L. (1994), *On the Derivation of Coseismic Displacement Fields Using Differential Radar Interferometry: The Landers Earthquake*, J. Geophys. Res. *99*, 19,617–19,634.

(Received February 22, 2002, revised December 13, 2002, accepted April 16, 2003)

 To access this journal online:
http://www.birkhauser.ch

Pure appl. geophys. 161 (2004) 1359–1377
0033–4553/04/071359–19
DOI 10.1007/s00024-004-2509-2

© Birkhäuser Verlag, Basel, 2004

❙ Pure and Applied Geophysics

GPS Monitoring in the N-W Part of the Volcanic Island of Tenerife, Canaries, Spain: Strategy and Results

J. Fernández[1,*], F. J. González-Matesanz[2], J. F. Prieto[3],
G. Rodríguez-Velasco[1], A. Staller[3],
A. Alonso-Medina[4], and M. Charco[1]

Abstract — This paper describes design, observation methodology, results and interpretation of the GPS surveys conducted in the areas of the N-W of Tenerife where deformation was detected using InSAR. To avoid undesirable antenna positioning errors in the stations built using nails, we designed and used calibrated, fixed-length metal poles, allowing us to guarantee that the GPS antenna was stationed with a height repeatability of the order of 1 mm and of less than 3 millimeters on the horizontal plane. The results demonstrate that this system is ideal for field observation, especially to detect small displacements that might be masked by accidental errors in height measurements or centering when observed with a tripod. When observations were processed, we found that using different antenna models in the same session sometimes causes errors that can lead to rather inaccurate results. We also found that it is advisable to observe one or two stations in all the sessions. The results have reconfirmed the displacement in the Chío deformation zone for the period 1995–2000 and indicate a vertical rebound from 2000 to 2002. They also confirm that the subsidence detected by InSAR to the south of the Garachico village has continued since 2000, although the magnitude of the vertical deformation has increased from around 1 cm to more than 3 cm a year. Detected displacements could be due to groundwater level variation throughout the island. A first attempt of modelling has been made using a simple model. The results indicate that the observed deformation and the groundwater level variation are related in some way. The obtained results are very important because they might affect the design of the geodetic monitoring of volcanic reactivation on the island, which will only be actually useful if it is capable of distinguishing between displacements that might be linked to volcanic activity and those produced by other causes. Even though the study was limited to a given area of Tenerife, in the Canary Islands, some conclusions apply to, and are of general interest in similar geodynamic studies.

Key words: Tenerife, Canary Islands, geodetic volcano monitoring, GPS, InSAR.

[1] Instituto de Astronomía y Geodesia (CSIC-UCM). Facultad de Ciencias Matemáticas, Universidad Complutense de Madrid, Ciudad Universitaria, 28040-Madrid, Spain.
E-mail: jose_fernandez@mat.ucm.es, gema_rodriguez@mat.ucm.es, charco@mat.ucm.es
[2] Instituto Geográfico Nacional, Ministerio de Fomento, C/ General Ibáñez Íbero 3, 28003-Madrid, Spain. E-mail: fjgmatesanz@mfom.es
[3] Dpto. Ingeniería Topográfica y Cartografía, EUIT Topográfica, Universidad Politécnica de Madrid. E-mail: jprieto@euitto.upm.es, astaller@euitto.upm.es
[4] Departamento de Física Aplicada, Esc. Universitaria de Ing. Técnica Industrial, Universidad Politécnica de Madrid, C/ Ronda de Valencia 3, 28012-Madrid. E-mail: aurelia@fais.upm.es
* Corresponding author

1. Introduction

Although geodetic monitoring techniques have been widely used in areas of seismic or volcanic activity, the difficulty inherent to their discrete nature means that they must be deployed carefully to ensure the best possible detection or sensitivity of these points (see e.g., BALDI and UNGUENDOLI, 1987; JOHNSON and WYATT, 1994; SEGALL and MATTHEWS, 1997; YU et al., 2000). In many cases, a more global monitoring method, is required yet at the same time one that offers the highest level of sensitivity which enables detection of the phenomenon. Interferometry radar (InSAR) techniques have been shown to play an important role in seismic and volcanic monitoring because they cover large areas (100 × 100 km) and can be easily systematized in monitoring (see e.g., MASSONNET and FEIGL, 1998; BÜRGMANN et al., 2000; MASSONNET and SIGMUNDSON, 2000; HANSSEN, 2001). The limitations inherent to the GPS and InSAR techniques (mainly observations at discrete surface points in the case of GPS and existence of non-coherent areas and the fact that, at present, the three displacement components cannot be obtained in SAR interferometry) can be overcome by using them together or other techniques (e.g., PUGLISI and COLTELLI, 2001; RODRÍGUEZ-VELASCO et al., 2002; FERNÁNDEZ et al., 2003).

The Canary Islands are seven islands and several islets of volcanic origin located between parallels 27 ° and 30 ° latitude North and between the parallels 13 ° and 19 ° longitude West. Located less than 100 km from the African continent, the Canary Islands were formed on the continental rise and slope as independent structures (ARAÑA and ORTÍZ, 1991). Despite being on a passive margin, they form a volcanic archipelago with a long-standing history of volcanic activity that began more than 40 m.y. ago (ANCOCHEA et al., 1990). At least a dozen eruptions occurred on the islands of Lanzarote, Tenerife, and La Palma between 1500 and 1971. Tenerife, the biggest island of the Archipelago, and its eruptive system are dominated by the Teide (3718 m) (ABLAY and MARTÍ, 2000) and Las Cañadas Caldera (MARTÍ et al., 1994). At the northern border of this caldera the Teide strato-volcano has been formed over the last one hundred and fifty thousand years. One of the most important eruptions took place in 1706 (Arenas Negras volcano) and destroyed part of Garachico village. Las Cañadas Caldera and the Teide are the areas where most volcanic research, in particular geodetic measurements (e.g., SEVILLA and ROMERO, 1991; CAMACHO et al., 1991), has been performed without any significant displacement or gravity changes having been detected to date.

Interferometry Synthetic Aperture Radar (InSAR) has been used to monitor the island of Tenerife during the period 1992–2000 (CARRASCO et al., 2000; FERNÁNDEZ et al., 2002). This technique can be used to cover the whole island in a single interferogram and obtain fairly comprehensive information about possible deformation in areas with a high level of coherence. The analysis of the interferograms revealed two areas with deformation: subsidence (see e.g., FERNÁNDEZ et al., 2002), located to the south of Garachico (an area of approximately 15 km^2 with a maximum subsidence of 10 cm from 1993 to 2000) and the Pinar de Chío zone (an area of approximately 8 km^2 with a maximum subsidence of 3 cm from 1993 to 2000, with a faster displacement velocity since 1997, see CARRASCO *et al.* (2000)). The location of these areas is displayed in Figure 1. From now on, we will refer to the two areas as Deformation Zones (DZs) and DZ1 and DZ2, respectively. However, they were unable to obtain information about possible displacements throughout the island, because although the whole island fits in one SAR image, the level of coherence is not good over the entire surface. In fact, in some areas there is no coherence at all (FERNÁNDEZ *et al.*, 2002, 2003). Therefore another technique, GPS, had to be used to monitor these areas, and to validate the deformations detected. We defined a GPS network that covered the entire island, which was densified in the areas of

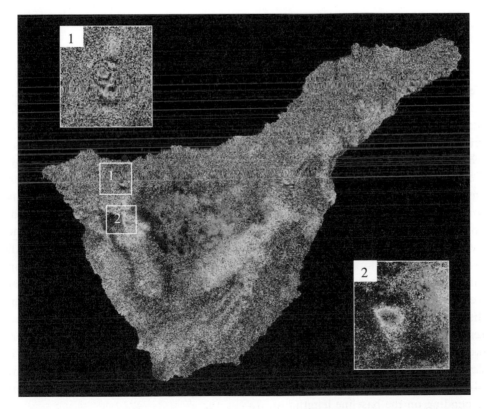

Figure 1

Differential interferogram from Tenerife island corresponding to 2 Aug. 1996–15 Sept. 2000. No fringe can be seen in the Las Cañadas area, so there is no deformation from 2 Aug. 1996–15 Sept. 2000 1) and 2) represent Garachico and Chío subsidence, respectively, from 20 Jul. 1993–15 Sept. 2000 differential interferogram ($B_\perp = 180$ m, $\Delta d = 2614$ days). The Garachico subsidence has 3 fringes, in other words, about 9 cm of ground subsidence from 20 Jul. 1993–15 Sept. 2000; the Chío subsidence has 1 fringe, that is to say, about 3 cm of ground subsidence from 20 Jul. 1993–15 Sept. 2000. ◆ indicates the location of Garachico village.

deformation detected with InSAR. This network was observed in 2000 and the obtained results (RODRÍGUEZ-VELASCO *et al.*, 2002; FERNÁNDEZ *et al.*, 2003) allowed us to validate the deformation found in Pinar de Chío, but not the one found to the south of the town of Garachico. Furthermore, certain antenna-height measuring problems were encountered in certain stations observed with tripods. To solve these two aspects wc designed a micro-GPS network covering both deformation areas located in the N-W part of the Volcanic Island of Tenerife and modified the observation methodology.

This paper briefly describes the design of the campaigns, the observation methodology used, the results of the campaigns conducted in 2000, 2001 and 2002, the conclusions obtained when these results were compared with the results of a previous campaign and from InSAR, as well as a first tentative interpretation.

2. *2000 GPS campaign*

Even though one of the characteristics associated with volcanic activity in Tenerife is that eruptions normally have not occurred more than once in the same volcanic structures, the geodetic monitoring conducted until 2000 was focused mainly on the Caldera de Las Cañadas, where a geodetic micronetwork and a levelling profile have been installed (SEVILLA and ROMERO, 1991; SEVILLA *et al.*, 1996). This is because the biggest volcanic risk on the island is associated with the Teide stratovolcano located in the Caldera. A sensitivity test of this geodetic network showed a clear need to extend the existing geodetic network to cover the full island for volcano monitoring purposes using GPS techniques (YU *et al.*, 2000). This conclusion, together with the detection of two unexpected movements on the island, by the processing of Synthetic Aperture Radar (SAR) data that were beyond the scope of the traditional geodetic network installed on the island due to coverage limitations, prompted us to design and observe a GPS network covering entirely Tenerife in 2000, mainly for the purposes of the geodetic monitoring of possible displacements associated with volcanic reactivation and corroborating the obtained results by InSAR. This GPS network has a station in DZ2 and a densification in DZ1 located to the south of the village of Garachico (Fig. 2). The network and its densification was observed in August 2000 using 5 receivers and relative static observations (HOFMANN-WELLENHOF *et al.*, 1992) during periods of 5 and 2 hours, depending on the baseline length.

The data collected for the global network and DZ1 densification during the 2000 survey were processed separately using Bernese 4.2 software (BEUTLER *et al.*, 2001) with precise ephemeredes. To fully guarantee that the results were included in the REGCAN-95 reference system (CATURLA, 1996), based on the ITRF93 date 94.9 (where we have previous values of coordinates for many of the stations), Chinobre, Teide and Bocinegro stations were considered fixed in the GPS observation process.

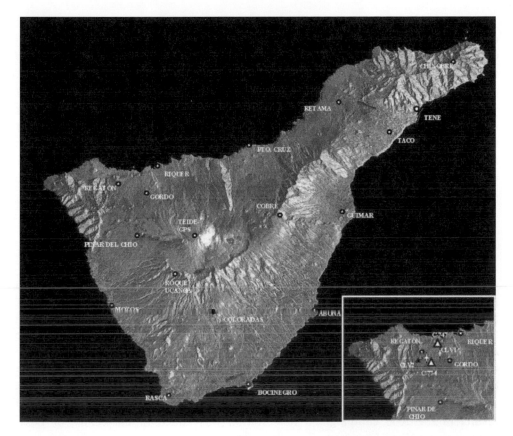

Figure 2
Global GPS network: 17 stations from REGCAN95 together with the permanent station TENE, marked with white circles. A densification network sketch appears on the bottom right. It is formed by 2 fourth-order stations (white triangles) and 2 benchmarks (white stars) (after FERNÁNDEZ et al., 2003).

Regatón station, near DZ1, was used as a fixed reference station in that zone's subnetwork. The precision of the obtained results is within one centimeter in height and several millimeters in horizontal coordinates. For more details see RODRÍGUEZ-VELASCO et al. (2002) and FERNÁNDEZ et al. (2003). Significant coordinate variations were obtained in several stations distributed throughout the island, showing a subsidence of station Pinar de Chío, of the same order as obtained using InSAR in DZ2. Therefore the two techniques have confirmed one another. Some coordinate variations were detected in areas where the band C radar observation cannot be used for deformation monitoring on account of the very low coherence, due to the existence of vegetation. This demonstrates that the two techniques complement one another, and that they should be combined in any monitoring methodology that is to be efficient in detecting possible deformations associated with volcanic reactivation on the island.

The obtained results in the DZ1, to the south of the town of Garachico, were not definitive enough to confirm the displacements detected using InSAR (see RODRÍ-GUEZ-VELASCO *et al.*, 2002; FERNÁNDEZ *et al.*, 2003). Future observations were considered necessary and more stations would have to be included in this DZ1, preferably in the center where the largest deformation occurs and is therefore easier to detect. Furthermore, it must be taken into account that several stations in the deformation zone were observed with a tripod, and therefore the antenna height measurements and centering might contain inaccuracies that should be eliminated as far as possible, considering the magnitude of the vertical displacement (of the order of one cm, as shown by InSAR results for the 1993–2000 period) that is to be detected.

3. 2001 GPS Campaign

In an attempt to solve the aforementioned problems, we designed and observed a micro-GPS network limited to the DZs. This micro-network is formed by four vertices of the REGCAN95, run by the National Geographic Institute of Spain (IGN) (CATURLA, 1996), two benchmarks of the fourth-order network and three nails located in the deformation zone, Figure 3a. Geodetic vertices of the REGCAN95 network Riquer, Regatón and Roque de Ucanca were selected as reference stations because they were regarded as located outside the DZs. When this micro-network was being designed, and taking into consideration the obtained results in the 2000 campaign, a new station was added in the center of DZ1, CLAVO3 (CLV3 in Fig. 3a).

For the nails we designed fixed-length poles which were perfectly numbered and calibrated, and secured them with steel cable by means of a turnbuckle and a galvanized flat (Fig. 4). The aim was to avoid having to measure the antenna height in this and future campaigns, thus eliminating a frequent source of errors. Using these poles let us assure a height repeatability of the order of 1 mm. Each pole is formed by a stainless steel pipe 1.125 meters high and 3 cm in diameter. The accidental errors in component x and y caused by levelling the pole remain below 3 millimeters due to the use of a 10' level and to the fact that the pole is fastened with steel guys (see Fig. 4).

The GPS observation of the micro-network conducted in 2001 took account of the following aspects: minimum observation period of 4 hours in morning and afternoon sessions for baselines between 0.8 and 20 km; use of 4 simultaneous receivers; CLAVO 3 station, located in the center of the deformation zone, was observed in all the sessions (except in the first, due to an antenna problem). The observation sessions are summarised in Table 1.

In order to facilitate the comparison of results, the data collected were processed as 2000 data, using Bernese 4.2 software. To guarantee that our results were included

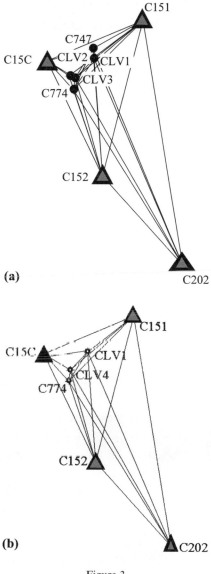

Figure 3
(a) 2001 and (b) 2002 GPS micro-networks.

in the REGCAN-95 reference system, in the GPS observation process we established the Regatón station (C15C), near DZ1, as a fixed station to the coordinates determined in the 2000 GPS campaign (FERNÁNDEZ *et al.*, 2003).

The data analysis demonstrated that a set of GPS vectors were obtained at the C151 Riquer station grouped in the three sessions in which the GPS antenna was located at that station throughout the campaign. The values of the altitudes assigned

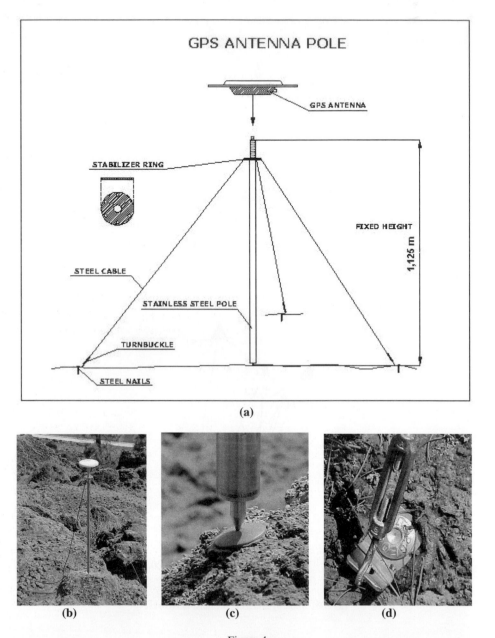

(a)

(b)   **(c)**   **(d)**

Figure 4
(a) Scheme of the steel pole for observation in benchmarks, (b) steel pole at benchmark 774, (c) pole end, and (d) hook and eye turnbuckle.

to Riquer by the vectors of the first two sessions were grouped around values that only differ by a few mm, but differing from the value of the altitude assigned by the third session by around eight centimeters. Different tests were conducted to study its

Table 1

(a) Surveys carried out in 2001 and 2002 of the geodetic GPS network in Tenerife; (b) distribución de stations observed at the different sessions

(a)

Year	From	To	Sessions	Duration	Receivers	Stations	Baselines	Parameters	Observations
2001	26 Jul.	30 Jul.	5	4 or 8 hours	4	9	19	366	8808
2002	25 Jul.	29 Jul.	5	6 hours	5	7	20	394	13932

(b)

	OBSERVATION OF STATIONS									
STATION	2001 SESSIONS					2002 SESSIONS				
	1	2	3	4	5	1	2	3	4	5
C202	X	X			X	X		X	X	
C774	X	X		X	X	X	X		X	
C152	X	X		X	X	X		X		X
C747		X	X		X					
C151		X	X		X		X		X	X
CLV3		X	X	X	X					
C15C			X	X		X	X	X	X	X
CLV2			X	X						
CLV1					X		X	X		X
CLV4						X	X	X	X	X

effect: in (a), we processed the data considering the Riquer observations in the three sessions in which the station was used; in (b) we only considered the first two sessions in Riquer, observed with the same antenna; in (c) we only considered the last day and in (d) we used none of the observations in this station. The obtained results in the four cases using Bernese 4.2 were similar (in terms of level of precision) for all the points except Riquer. When these results are compared with the rest of the GPS network in a constraint-free adjustment (LEICK, 1990), and conducting the pertinent tests to detect gross errors, the observations made in the first two sessions are rejected, although it is significant to note that both are grouped around values that only differ by a few millimeters, as mentioned above. The network adjustment indicates that the most coherent solution for station C151 is the one provided by the third session, case (c), yet the fact that the rejected sessions are grouped independently and separately inclines one to think that this station's results should not be considered definitive until a redundancy of results higher than the one found in the 2001 campaign is obtained for this station.

After the 2001 campaign we concluded that further campaigns should be conducted with the same methodology, although using the same model of geodetic antennae in all the stations and sessions to avoid causing any noise in the level of detection required in this work, close to the precision that the technique can provide without permanent observation.

4. 2002 GPS Campaign

The 2002 campaign was based on the main conclusions drawn from the results of the 2001 campaign. However, during the inspection prior to the observations, we found that two benchmarks, CLV2 and CLV3, had been destroyed. In an attempt to solve the problem caused by the disappearance of station CLV3, located in the middle of DZ1, and considering future campaigns, we installed another nail nearby at a location that was named CLAVO4, CLV4.

In the observation we used the same geodetic antenna model for all stations and sessions. Station C15C, considered fixed in the data processing, and the new station CLV4 were observed in all the sessions. Figure 3b is a diagram of the micro-network. The observation characteristics are also given in Table 1. The data processing was performed in exactly the way as in the 2000 and 2001 campaigns. The obtained results are given in Table 3.

5. Discussion of Results and Interpretation

The results of the 2001 and 2002 surveys can be compared with those obtained in 1995 and 2000 (see FERNÁNDEZ *et al.*, 2003) at the coinciding points. The differences are shown graphically in Figure 5. The comparison of the 2001 coordinates recorded at the Pinar de Chío station, C152, to the 1995 coordinates

Table 2

Coordinates of the DZs micro-network stations observed during the 2001 campaign, determined in the REGCAN95 system. Precision in meters. The Regatón station, whose coordinates have not been assigned a level of precision, has been considered fixed. See text for more details

STATION		LATITUDE	LONGITUDE	ELLIPSOIDAL HEIGHT
NAME	CODE	(° ' ")		(m)
RIQUER	C151	28 22 48.04568 ± 0.001	−16 43 02.53895 ± 0.001	209.999 ± 0.006
PINAR DE CHIO	C152	28 16 08.36172 ± 0.001	−16 44 58.32698 ± 0.001	1641.354 ± 0.003
ROQUE DE UCANCA	C202	28 12 19.98654 ± 0.001	−16 41 00.29192 ± 0.001	2155.187 ± 0.004
REGATÓN	C15C	28 21 02.65978	−16 47 39.10352	844.727
747	C747	28 21 34.86858 ± 0.001	−16 45 27.61258 ± 0.001	652.729 ± 0.004
774	C774	28 19 48.21827 ± 0.001	−16 46 21.66942 ± 0.001	1129.869 ± 0.004
CLAVO 1	CLV1	28 21 08.98676 ± 0.001	−16 45 23.28204 ± 0.001	766.721 ± 0.004
CLAVO 2	CLV2	28 20 24.72778 ± 0.001	−16 46 35.23842 ± 0.001	954.600 ± 0.004
CLAVO 3	CLV3	28 20 18.64512 ± 0.001	−16 46 17.08318 ± 0.001	990.585 ± 0.003

Table 3

Coordinates of the DZs micro-network stations observed during the 2002 campaign, determined in the REGCAN95 system. Precision in meters. The Regatón station, whose coordinates have not been assigned a level of precision, has been considered fixed. See text for more details

STATION		LATITUDE	LONGITUDE	ELLIPSOIDAL HEIGHT
NAME	CODE	(° ' ")		(m)
RIQUER	C151	28 22 48.04542 ± 0.001	−16 43 02.53866 ± 0.001	209.975 ± 0.003
PINAR DE CHIO	C152	28 16 08.36122 ± 0.001	−16 44 58.32737 ± 0.001	1641.383 ± 0.003
ROQUE DE UCANCA	C202	28 12 19.98621 + 0.001	−16 41 00.29171 ± 0.001	2155.185 ± 0.003
REGATÓN	C15C	28 21 02.65978	−16 47 39.10352	844.727
774	C774	28 19 48.21778 ± 0.001	−16 46 21.66912 ± 0.001	1129.812 + 0.003
CLAVO 1	CLV1	28 21 08.98671 ± 0.001	−16 45 23.28239 ± 0.001	766.699 ± 0.003
CLAVO 4	CLV4	28 20 18.56635 ± 0.001	−16 46 15.20997 ± 0.001	996.058 ± 0.003

confirmed the subsidence obtained results by InSAR. The difference is larger, albeit only slightly, than 2σ. We also obtained horizontal displacements of around one cm in the N-W direction. The vertical deformation obtained results upon comparing the values of 2000 and 2001 are not significant, as was to be expected, taking into account the InSAR results (approximately 3 cm of deformation in eight years). In this second case of comparison, the horizontal deformation results are significant again in the N-W direction. However the real magnitude of the horizontal displacement might be smaller due to the possibility of errors in the antenna centering, because in 2000 this station was observed without a tribrach due to availability problems on the observation day. Comparing the 2002 and 1995 coordinates shows the height values to be quite identical, with differences below precision level. Horizontal displacement is still N-W, increasing in magnitude to the N and decreasing to the W. If one compares 2002 with 2000, the horizontal displacement is still N-W, but compared to the results for 2001 the displacement is N-E. In short, vertical displacement at the Pinar de Chio station, see Figure 5, indicates clear subsidence from 1995 to 2000 (coinciding with InSAR results) and elevation from 2000 to 2002.

At the Roque de Ucanca station, when the 1995 coordinates were compared with the 2001 coordinates, the vertical and horizontal differences were around 2σ, and could not be considered clearly significant. The vertical differences found upon comparing 2000 and 2001 are not significant either, although the horizontal ones are. Comparing the 2002 height with the values for 1995, 2000 and 2001 gives similar results, a tendency showing subsidence, though not significant over the noise level. The horizontal components display a N-W displacement, as in Pinar de Chío. Therefore new campaigns must be conducted in order to look for possible displacement at this station, which is next to C152 and could also be starting to display displacements. Therefore this point must be studied very carefully in future campaigns.

In the fourth order pin C774, the vertical coordinate variations in the comparison of 2000, 2001 and 2002 with respect to 1995 were not significant because the 1995 coordinates were not very accurate. Comparison between 2000 and 2001 results are not significant either, but comparing 2002 with 2000 and 2001 values produces clearly significant subsidence. Comparing the 2000 coordinates with 2001 and 2002 reveals horizontal displacement in the S-W direction, although one must remember that the 2000 observation might contain a centering error. Comparing the 2002 and 2001 reveals N-W displacement.

At pin 747 the vertical coordinate differences include an uplifting of 5.7 cm in the 95–00 period, an uplifting of 1.8 cm in the period 95—01, and a subsidence of 3.9 cm in the period 00-01. Comparing the 1995 and 2000 coordinates reveals E-W displacement that is also observed in the period 2000-2001. The station is located on the roof of a small water tank, the height of which more than probably varies in line with the level of water in the tank. In fact, at first it was thought that the tank was no longer in use, and it was chosen due to its proximity to the DZ and because it offered previous coordinates with which to make comparisons. The obtained results, together with the inspections carried out subsequently (FERNÁNDEZ *et al.*, 2003) have shown that the tank is still being used and that the level of water inside is far from constant. Therefore, and also taking into account that when the station is accurately located, it is clearly outside the DZ, we can conclude that the choice of the point was unfortunate and the variation of its coordinates is far from significant. This benchmark was removed from the network in the 2002 observations.

At stations Clavo1 and Clavo2, CLV1 and CLV2, respectively, if we compare the 2001 and 2000 coordinates, the vertical coordinate differences are not significant as they are less than 2σ or very close, and we must consider the more than probable existence of accidental errors in the antenna height-measuring in the 2000 observation with tripods. Due to its disappearance, CLV2 was not observed in 2002. If we compare the 2002 coordinates of CLV1 with the 2000 and 2001 values, see Figure 5, we observe a clear and significant subsidence, and if we do not consider horizontal coordinates from 2000, which might contain centring errors due to the use of tripods, there is also clear horizontal displacement in the E direction.

At station Riquer, C151, without considering the results for 2001, due to the aforementioned problem, a clear subsidence and a N–W displacement is observed, see Table 2.

Apart from the need to check certain stations, one question that remains is to determine the possible causes of the subsidence detected. These causes could differ widely from one area to another: compacting of the lava emitted during the last eruptions, subduction post eruptions caused by the emptiness of the magma chambers, volcanic reactivation, this subsidence could be caused by a dyke (YU *et al.*, 2000), or the extraction of water from water tables. It must be stressed that numerous tunnels have been drilled in the subsoil of most of the island to draw drinking water for the island's inhabitants (GOBIERNO DE CANARIAS and CABILDO INSULAR DE

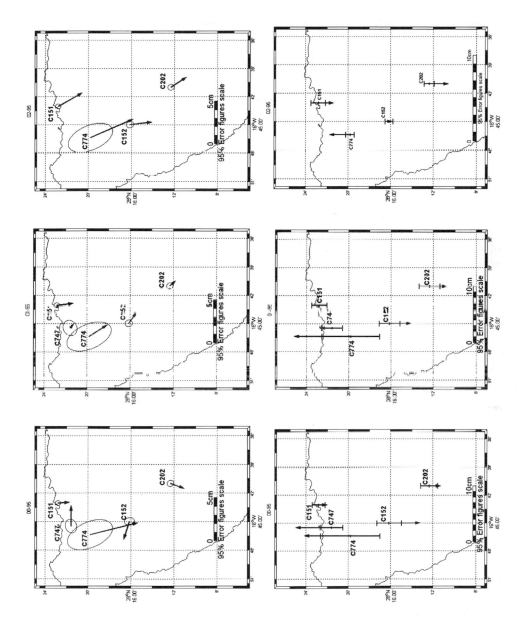

Figure 5

Horizontal and vertical displacement with errors. Vectors with error ellipses for horizontal and with error bars for vertical motions. Differences, given in cm, are computed by comparison of coordinates measured in 1995, 2000, 2001 and 2002 for the repeated stations of the DZs micro-network. For example, 00–95 indicates differences between 2000 and 1995 coordinates.

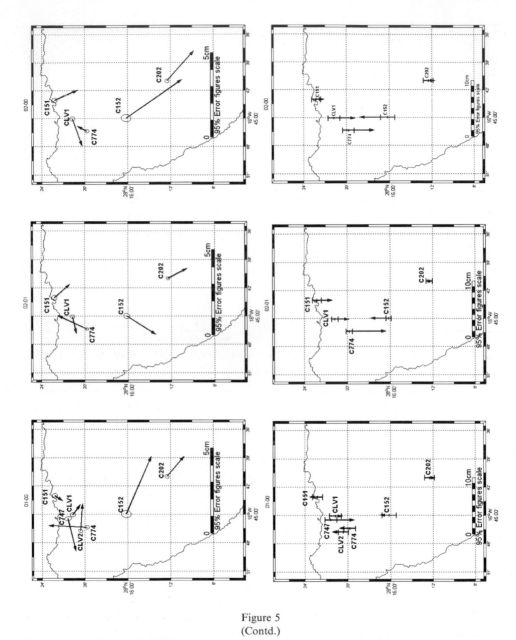

Figure 5
(Contd.)

TENERIFE, 1989). At present it is difficult to distinguish the real cause from the others with certainty. Different geodetic and geophysical experiments are being prepared to determine the origin and evolution of the subsidence (FERNÁNDEZ *et al.*, 2002).

The deformations found are located in areas where large amounts of groundwater have been drawn since the end of the 19th century (Government of Canary Islands and CABILDO INSULAR DE TENERIFE, 1989). Furthermore, since 2000, there has been a clear increase in rainfall on the island (I. Farrujia, personal communication, 2002), as a result of which groundwater levels have risen in some areas of the island, or are dropping more slowly.

This leads us to think that the groundwater decrease/recharge may be at least partly the cause of the displacements observed, as has been seen elsewhere (KAUFMANN and AMELUNG, 2000; HOFFMANN et al., 2001; LE MOUÉLIC et al., 2002; WATSON et al., 2002). In order to test this hypothesis, we will endeavor to interpret the displacements using the model developed by GEERTSMA (1973) and used in similar problems by XU et al. (2001) and LE MOUÉLIC et al. (2002). We will only consider the vertical component of the displacement in our study. A future, more complete interpretation should also consider the horizontal components. The model assumes that the reservoir is a circular disk of radius R and height h buried parallel to the flat earth surface at depth D. The elastic earth is treated as an elastic half-space with Poisson's ratio δ. Then the vertical component of the displacement of the surface u is

$$ u - (2\delta - 2)\Delta h R \int_0^\infty e^{-D\alpha} J_1(\alpha R) J_0(\alpha r) d\alpha = (2\delta - 2)\Delta h A \tag{1} $$

where r is the radial coordinate along the surface with origin above the center of the reservoir; Δh is the height change of the reservoir due to its recharge or compaction, and α is the integration variable.

For the elastic parameters we could consider those determined by FERNÁNDEZ et al. (1999). They give as properties for the 3.5 km of the crust in their cortical model (from 2 km over sea level to 1.5 km below it) a density of 2100 kg m^{-3}, and Lamé parameters $\lambda = 8 \cdot 10^9$ Pa and $\mu = 7 \cdot 10^9$ Pa. Therefore the Poisson's ratio is $\delta = 0.27$. Furthermore, the information available (CONSEJO INSULAR DE AGUAS DE TENERIFE, 2002) can be used to obtain an approximation of the depth and radius values of the underground aquifers. We will annotate the values of DZ1 with subindex 1 and the value of DZ2 with subindex 2. We derive $D_1 = 0.4$ km, $R_1 = 2$ km, $D_2 = 1.2$ km and $R_2 = 3$ km. Using the values tabulated by GEERTSMA (1973) of parameter A, see equation (1), as a function of $\rho = r/R$ and $\eta = D/R$, we obtain for $\rho_1 = \rho_2 = 0$, $\eta_1 = 0.2$ and $\eta_2 = 0.4$, the values $A_1 = 0.8039$ and $A_2 = 0.6286$, which are the maximum vertical deformation u_{max} in $r = 0$ km in DZ1 and DZ2, respectively. Finding the value of (1), we obtain the expression of the water table variation as a function of the deformation in $r = 0$, u_{max}, δ and A:

$$ \Delta h = \frac{u_{max}}{(2\delta - 2)A}. \tag{2} $$

Replacing the maximum subsidence values in the period 1993–2000 in DZ1 and DZ2, which are around −10 and −4 cm, respectively (values obtained from the InSAR and GPS results, averaging values in the case of DZ2), we obtain the respective water table level variation values for that period $\Delta h_1 = -8.5$ cm and $\Delta h_2 = -4.4$ cm. However, if we consider the period 2000–2002, the deformation values in DZ1 and DZ2 would now be around -6 and $+4.4$ cm, respectively, producing water table variations $\Delta h_1 = -5$ cm and $\Delta h_2 = 4.8$ cm.

6. Conclusions

This paper describes the design, observation methodology used and the results of the GPS observation campaign conducted in 2001 in the areas of the N-W of Tenerife, where deformation was detected using InSAR (CARRASCO *et al.*, 2000; FERNÁNDEZ *et al.*, 2002) and discusses possible interpretations.

To avoid undesirable antenna stationing errors in the stations built with nails, we designed and used perfectly numbered and calibrated, fixed-length metal poles, secured with steel cable by means of a turnbuckle and a galvanized flat. This guaranteed that the GPS antenna could be stationed on the control nail with a height repeatability of the order of 1 mm and below 3 mm on the horizontal plane. The obtained results guarantee that this system is ideal for field observation, and considerably better than using tripods, especially in areas where, as in Tenerife, one wants to detect small displacements that otherwise might be masked by accidental errors in measuring heights or centering, as was inferred in the 2000 campaign (FERNÁNDEZ *et al.*, 2003).

When the 2001 observations were processed, we found that using different antenna models in the same session can cause errors that can lead to inaccurate results. Therefore the same antenna model was used at all stations in the 2002 campaign. We also observed that, in order to make the data processing and global network adjustment more reliable, one or two stations should be observed in all the sessions. Therefore, it was decided that the fixed station, Regatón, and central station in DZ1, where the highest deformation value is expected, will always be observed in the next campaigns. This was the procedure followed in the 2002 campaign.

The results have reconfirmed the displacement in the Chío deformation zone for the period 1995–2000 and indicate that vertical coordinate increased again from 2000 to 2002. They have drawn attention to the existence of possible displacements at station C202, next to this zone. The results also confirm that the subsidence detected by InSAR to the south of the Garachico have continued since 2000, and also indicate that the magnitude of the vertical deformation to be expected in the one-year interval has increased from around one cm a year to more than 3 cm a year. These values indicate the need to continue monitoring the displacements.

One of the possible reasons for the detected displacements could be the groundwater level variation. A first attempt has been made to model this with a simple model (GEERTSMA, 1973) that can only be regarded as an approximation to the real problem on the Island of Tenerife, due to both to the type of reservoir and the medium considered (homogeneous and without fractures). However, the results do not contradict the groundwater level decrease values (GOBIERNO DE CANARIAS and CABILDO INSULAR DE TENERIFE, 1989) and, in Pinar de Chío are coherent with the increase in rainfall in the recent years (I. FARRUJIA, personal communication, 2002). These results seem to indicate that the observed deformation and the groundwater level variation are somehow related. Also, in the case of Pinar de Chío, the deformation measured could also be affected by other factors such as the system of fractures existing in the zone (I. FARRUJIA and J. COELLO, personal communication, 2002). Consequently, there is clearly a need to continue with the described research, with further monitoring of the displacements in order to study their evolution; further interpretation using more and different data (gravity, gases, etc.); real, detailed and up-to-date information about the evolution of the ground waters; and the development of models that represent the island's characteristics more accurately. It should also be noted that this research is also very important, not only for studying the displacements that might be associated, at least partly, to the extraction of groundwater, but must be capable of clearly distinguishing them from other displacements on the island's surface that might be linked to a future magmatic reactivation. Once effects of water level variations have been removed or have been shown to be irrelevant, then the continuous monitoring of ground deformation becomes a useful tool for volcanic activity monitoring. When combined with seismic, hydrologic and fumarolic activity, it provides the means to evaluate the physical processes at work, at greater depth.

Acknowledgements

This research has been funded through the Ministry of Science and Technology project AMB99-1015-C02, ESA-ESRIN Contract No.13661/99/I-DC and the MEDSAR contract with INDRA Espacio SA. We thank the Consejo Insular de Aguas de Tenerife for information regarding groundwater extraction and rainfall. We thank S. Le Mouélic, K.F. Tiampo, I. Farrujia and J. Coello for their useful comments. We are thankful for the review by A. Donnellan, J. Fonseca and F. Cornet which improved this paper.

REFERENCES

ABLAY, G. J. and MARTÍ J, (2000), *Structure, Stratigraphy, and Evolution of the Pico Teide-Pico Viejo Formation, Tenerife, Canary Islands*, J. Volcanol. Geotherm. Res. *103*, 175–208.

ANCOCHEA, E., FÚSTER J. M., IBARROLA E., CENDRERO A., COELLO J., HERNÁN F., CANTAGREL J. M., and JAMOND C, (1990), *Volcanic Evolution of the Island of Tenerife (Canary Islands) in the Light of New K-Ar Data*, J. Volcanol. Geotherm. Res. *44*, 231–249.

ARAÑA, V. and ORTIZ R, *The Canary Islands: Tectonics, magmatism and geodynamic framework*, In *Magmatism in Extensional Structural Settings. The Phanerozoic African Plate* (A. B. Kampunzu, and R. T. Lubala, eds.) (Springer-Verlag, Berlin 1991) pp. 209–249.

BALDI, P. and UNGUENDOLI, M., *Geodetic networks for crustal movements studies.* In (Stuart Turner, (ed.), Lecture Notes in Earth Sciences, 12. Applied Geodesy (Springer-Verlag 1987) pp. 135–161.

BEUTLER, G., BOCK, H., BROCKMANN, E., DACH, R., FRIDEZ, P., GURTNER, W., HUGENTOBLER, U., INEICHEN, D., JOHNSON, J., MEINDL, M., MERVANT, L., ROTHACHER, M., SCHAER, S., SPRINGER, T., and WEBER, R. (2001) *Bernese GPS Software Version 4.2.* (eds. U. Hugentobler, S. Schaer, and P. Fridez.) Astronomical Institute, University of Berne, 515 pp.

BÜRGMANN, R., ROSEN, P. A., and FIELDING, E. J. (2000), *Synthetic Aperture Radar Interferometry to measure Earth's Surface Topography and its Deformation*, Ann. Rev. Eath Planet. Sci. *28*, 169–209.

CAMACHO, A. G., VIEIRA, R., and TORO, C. (1991), *Microgravimetric Model of the Las Cañadas Caldera (Tenerife)*, J. Volcanol. Geotherm. Res. *47*, 75–88.

CARRASCO, D., FERNÁNDEZ, J., ROMERO, R., ARAÑA, V., MARTÍNEZ, A., MORENO, V., APARICIO, A., and PAGANINI, M. (2000), *First Results from Operational Volcano Monitoring in the Canary Islands*, ESA, SP-461, ERS-ENVISAT Symposium, Gothenburg, Sweden 16–20/10/2000. CD-ROM.

CATURLA, J. L. (1996), *REGCAN95, Nueva Red Geodésica de las Islas Canarias.*, Instituto Geográfico Nacional, Area de Geodesia. Internal Report.

CONSEJO INSULAR DE AGUAS DE TENERIFE (2002), http://www.aguastenerife.org/sup.html.

FERNÁNDEZ, J., CARRASCO, J. M., RUNDLE, J. B., and ARAÑA, V. (1999), *Geodetic Methods for Detecting Volcanic Unrest: A Theoretical Approach*, Bull. Volcanol. *60*, 534–544.

FERNÁNDEZ, J., ROMERO, R., CARRASCO, D., LUZÓN, F., ARAÑA, V. (2002), *InSAR Volcano and Seismic Monitoring in Spain. Results for the Period 1992–2000 and Possible Interpretations*, Optics and Lasers in Engin. *37*, 285–297.

FERNÁNDEZ, J., YU, T.-T., RODRÍGUEZ-VELASCO, G., GONZALEZ-MATESANZ, F. J., ROMERO, R., RODRÍGUEZ, G., QUIRÓS, R., DALDA, A., APARICIO, A., and BLANCO, M. J. (2003), *New Geodetic Monitoring System in the Volcanic Island of Tenerife, Canaries, Spain. Combination of InSAR and GPS Techniques*, J. Volcanol. Geotherm. Res. *124*, 241–253.

GEERTSMA, J. (1973), *Land Subsidence above Compacting Oil and Gas Reservoirs.*, J. Pet. Tech. 734–744.

GOBIERNO DE CANARIAS, CABILDO INSULAR DE TENERIFE (1989), *Plan hidrológico Insular de Tenerife.* Avance: bases para el planeamiento hidrogeológico. Ed. Cab. Insul. de Tenerife, 133 pp.

HANSSEN, R. F., *Radar Interferometry: Data Interpretation and Error Analysis* (Kluwer Academic Publishers, The Netherlands, 2001) 308 pp.

HOFFMANN, J., ZEBKER, H. A., GALLOWAY, D. L., and AMELUNG, F. (2001), *Seasonal Subsidence and Rebound in Las Vegas Valley, Nevada, Observed by Synthetic Aperture Radar Interferometry*, Water Resources Res. *37*, 1551–1566.

HOFMANN-WELLENHOF, B., LICHTENEGER, H., and COLLINS, S. J., *GPS Theory and Practice*, 2nd edi. (Springer Verlag Wien New York, 1992) 326 pp.

JOHNSON, H. O. and WYATT, F. K. (1994), *Geodetic Network Design for Fault-mechanics Studies*, Manuscripta geodaetica *19*, 309–323.

KAUFMANN, G. and AMELUNG, F. (2000), *Reservoir-induced Deformation and Continental Rheology in Vicinity of Lake Mead, Nevada*, J. Geophys. Res. 16,341–16,358.

LE MOUÉLIC, S., RAUCOULES, D., CARNEC, C., and KING, C. (2002), *A Ground Uplift in the City of Paris (France) Detected by Satellite Radar Interferometry*, Geophys. Res. Lett. *29*, 1853, doi:10.1029/2002GL015630.

LEICK, A., *GPS Satellite Surveying.* (Wiley, New York Chichester Brisbane Toronto Singapore 1990).

MARTÍ, J., MITJAVILA, J., and ARAÑA, V. (1994), *Stratigraphy, Structure and Geochronology of the Las Cañadas Caldera (Tenerife, Canary Islands)*, Geol. Mag. *131*, 715–727.

MASSONET, D. and FEIGL K. L. (1998), *Radar Interferometry and its Application to Changes in the Earth's Surface*, Rev. Geophys. *36*, 441–500.

MASSONNET, D. and SIGMUNDSSON, F. (2000), *Remote sensing of volcano deformation by radar interferometry from various satellites.* In *Remote Sensing of Active Volcanism*, Geophysical Monograph *116*, AGU, pp. 207–221.

PUGLISI, G. and COLTELLI M, (2001), *SAR Interferometry Applications on Active Volcanism: State of the Art and Perspectives for Volcano Monitoring*, Il Nuovo Cimento *24*, 133–145.

RODRÍGUEZ-VELASCO, G., ROMERO, R., YU, T.-T., GONZÁLEZ-MATESANZ, J., QUIRÓS, R., DALDA, A., CARRASCO D., and FERNÁNDEZ, J. (2002), *Introducción de técnicas espaciales a los sistemas de vigilancia geodésica en Tenerife (Islas Canarias). Asamblea Hispano-Portuguesa de Geodesia y Geofísica. Valencia (España) 4–8 Febrero 2002. Proceedings Tomo II*, Editorial UPV, pp. 746–750.

SEGALL, P. and MATTHEWS, M. (1997), *Time-dependent Inversion of Geodetic Data*, J. Geophys. Res. *102*, 22,931–22,409.

SEVILLA, M. J. and ROMERO, P. (1991), *Ground Deformation Control by Statical Analysis of a Geodetic Network in the Caldera of Teide*, J. Volcanol. Geotherm. Res. *47*, 65–74.

SEVILLA, M. J., VALBUENA, J. L., RODRÍGUEZ-DÍAZ, G. and VARA, M. D. (1996), *Trabajos altimétricos en la Caldera del Teide*, Física de la Tierra *8*, 117–130.

WATSON, K. M., BOCK, Y., and SANDWELL, D. (2002), *Satellite Interferometric Observation of Displacements Associated with Seasonal Groundwater in the Los Angeles Basin*, J. Geophys. Res. *107*, B4, 10.1029/2001JB000470.

XU, H., DVORKIN, J., and NUR, A. (2001), *Linking Oil Production to the Surface Subsidence from Satellite Radar Interferometry*, Geophys. Res. Lett. *28*, 1307–1310.

YU T.-T., FERNÁNDEZ, J., TSENG, C.-L., SEVILLA, M. J., and ARAÑA V. (2000), *Sensitivity Test of the Geodetic Network in Las Cañadas Caldera, Tenerife, for Volcano Monitoring*, J. Volcanology and Geothermal Res. *103*, 393–407.

(Received March 30, 2002, revised March 24, 2003, accepted April 28, 2003)

 To access this journal online:
http://www.birkhauser.ch

Pure appl. geophys. 161 (2004) 1379–1397
0033–4553/04/071379–19
DOI 10.1007/s00024-004-2510-9

© Birkhäuser Verlag, Basel, 2004

| Pure and Applied Geophysics

Far-field Gravity and Tilt Signals by Large Earthquakes: Real or Instrumental Effects?

Giovanna Berrino[1] and Umberto Riccardi[2,°]

Abstract — A wide set of dynamics phenomena (i.e., Geodynamics, Post Glacial Rebound, seismicity and volcanic activity) can produce time gravity changes, which spectrum varies from short (1... 10 s) to long (more than 1 year) periods. The amplitude of the gravity variations is generally in the order of $10^{-8}...10^{-9}$ g, consequently their detection requires instruments with high sensitivity and stability: then, high quality experimental data. Spring and superconducting gravimeters are intensively used with this target and they are frequently jointed with tiltmeters recording stations in order to measure the elasto-gravitational perturbation of the Earth. The far-field effects produced by large earthquakes on records collected by spring gravimeters and tiltmeters are investigated here. Gravity and tilt records were analyzed on time windows spanning the occurrence of large worldwide earthquakes; the gravity records have been collected on two stations approximately 600 km distant. The background noise level at the stations was characterized, in each season, in order to detect a possible seasonal dependence and the presence of spectral components which could hide or mask other geophysical signals, such as, for instance, the highest mode of the Seismic Free Oscillation (SFO) of the Earth. Some spectral components (6.5'; ~8'; ~9'; ~14', ~20', ~51') have been detected in gravity and tilt records on the occasion of large earthquakes and the effect of the SFO has been hypothesized. A quite different spectral content of the EW and NS tiltmeter components has been detected and interpreted as a consequence of the radiation pattern of the disturbances due to the earthquakes. Through the analysis of the instrumental sensitivity, instrumental effects have been detected for gravity meters at very low frequency.

Key words: Gravimeters, tiltmeters, earthquakes, seismic free oscillation.

1. Introduction

Spring and superconducting gravimeters are intensively used to record time gravity changes at the Earth's surface caused by many geophysical phenomena. Moreover, tiltmeters and GPS recording stations are frequently coupled with gravimeters in order to measure the elasto-gravitational perturbation of the Earth. A wide set of geophysical phenomena can affect the elasto-gravitational equilibrium:

---

[1] Istituto Nazionale di Geofisica e Vulcanologia - Osservatorio Vesuviano, Via Diocleziano, 328, 80124 Naples, Italy. E-mail: berrino@ov.ingv.it

[2] Dipartimento di Geofisica e Vulcanologia, Università «Federico II» di Napoli, L.go S.Marcellino, 10, 80138 Naples, Italy.

° F.S.E. (Fondo Sociale Europeo -European Community -)

e.g., Seismic Free Oscillations (SFO), luni-solar tides, Free Core Nutations (FCN), Earth's rotation changes, atmospheric and oceanic loading, Plate Tectonics and related underground mass redistribution due to volcanic and seismic processes and ground water dynamics. The spectrum of the changes of the Earth's figure due to these phenomena varies from short periods, 1...10 s (microseismic noise), to periods longer than 1 year (Geodynamics, Post Glacial Rebound, Polar motion) (CROSSLEY and HINDERER, 1995; HINDERER and CROSSLEY, 2000). The amplitude of the gravity changes is generally in the order of $10^{-8}...10^{-9}$ g, thus their detection requires instruments with high sensitivity and stability, as well as a very low instrumental drift; subsequently, high quality experimental data. Several natural and man-made sources can affect geophysical measurements reducing the Signal to Noise Ratio (SNR). The main natural noise is the seismic activity. Other authors have shown that disturbances induced by earthquakes can affect spring gravimeters (TORGE, 1989; MELCHIOR, 1983; RICCARDI et al., 2002, ZÜRN et al., 2002). This can produce significant changes in the instrumental sensitivity because of spring gravimeters mechanical perturbations due to some characteristic dominant frequencies of the noise at the station on the occasion of large earthquakes. Therefore, the stacking of instruments is strongly recommended to improve the quality of the collected signals and concerted efforts are on going to develop standards for low noise geophysical recording stations.

The far-field effects produced by large earthquakes on some geophysical records, mainly those collected by spring gravimeters and tiltmeters, are investigated. This study is developed by adopting gravimeters and tiltmeters in a most standard featuring of data acquisition, with a typical sampling frequency of $16.7 \cdot 10^{-3}$ Hz (1 datum per minute), suitable for tidal research and the other previously quoted geophysical investigations. Obviously this sampling rate is too low for seismological purposes, for the "low frequency" seismology (concerning the SFO) too; nevertheless, it is suitable for the present study intended to improve the knowledge of the transfer function of gravimeters and tiltmeters in the frequency range of the disturbances produced by large worldwide earthquakes.

2. Data Acquisition and Analyses

a) Instrumental Set-up

The far-field effects produced by large earthquakes have been evaluated on the signals acquired by two LaCoste and Romberg gravity meters — the D-126 and G-1089, both equipped with feedback systems — the GWR superconducting SG-C023, and two Applied Geomechanics biaxial tiltmeters AGI-710, indicated as T1 and T2.

The D-126 continuously operates since 1987 on Mt. Vesuvius, an active volcano near Napoli (Southern Italy), is placed in an underground laboratory 20 m deep (BERRINO et al., 1997) at the Osservatorio Vesuviano Old Building (φ: 40.82° N,

λ: 14.40° E; h: 608 m). The meter is equipped with a Maximum Voltage Retroaction (MVR) feedback (VAN RUYMBEKE, 1991) implemented at the Royal Observatory of Belgium (ROB) in Brussels and upgraded in 1994. The data acquisition is provided by DAS or μDAS systems developed at the ROB (VAN RUYMBEKE et al., 1995). The G-1089 operates since October 2000 in a ground level laboratory at the Istituto di RadioAstronomia (IRA) of the National Research Council in Medicina (Bologna-North Italy) (φ: 44.52° N, λ: 11.65° E; h: 50 m). The G-1089 is equipped with a MVR feedback by LaCoste and Romberg; it is settled on a concrete pillar a few meters away from the superconducting meter SG-C023, owned by the Bundesamt für Kartographie und Geodäsie (BKG-Frankfurt am Main) (SCHWAHN et al., 2000). The two AGI-710, T1 and T2, are respectively operating since 1999 on the pillars occupied by the spring gravity meter and the SG, in order to monitor the change of the ground tilt. The data acquisition system for both G-1089 and AGI tiltmeters consists of a digital card by National Instruments (NIDAQ700) installed in a laptop PC. Data (1 sample/min) are stored on site and the station is managed via Internet from the Dipartimento di Geofisica e Vulcanologia in Napoli, where signals are processed. The Medicina SG station belongs to global and European networks for checking and validation of space geodetic observations devoted to global geodynamics studies and to defining a more detailed pattern for the Africa-Eurasia plate interaction in the Mediterranean basin (ZERBINI et al., 2000).

The location of the stations is shown in Fig. 1.

b) Analysis of the Background Noise Level

Before analyzing gravity and tilt signals recorded during and after large earthquakes, the noise level at each station was characterized. Several time windows enduring about 1 week were selected in each season, aimed at investigating a possible seasonal dependence of the noise features at the station. The amplitude and spectral content of the noise on 1 minute sampling gravity and tilt data have been analyzed. The normal gravity tide was removed from the gravity signals; the former is defined as the theoretical tide amplified accounting for tidal parameters (amplitude and phase) obtained through the harmonic analyses on extended series of gravity records collected at each station. A trend computed by least-square approximation was subtracted from tilt records. The obtained residual signals were convolved with a Hanning window to reduce "truncation" effects and then Fourier analyses were applied. The power spectra of the residual signals are shown in Fig. 2. For the spring gravimeters (D-126 and G-1089) power spectra have been computed for each season, while for the SG-C023, the available data were collected only during autumn and winter. For the tiltmeters an example of noise spectra, computed on two weekly samples recorded during autumn, is plotted for one station. The results of these spectral analyses show a flat trend in the analyzed spectral band according to the standard New Low Noise Model (NLNM) (Fig. 2c) (PETERSON, 1993).

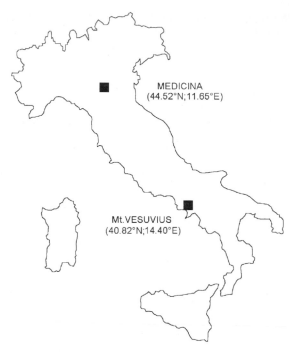

Figure 1
Location map of the recording gravity and tilt stations.

As regards the SG-C023 meter, a steady strong peak at $3.5 \cdot 10^{-3}$ Hz (4.8') is detected. This component seems to be a leading wave which modulates the output signal of the meter so it can be considered as due to instrumental setting. The meter's owner also shared this idea (Richter, personal communication). Excepting this component, the shape of the spectrum is quite similar to that for the G-1089 operating in the same laboratory, although with an expected lower amplitude; this is true also for tiltmeters.

c) Gravity and tilt records

Gravity and tilt records collected during and after large earthquakes were analyzed. The considered worldwide earthquakes which occurred between 2000 and 2001 are listed in Table 1; seismic parameters were taken from the USGS (USGS-NEIC Web site). The seismic events detected by the main part of the stations and with a useful SNR are highlighted in bold characters and with a grey background. In reason of the best quality of the available data, records acquired on the occasion of three seismic events among those listed in Table 1 are presented and discussed (Figs. 3 and 4).

The first seismic event in Table 1, which is the strongest one (magnitude: 8.0), occurred on 16th November 2000, was recorded by the SG-C023 gravimeter and that

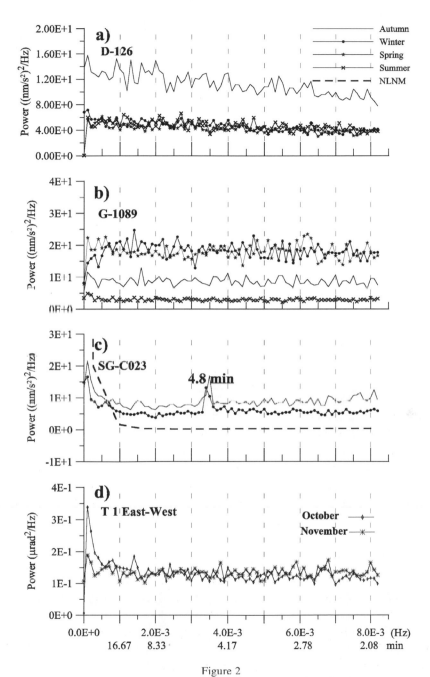

Figure 2

Power spectra of the noise signal computed on weekly data sets collected in different seasons for D-126 (a), G-1089 (b), SG-C023 (c) and AGI-710 (d), according to the available data, with indication of the standard New Low Noise Model (NLNM) as reference.

Table 1

List of some large worldwide earthquakes which occurred between 2000 and 2001. The three earthquakes selected and discussed in the paper are highlighted in bold characters and with a grey background. List of the abbreviations adopted in the table: Magn. = Magnitude; N.A. = Not Available; N.D. = Not Detected; S = Suitable SNR; Ant. = Antipodes.

DATE Time	LOCATION	Lat. (°) Long. (°)	Magn.	Depth (km)	Distance (km)	Azimut	AGI-710	G-1089	SG-C023	D-126
16 /11/00 04:55	**Papua-New Guinea**	**3.9S 152E**	**8.0**	**33**	**16100**	**SE**	**T1EW-NS T2EW-NS**	**Out of range**	**S**	**S**
17/11/00 21:02	Papua-New Guinea	5.4S 151.7E	7.6	33	16100	SE	T1EW-NS T2EW-NS	S	S	N.D.
6/12/00 17:16	Turkmenistan SE Caspian Sea	39.6N 54.8 E	7.2	30	4600	ESE	T1EW-NS T2EW-NS	Out of range	S	Out of range
10/01/01 16:02	Kodiak island (Alaska)	57.0N 153.6W	6.7	33	15640	Ant.	NO	S	S	S
13/01/01 17:33	El Salvador	12.8N 88.8W	7.6	39	11615	WSW	T1EW-NS T2EW-NS	Out of range	S	Out of range
26/01/01 03 :16	**Southern India**	**23.4N 70.3E**	**7.9**	**24**	**6900**	**SE**	**T1EW-NS T2EW-NS**	**Out of range**	**S**	**Out of range**
13/02/01 14:22	El Salvador	13.6N 88.9W	6.6	13	11500	WSW	N.D.	S	S	S
13/02/01 19:28	**South Sumatra (Indonesia)**	**5.0S 102.4E**	**7.3**	**33**	**11500**	**SE**	**N.D.**	**S**	**S**	**S**
24/02/01 07:23	Northern Molucca Sea (Sulawesi)	1.27N 126.0E	6.6	35	13570	SE	GAP	GAP	S	N.D.
28/02/01 18:54	Seattle (Washington) USA	47.15N 122.7W	6.8	53	13800	WNW	N.D.	S	N.A.	N.D.
24/03/01 06:28	West Honshu (Japan)	34.1N 132.6E	6.5	33	13340	ESE	N.D.	S	N.A.	S
09/04/01 17:38	Albania-Greece		4.6		1035	SE	N.D.	S	N.A.	S
28/04/01 04:49	Fidji Island	18.0S 177W	6.9	352	19780	Ant.	N.D.	S	N.A.	S
25/05/01 00:40	Kurili Islands	44.3N 148.4E	6.7	33	14490	E	N.D.	S	N.A.	S

Date/Time	Location	Coordinates								
23/06/01 20:33	Perù	16.1S 73.4W	8.1	33	11615	SW	T1EW-NS T2EW-NS	Out of range	N.A.	S Out of range
07/07/01 09:39	Perù	17.4S 71.8W	7.6	33	11615	SW	T2EW	Out of Range	N.A.	S
17/07/01 15:06	Val Venosta Merano (BZ-Italy)	46.7N 11.4E	4.8	10	276	NE	T1EW-NS T2EW-NS	N.D.	N.A.	N.D.
24/07/01 5:00	Northern Chile	19.3S 69.W	5.9	33	11345	SW	N.D.	S	N.A.	N.D.
26/07/01 00 :24	Aegean Sea	39.0N 24.2E	6.3	10	1495	SE	T1EW-NS T2EW-NS	Out of Range	N.A.	Out of range
28/07/01 07:33	Southern Alaska	59.1N 155.1W	6.5	139	15000	Ant.	N.D.	S	N.A.	Out of range
02/08/01 23:52	Kamchatka	56.4N 163.6E	6.2	27	15180	Ant.	N.D.	S	N.A.	S

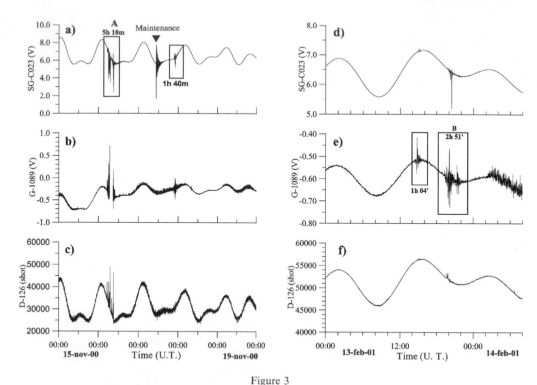

Figure 3

Gravity record related to the Papua-New Guinea (a, b, c) and Sumatra Island (d, e, f) earthquakes (ref. Table 1). The seismic events are respectively highlighted in box A and box B, in which the length of the analyzed time section is shown.

on Mt. Vesuvius, while the G-1089 was out of range. The tiltmeter records have a bad SNR. For this event a signal of about 5 hours recorded during the seismic event was analyzed. The records are shown in Figure 3 and the analyzed section of the signal is highlighted in the box A (Fig. 3a). The results of the spectral analyses are shown in Figs. 5, 6,7. In order to interpolate the wide gap of the G-1089 recording (ref. Fig. 3b), the time derivative of the signal has been computed; this is the reason why the spectrum does not contain lower frequencies (Fig. 5c). The lower frequencies $1.3 \cdot 10^{-4}$ Hz and $1.6 \cdot 10^{-4}$ Hz (128' and 102') are certainly due to an inefficient trend removal (Figs. 6a,b). The component at $2.92 \cdot 10^{-4}$ Hz (57') is close to the fundamental spheroidal SFO, whereas the peak at $8.3 \cdot 10^{-4}$ Hz (20') is coincident with the $_0S_5$ component, as well as the peak at $1.17 \cdot 10^{-3}$ Hz (14') is coincident with the $_0S_6$. In Figure 7 the higher frequencies of the spectra are plotted. A peak at $2.56 \cdot 10^{-3}$ Hz (6.5') is well evident in spring gravity meters' records. This is close to the $_0S_{16}$ - $_0S_{17}$. In Figure 5c peaks around 3–4 min are also evident; as suggested by some authors they could be excited by long-period Rayleigh waves (LR) (EKSTRÖM, 2001; KANAMORI and MORI, 1992; ZÜRN et al., 2002) or by a close $_0S_{30}$ SFO (LAY and

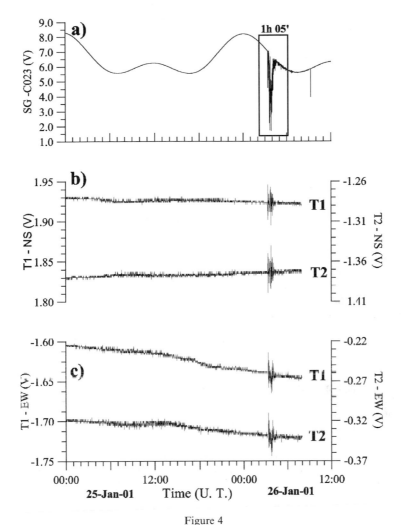

Figure 4
Signals related to the India earthquake (ref. Table 1) by means of the SG-C023 (a) and the two tiltmeters AGI-710 (b – NS components, c – EW components). The seismic event is highlighted in a box on which the length of the analyzed time section is shown.

WALLACE, 1995). A cross-spectrum between D-126 and SG-C023 has been computed and the coherence function is plotted in Fig. 8. The main component at very low frequency $(0.13 \cdot 10^{-3}$ Hz) reflects the incomplete modelling and removing of the trend in the gravity records. Otherwise, the component at $0.33 \cdot 10^{-3}$ Hz (51') is close to the fundamental spheroidal mode (Eulerian eigenperiod). The higher coherence (c = 0.94) can be stressed for the component at about $2.0 \cdot 10^{-3}$ Hz (8') which is close to the $_0S_{13}$ (Fig. 8).

The seismic event which occurred on 13th February, 2001 in the Sumatra Island was recorded by all gravimeters and highlighted in box B (ref. Figs. 3d,e,f). The

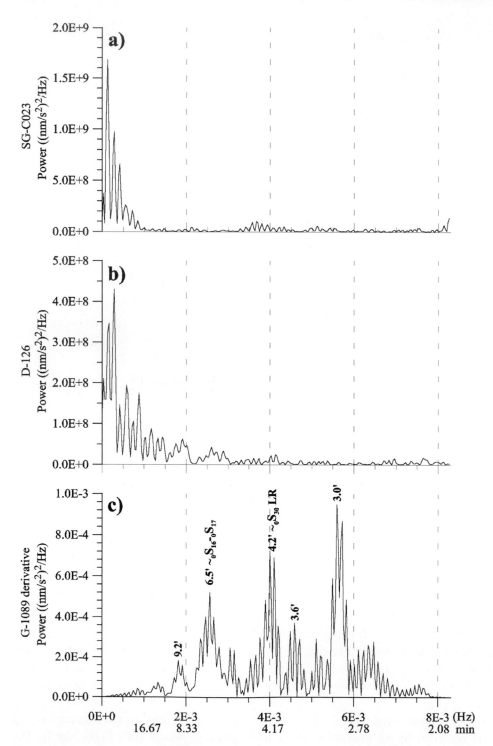

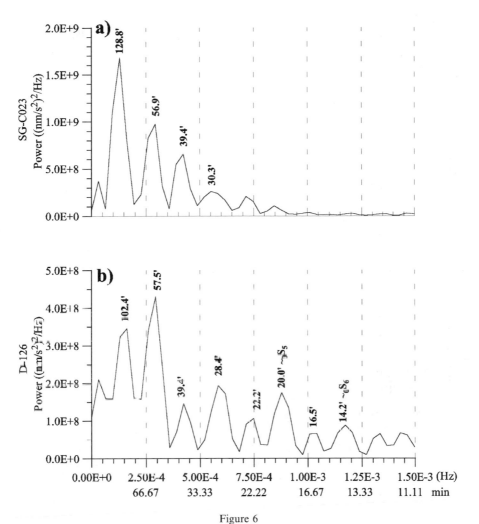

Figure 6

Lower frequency part of the power spectra shown in Figure 5 for SG-C023 (a) and D-126 (b). The period of the main spectral peaks is labelled with the indication of the correspondent SFO's components.

spectra are depicted in Figure 9. They show a well evident component in the G-1089 record at the frequency $2.22 \cdot 10^{-3}$ (7.5') close to the SFO $_0S_{14}$.

Concerning the seismic event which occurred in India on 26th January 2001, a time section of 65' has been analyzed; this event has a good SNR on the

◄

Figure 5

Power spectra computed on gravity recording collected during the Papua-New Guinea earthquake (event "A" in Fig. 3a) from SG-C023 (a), D-126 (b) and G-1089 (c). For the G-1089 the period of the main spectral peaks is labelled with the indication of the correspondent SFO's components.

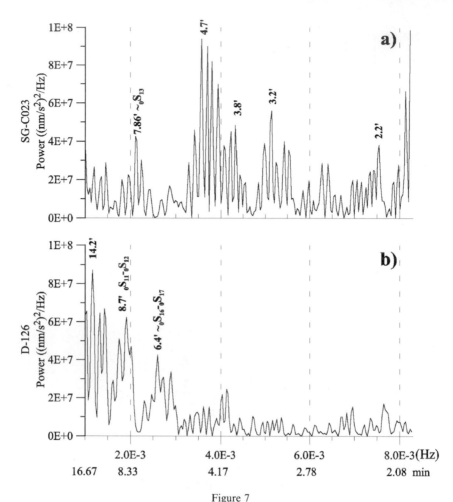

Figure 7

Higher frequency part of the power spectra shown in Figure 5 for SG-C023 (a), D-126 (b). The period of the main spectral peaks is labelled with the indication of the correspondent SFO's components.

superconducting meter and tiltmeters. The records are shown in Figure 4, while the spectra are plotted in Figure 10. A "typical" spectral content has been detected in the records collected with tiltmeters. It is characterized by different dominant harmonics for E-W and N-S tilt components. This feature has been detected every time when a seismic event is recorded by means of tiltmeters. Interesting harmonics have been observed close to the $_0S_{11}$ and $_0T_{12}$. A satisfactory correlation can be underlined

▶

Figure 8

Coherence function (a) and cross-power spectra, both in linear (b) and logarithmic scale (c), computed on gravity recording from SG-C023 and D-126 during the Papua-New Guinea earthquake (referred in Fig. 3a as "A"). The period of the main spectral peaks is labelled with the indication of the correspondent SFO's components.

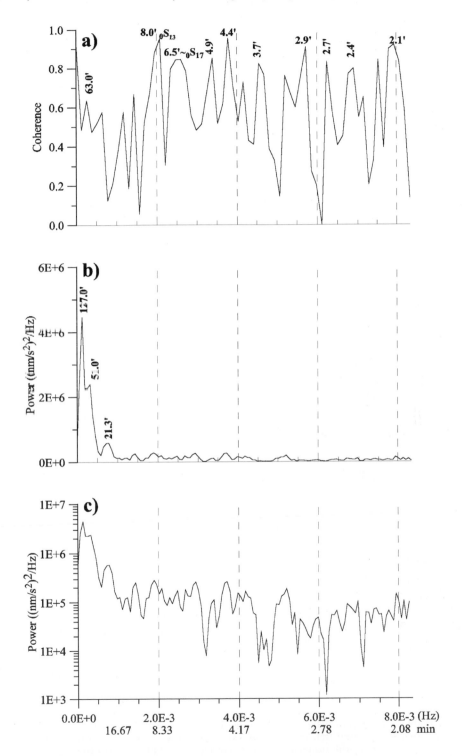

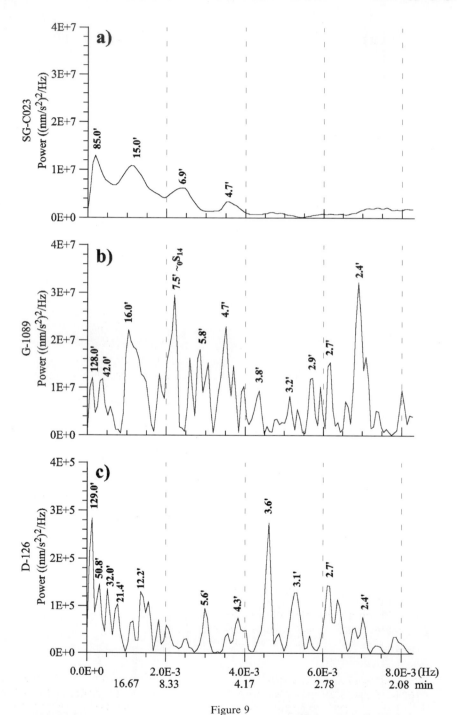

Figure 9

Power spectra computed on gravity recording from SG-C023 (a), G-1089 (b), and D-126 (c) during the Sumatra Island seismic event (ref. Figs. 3d, e, f, event "B"). The period of the main spectral peaks is labelled with the indication of the correspondent SFO's components.

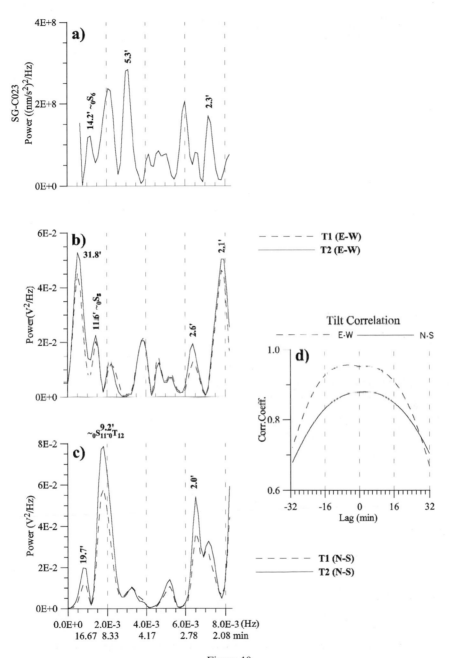

Figure 10

Power spectra computed on recording from SG-C023 (a) and the two AGI-710 (b – EW components, c – NS components) during the India earthquake (ref. Fig. 4) and correlation analysis between North-South and East-West tilt components (d). The period of the main spectral peaks is labelled with the indication of the correspondent SFO's components.

between signals collected on the axis of the tiltmeters with the same orientation. This could indicate that these spectral features are due to the radiation pattern of the ground movement during the seismic event.

3. Discussion

The analysis of the noise at each station showed, for both gravity and tilt stations, a flat trend in the analyzed spectral band in agreement with the standard New Low Noise Model (NLNM) and New High Noise Model (NHNM). It means that real geophysical signals inside this frequency band can be detected. A quantitative study of the lowest and highest mode of the SFO of the Earth might be very hard to pursue, therefore a data acquisition with a higher sampling rate would be needed. Some spectral components (6.5'; ~8'; ~9'; ~14', ~20', ~51') have been detected at a suitable level of reliability and the effect of the Earth's SFO (spheroidal component for gravimeters and torsional and spheroidal components for tiltmeters) can be hypothesized. The tiltmeters response, in the frequency domain, is quite different for the EW and NS components, probably as a consequence of the radiation pattern of the disturbances produced by earthquakes.

In order to investigate the effect of large earthquakes on gravity meters, at a very low frequency also, the time evolution of the instrumental sensitivity on the occasion of the seismic events has been studied. A theoretical value of the instrumental sensitivity was computed and compared with the experimental calibration factors periodically monitored "on site"; the latter are obtained inducing changes in the spring length through known "dial" turning and fitting this, in the least-square sense, against the instrumental output. Detailed information on the different calibrations carried out at these stations is given in RICCARDI et al. (2002). The theoretical instrumental sensitivity was determined by a regression analysis between the meter's output signal and the "normal" gravity tide. The signals were pre-processed interpolating the main outliers (step, spike or gap), some of which originating from seismic events or increase of noise level at the station, subsequently the regression coefficients were computed. This way, a set of weekly theoretical values of calibration factor was obtained and compared with the results from the repeated calibrations. This comparison allows for an investigation of whether the calibration factor truly reflects change in the instrumental response or is merely due to the adopted "on site" calibration procedures. A time interval of contemporary records by D-126, G-1089 and SG-C023 was selected (Fig. 11). A good agreement between the time evolution of the theoretical factors and those obtained by the "on site" calibration was detected and a similar trend can be envisaged among the theoretical sensitivity time evolution for the gravimeters. As regards the SG-C023, the change of sensitivity is very small, always within 1%, and meaningless (Fig. 11b). Otherwise, the change of sensitivity of G-1089 shows a similar trend although with an amplitude larger than the D-126's.

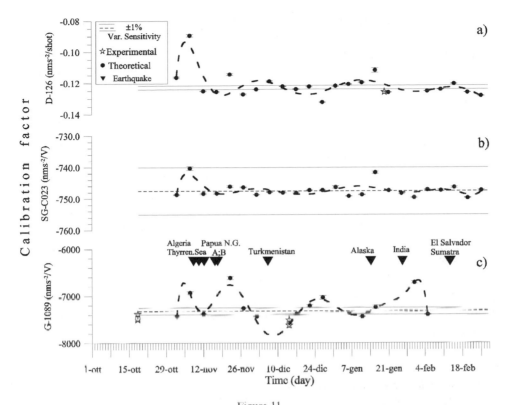

Figure 11

Theoretical calibration factors time evolution computed on weekly data sets for D-126 (a), SG-C023 (b), and G-1089 (c). Comparison with the experimental calibrations and time occurrence of some worldwide earthquakes (c). The dashed curves represent data interpolation. The ±1% range is the acceptance of the instrumental sensitivity changes.

These trends have been plotted against the time occurrence of certain worldwide large earthquakes (Fig. 11c). A time correlation between the higher changes of instrumental sensitivity and the occurrence of seismic events can be observed. More detailed discussion concerning the instrumental sensitivity changes on the occasion of large earthquakes is given in RICCARDI et al. (2002); they also suggest that the highest part of the SFO frequency band includes the fundamental mode of oscillation (T_0: 15 – 20 s) of the spring sensor for both LaCoste and Romberg models D and G (TORGE, 1989). In light of these results, a mechanical perturbation of the sensor device, due to some dominant frequencies of the noise at the station on the occasion of large earthquakes, may be hypothesized. As inferred by the results from the monitoring of the sensitivity, these instrumental disturbances can last several weeks. Changes in time of calibration factors for different kinds of mechanical gravimeters have been detected by other authors (e.g., BONVALOT et al., 1998; BUDETTA and CARBONE, 1997), however there is insufficient information to deduce if mechanical perturbations are typical of all spring gravimeters or only those of LaCoste and Romberg.

4. Conclusion

The far-field effects produced by large earthquakes on records collected by LaCoste and Romberg gravimeters and tiltmeters have been investigated. Gravity and tilt records were analyzed on time windows spanning the occurrence of large worldwide earthquakes.

A mechanical perturbation of the gravity sensors, due to some dominant frequencies of the noise on the occasion of large earthquakes, may be hypothesized for spring gravimeters. The observed change of the gravimeters instrumental sensitivity shows that these disturbances can last several weeks. However, some geophysical signals were also detected in gravity and tilt records acquired in the event of large earthquakes. Several spectral components (6.5'; ~8'; ~9'; ~14', ~20', ~51') have been recognized due to the effect of the Earth's SFO (spheroidal component for gravimeters and torsional and spheroidal components for tiltmeters). Both tiltmeters have shown a quite different response, in the frequency domain, for the EW and NS components, probably as a consequence of the radiation pattern of the disturbances produced by earthquakes.

In conclusion, it can be stressed that large worldwide earthquakes can produce far-field effects, detectable by tilt and gravity meters, including both real geophysical and instrumental signals.

Acknowledgements

The authors are indebted to S. Bonvalot and L. Rivera for their critical reviews of the paper; they are very grateful to Prof. Gennaro Corrado from the Dipartimento di Geofisica e Vulcanologia (University "Federico II" of Napoli), for his highly valued suggestions and stimulating discussions, which improved this research.

The authors also thanks Dr. Bernd Richter, from the Bundesamt für Kartographie und Geodäsie (BKG-Frankfurt am Main), who furnished data of the superconducting gravity meter SG-C023.

This research has been partially supported by the Agenzia Spaziale Italiana, Grant ASI I/R/144/01.

REFERENCES

BERRINO, G., CORRADO, G., MAGLIULO, R., and RICCARDI, U. (1997), *Continuous Record of the Gravity Changes at Mt. Vesuvius*, Annali di Geofisica XL, N5, 1019–1028.
BONVALOT, S., DIAMENT, M., and GABALDA, G. (1998), *Continuous Gravity Recording with Scintrex CG-3M Meters: a Promising Tool for Monitoring Active Zones*, Geophys. J. Int. *135*, 470–494.

BUDETTA, G. and CARBONE, D. (1997), *Potential Application of the Scintrex CG-3M Gravimeter for Monitoring Volcanic Activity: Results of Field Trials on Mt. Etna, Sicily*, J. Volcano. Geotherm. Res. *76*, 199–214.

CROSSLEY, D. and HINDERER, J. (1995), *Global Geodynamics Project –GGP*, Cah. Cent. Eur. Géodyn. Séismol. *11*, 244–271.

EKSTRÖM, G. (2001), *Time Domain Analysis of Earth's Long-period Background Seismic Radiation*, J. Geophys. Res. *106, B11*, 26483–26493.

HINDERER, J. and CROSSLEY, D. (2000), *Time Variations and Inferences on the Earth's Structure and Dynamics*, Surveys in Geophys. *21*, 1–45.

KANAMORI, H. and MORI, J. (1992), *Harmonic Excitation of Mantle Rayleigh Waves by the 1991 Eruption of Mount Pinatubo, Philippines*, Geophys. Res. Lett. *19*, 721–724.

LAY, T. and WALLACE, T. C., *Modern Global Seismology* (Academic Press, S. Diego, California 1995).

MELCHIOR P., *The Tides of the Planet Earth* (Pergamon Press, Oxford 1983).

PETERSON, J. (1993), *Observations and Modelling of Seismic Background Noise*, Open File Report *93–322*, U.S. Department of Interior Geologica Survey, Albuquerque, New Mexico.

RICCARDI, U., BERRINO, G., and CORRADO, G. (2002), *Changes in the Instrumental Sensitivity for same Feedback Equipping LaCoste and Romberg Gravity Meters*, Metrologia, *39*, 509–515.

SCHWAHN, W., BAKER, T., FALK, R., JEFFRIES, G., LOTHAMMER, A., RICHTER, B., WILMES, H., and WOLF, P. (2000), *Long-term Increase of Gravity at the Medicina Station (Northern Italy) Confirmed by Absolute and Superconducting Gravimetric Time Series*, Cah. Cent. Eur. Géodyn. Séismol. *17*, 145–168.

TORGE, W., *Gravimetry* (de Gruyter, Berlin, New York 1989).

USGS-NEIC (National Earthquakes Information Center) Web Site: http:// neic.usgs.gov

VAN RUYMBEKE, M. (1991), *New Feedback Electronics for LaCoste and Romberg Gravimeters*, Cah. Cent. Eur. Géodyn. Séismol. *4*, 333–337.

VAN RUYMBEKE, M., VIEIRA, R., d'OREYE, N., SOMERHAUSEN, A., and GRAMMATIKA, N. (1995), *Technological Approach from Walferdange to Lanzarote. The EDAS Concept*. In Proceeding 12th Int. Symp. on Earth tides, Science press (Beijing, China), pp. 53–62.

ZERBINI, S., PLAG, H. P., and RICHTER, B. (eds.) (2000), *Wegener: Observations and Models*, J. Geody. (Special Issue), *30*, 120 pp.

ZÜRN, W., BAYER, B., and WIDMER, R. (2002), *A 3.7 mHz Signal on June 10, 1991*, Bulletin d'Information des Marées Terrestres, *135*, 10717–10724.

(Received March 22, 2002, revised March 24, 2003, accepted April 3, 2003)

 To access this journal online:
http://www.birkhauser.ch

Pure appl. geophys. 161 (2004) 1399–1413
0033–4553/04/071399–15
DOI 10.1007/s00024-004-2511-8

© Birkhäuser Verlag, Basel, 2004

Pure and Applied Geophysics

Study of Volcanic Sources at Long Valley Caldera, California, Using Gravity Data and a Genetic Algorithm Inversion Technique

M. Charco[1], J. Fernández[1], K. Tiampo[2], M. Battaglia[3], L. Kellogg[4], J. McClain[4], and J. B. Rundle[4,5]

Abstract — We model the source inflation of the Long Valley Caldera, California, using a genetic algorithm technique and micro-gravity data. While there have been numerous attempts to model the magma injection at Long Valley Caldera from deformation data, this has proven difficult given the complicated spatial and temporal nature of the volcanic source. Recent work illustrates the effectiveness of considering micro-gravity measurements in volcanic areas. A genetic algorithm is a problem-solving technique which combines genetic and prescribed random information exchange. We perform two inversions, one for a single spherical point source and another for two-sources that might represent a more spatially distributed source. The forward model we use to interpret the results is the elastic-gravitational Earth model which takes into account the source mass and its interaction with the gravity field. The results demonstrate the need to incorporate more variations in the model, including another source geometry and the faulting mechanism. In order to provide better constraints on intrusion volumes, future work should include the joint inversion of gravity and deformation data during the same epoch.

Key words: Long Valley Caldera, gravity change, genetic algorithm, fitness function.

1. Introduction

Volcanic activity produces deformation and gravity changes that often can be used as precursors of future eruptions. Monitoring active zones by applying geodetic techniques involves interpreting the deformations and gravity changes that occur before and during volcanic eruptions, since these effects reveal the physics and characteristics of the magma reservoir. This task requires the use of mathematical deformation models. The basic model is assumed to be a strain source embedded

[1] Instituto de Astronomía y Geodesia (CSIC-UCM), Facultad de Ciencias Matemáticas, Ciudad Universitaria, 28040-Madrid, Spain. E-mail: charco@mat.ucm.es
[2] Cooperative Institute for Research in Environmental Sciences, University of Colorado, Boulder, CO 80389, USA.
[3] Department of Geophysics, Stanford University.
[4] Department of Geology, University of California at Davis.
[5] Distinguished Visiting Scientist, Jet Propulsion Laboratory, Pasadena, CA 91125, USA.

within an elastic medium. This model was proposed by Mogi (1958) for modeling volcanic inflation and deflation episodes. The success in explaining ground deformation in several volcanic areas by the simplest model (Mogi model) made it the most commonly used deformation model. However, this model presents difficulties in modeling horizontal displacements and gravity changes (e.g., Rymer, 1994). Also point-like source models are not acceptable if we want to study the deformation in the source neighbourhood (e.g., De Natale and Pingue, 1993). Recent work has proposed a set of increasingly complex models which considers various geometries as well as more complicated distribution of elastic properties (for summary of works, Folch *et al.*, 2000; Tiampo *et al.*, 2000; Fernández *et al.*, 2001a). Rundle (1980; 1982) presented an elastic-gravitational model that considers a stratified half-space of homogeneous layers. It has been shown that considering the self-gravitation of the medium can be fundamental for interpreting and explaining gravity changes measured in active volcanoes (e.g., Fernández *et al.*, 1997, 2001a, b).

Different techniques can be used in exploration geophysics to obtain information pertaining to volcanic source parameters. Traditionally this kind of inversion has been approached using local optimization techniques such as steepest descent, conjugate gradients or the simultaneous iterative reconstruction methods. Discrepancies observed among the solutions using these methods are related to the non-linearity of the inverse problem (local searches are prone to trapping in local minimum) and to the inherent ambiguity in the inversion. They indicate a need for global optimization methods to determine source parameters. Monte-Carlo technique is the common global method to solve nonlinear inverse problems. Recently, genetic algorithm (GA) methods have been proposed in geophysics (e.g., Beauducel *et al.*, 2000; Billings *et al.*, 1994; Boschetti *et al.*, 1996; Stoffa and Sen, 1991). Genetic algorithms, like Monte-Carlo methods, use random processes to locate a near optimal solution and require no derivative information. The ability of a genetic algorithm to use the information contained in successful models, while still exploring the model space, suggests it would be more efficient than Monte-Carlo methods in solving optimization problems (Gallagher *et al.*, 1991).

In this paper we use the elastic-gravitational deformation model to perform the inversion of the temporal free-air gravity changes between 1982–1998 in the Long Valley Caldera, California, USA. The inversion of the gravity data is complicated since it is not continuously distributed in the space. We have employed the genetic algorithm inversion technique because it is proven to be a powerful tool in many nonlinear and multi-dimensional geophysical problems.

2. Long Valley Caldera. Geology

The Long Valley Caldera, located to the east of the Sierra Nevada (California), extends 32 km eastwesterly and 17 km north to south, with an average elevation of

2,200 m (Fig. 1). The caldera itself was formed in an explosive eruption of rhyolitic magma 0.7 m.y. ago, producing about 500 km³ of Bishop tuff both within and outside the caldera and about 300 km³ of ash dispersed over much of the western United States (ABERS, 1985). BAILEY et al. (1976) present a detailed description of caldera development and post-caldera volcanism. Collapse of the roof of the magma chamber accompanied the eruption, with subsidence of about 2–3 km along a fracture zone. Immediately following the eruption of Bishop tuff, a resurgent dome began developing in the west-central caldera accompanied by extrusion of rhyolitic tuffs and flows onto the caldera floor. Over the last 0.5 m.y. there have occasionally been further eruptions of rhyolites and basalts, largely in the caldera, moat and rim, although some basalt eruptions occurred to the south and north of western caldera margin (ABERS, 1985). The most recent eruptions in the Long Valley regions occurred 500 to 600 years ago with the intrusion of a large, north-striking silicic dike that surfaced in three places to produce three new domes in the Inyo volcanic chain. Most of the magma feeding this dike was derived from a chamber beneath the Inyo Mono Craters volcanic chain north of Long Valley; the last major eruption fed by the Long Valley magma chamber occurred in the west moat of the caldera approximately 100,000 years ago (BAILEY et al., 1976; HILL, 1984).

Over the last 20 years there have been three episodes of rapid inflation of the central resurgent dome, accompanied by seismicity inside and around the caldera, with no eruptions between these episodes. In October of 1978 seismicity resumed in the Long Valley region with the occurrence of a $M_L = 5.7$ event near Wheeler Fault south caldera. Migrating northward over the next few years, the main activity

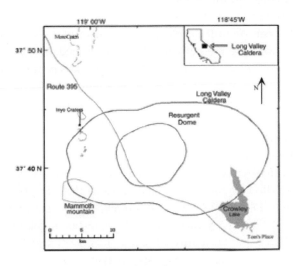

Figure 1

Long Valley Geology (TIAMPO et al., 2000). Much of Long Valley is covered by rocks formed during volcanic eruptions in the past two million years. A cataclysmic eruption 760,000 Ma formed Long Valley and ejected flows of hot glowing ash, which cooled to form the Bishop tuff.

concentrated in the south moat during May of 1980 with four earthquakes of M ≥ 6 and their subsequent aftershocks. In response to this seismicity, the California Division of Highways releveled Highway 395 in 1980, which runs from Tom's Place in the south to Lee Vining north of the caldera, and discovered 200–250 mm of uplift had occurred since 1975 (DENLINGER and RILEY, 1984). An extensive resurvey was carried out in 1982 and again in 1983 which demonstrated that uplift began sometime after 1975, probably in 1978 or 1979, and continued at an approximately constant rate through 1983 (CASTLE *et al.*, 1984). The second episode started in October 1989 after several years of relative calm (e.g., TIAMPO *et al.*, 2000). The seismicity increased again in conjunction with seismic activity under Mammoth Mountain, as extension rates across the caldera increased sharply to almost 9 mm/yr, followed by renewed activity under the south moat. Following the initial deformation, extension rates slowed to between 3 and 5 ppm/yr through 1991. Total uplift for the episode beginning in 1989 had reached 8 cm along the Route 395 leveling line, and 11 cm along the Rte 203 leveling traverse near Casa Diablo by 1992, before it decreased again to background levels of 1988 (LANGBEIN, 1989; LANGBEIN *et al.*, 1993, 1995). The most recent episode of inflation and uplift of the resurgent dome within the caldera occurred in 1997–1998.

3. Gravity Change

Areal gravity surveys are recognized as a valuable tool for mapping out the subsurface density distributions while temporal variations of the Earth's gravity field have been used to investigate the physical process involved in volcanic activity. Using these variations we can quantify the change in subsurface mass. During early June 1980 a high precision gravity network was established in Long Valley Caldera, California (ROBERTS *et al.*, 1988). The gravity network is centered near Tom's Place and extends from the Sierra Nevada west of Lee Vining, California, southeastward to a station in the White Mountains east of Bishop. The network was surveyed annually from 1980–1985. The early efforts at detecting and interpreting temporal and areal gravity changes over the interval 1980 to 1983 suggest some magma injection accompanied the uplift (JACHENS and ROBERTS, 1985). Unfortunately the uplift was so small compared to observational uncertainties in the gravity data (due to difficulty in correcting for shallow ground water fluctuations) that simple inflation cannot be completely ruled out. During the July 1998 survey, BATTAGLIA *et al.* (1999) measured 34 gravity stations. All gravity measurements were taken relative to the reference station of Tom's Place. Between July 1982 and July 1998 gravity within the caldera decreased substantially (the precise relative gravity measurements reveal a decrease in gravity of as much as -107 ± 6 μGal centered on the uplifting resurgent dome) whereas the control stations showed no substantial change. The stations with the largest gravity decrease are all located on the resurgent dome. After correcting by elevation differences using the free-air gradient and water table effect, the residual gravity field or free-air gravity change shows a peak of 64 $\pm$ 16 μGal centered on the resurgent dome. The location of the gravity stations and the free-air gravity change are shown in Figures 2 and 4.

We use the residual gravity anomalies to determine the parameters of the source or sources under surface. The inversion of geophysical data is complicated by the fact that observed data are contaminated by noise and are acquired at a limited number of observation points. Here we use a Genetic Algorithm (GA) technique to invert the positive residual gravity change using elastic-gravitational Earth models. A qualitative description of genetic algorithms is given by TIAMPO et al. (2000; 2004) and references included in both papers, therefore we refer to them for details. Although GAS have been adapted and applied to a variety of applications, they have proven to be an attractive global search tool suitable for irregular functions typically observed in nonlinear optimization problems in the physical sciences (e.g., STOFFA and SEN, 1991; BOSCHETTI et al., 1996; YU et al., 1998; BEAUDUCEL et al., 2000).

4. Forward Model

The goal of geophysics is to determine the properties of the Earth's interior or the characteristics of sources causing a local geodynamic phenomena, using large quantities of measured data. This constitutes the inverse problem. To understand how the data are affected by the Earth behavior we must be able to calculate synthetic data for an assumed Earth model. This constitutes the forward problem and involves deriving a mathematical relationship between data and model.

Here we have used a model where the Earth is represented by a layered homogeneous half-space possessing both elasticity and self-gravitation. RUNDLE (1980, 1982) obtained and solved the equations that represent the coupled elastic-gravitational problem for a stratified half-space of homogeneous layers, using the propagator matrix technique (HASKELL, 1953) to obtain the surface potential, gravity changes and deformation. The importance of including the gravity field for interpretations of measured gravity changes can be seen, for example, in FERNÁNDEZ et al. (1997) and FERNÁNDEZ et al. (2001b). It is shown that the gravitational effects are not significant in the displacements at the surface but the gravity field is important for modeling and interpreting the observed gravity changes. RUNDLE (1978) and WALSH and RICE (1979) showed that for a dilating point in a purely elastic media the gravity change is equal to the free-air gradient, i.e., a constant distortion gravity gradient. The inclusion of the intrusion mass and its interaction with the gravity field produces lower effects than estimated by other deformation models which only consider effects produced by a pressurized cavity in an elastic half-space (e.g., FERNÁNDEZ et al., 2001a,b,c). Therefore the change in volume of the magma chamber computed with the elastic models must be different than the volume

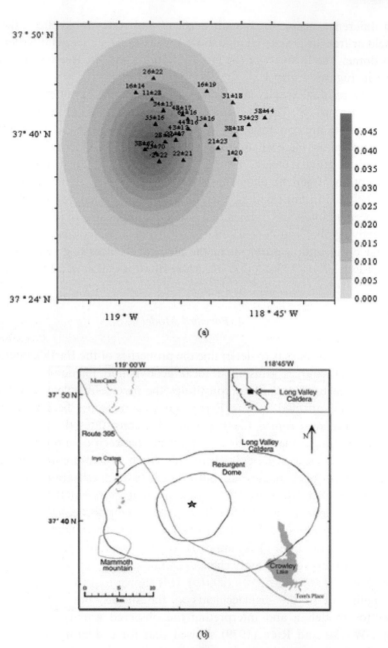

Figure 2
(a) Free-air gravity change in mGal predicted from the one-source inversion . The location of the gravity stations is marked by a triangle. The numbers indicate the residual gravity change in the network stations in μGal. (b) The star marks the source *x-y* location.

of magma which enters or leaves the chamber during an injection or removal process. The main advantage in using the full elastic-gravitational model is that it will yield accurate results for all elastic structures, even the "exotic" ones that might occasionally exist in volcanic regimes, e.g., with very low density fill in the magma chamber, unusual elastic modulus, etc. (e.g., RUNDLE, 1982; KISSLINGER, 1975).

Thus we use an elastic-gravitational model to get the parameters of the source. We assume a single point source located in a layered half-space with the properties shown in Table 1. The density and Lamé parameters have been obtained using the methodology described by FERNÁNDEZ and DÍEZ (1995) and geological information from HILL et al. (1985). The soft upper layer of 2-km thickness represents the deposit of ash and pumice which erupted during formation of Long Valley and the sediments and post-caldera volcanics. The granites underneath, which have a modulus of rigidity of 50 GPa, represent the prevolcanic basement in the area mainly formed by Mesozoic granitic rock of Sierra Nevada Batholitic and Paleozoic metasedimentary and Mesozoic metavolcanic rocks of Mt. Morrison, Gull Lake and Ritter range roof pendants (BAILEY, 1989). Bishop tuff is encountered at depth in the caldera and also appears as inclusions in intracaldera rhyolitic eruptives. ABERS (1985) assigned a density of 2.4 g/cm^3 to densely welded Bishop tuff. In the present work we assign a density of 2.56 g/cm^3 that is inside the density constraint limits computed by this author from the observed velocity ranges by using the Nafe-Drake velocity-density relation (e.g., GRANT and WEST, 1965).

The GA code we used is described by TIAMPO et al. (2004). This code tries to find the parameters (x and y location, depth, radius, mass and pressure increment of the magma chamber) that minimize the difference between the model response and the ground measurements. Thus the fitness function will be some appropriate function of the difference between the observed data values and the data values predicted from the model. In this work the value of chi-square is calculated using:

$$\chi^2 = \sum_k \frac{(y_k - f(x_k))^2}{\sigma_k^2}, \tag{1}$$

where y_k are the measured free-air gravity changes, $f(x_k)$ are the calculated gravity changes from the model, σ_k is the standard deviation for each measurement and k is the total number of measurements (TAYLOR, 1982; BEVINGTON and ROBINSON, 1992).

Table 1

Densities and Lamé constants for the crustal model of Long Valley Caldera

Layer	Thickness (km)	ρ (10^3 kg/m^3)	μ 10^{10} Pa	λ 10^{10} Pa
1	2.0	2.56	1.5	3.82
Half-space		2.65	5.0	2.7

This distribution is related to the metric distance concept. The chi-square value relates the data to the goodness of fit of model predictions.

The GA is a problem-solving method that seeks the fittest members of the population based upon maximum fitness value. Our optimization problem is to minimize the expression (1). This is equivalent to maximizing a function that we call the Fitness Value (FV):

$$FV = \frac{1}{\chi^2}.$$ (2)

This expression converts the fitness to a continuously increasing function.

5. *Inversion Results*

Although the numerous published models of deformation of Long Valley Caldera differ in detail, most have in common the following three sources: a primarily dilatational volume beneath the central part of the resurgent dome, centered at depth between 7–10 km, a secondary dilatational volume beneath the southern margin of the resurgent dome, centered at depth between 4–8 km, and a right lateral slip on a vertical west-northwest striking fault in the south moat.

First we estimated the *x-y* location, the depth, the pressure change, the radius and the mass increment assuming a simple spherical point source located in a layered elastic-gravitational half-space. The parameters obtained from the inversion are shown in Table 2 (a). Figure 2 shows the synthetic free-air gravity changes computed from the forward model using the parameters shown in Table 2 (a), together with the residual gravity in the network stations. The maximum predicted change of 50 μGal is located under the resurgent dome which has experienced uplift during the past 20 years. We estimated the parameters of the intrusion assuming a spherically symmetric point source embedded in the elastic-gravitational media described by Table 1. Figure 3 presents a comparison of observed and predicted residual gravity change (1982–1998) as a function of radial distance from the source. The gravity change comes from the intrusion mass. This is not surprising as the gravity measurements are not sensitive to pressure and radius parameters (TIAMPO *et al.*, 2004). This fact allows the inversion to discriminate between the role played by pressure changes and that played by mass displacements. Table 2 (a) and both Figures 2 and 3 show that it has been some magma recharge together with a certain increase in pressure inside the sub-caldera crust, although the pattern of the computed gravity change is different from the observed pattern (see Figure 3C in BATTAGLIA *et al.*, 1999).

Taking into account prior models used in Long Valley Caldera (e.g., HILL *et al.*, 1985; LANGBEIN *et al.*, 1995; FIALKO *et al.*, 2001; TIAMPO *et al.*, 2000; BATTAGLIA *et al.*, 2003), we have tried additional sources to model the free-air gravity change.

Table 2

(a) Source parameters obtained from the GA inversion for the single-source model and (b) parameters of the two-sources inversion

(a)

Parameter	Value
x	118° 53' 59"W
y	37° 41' 58" N
Depth c (km)	8.4
Increment magma pressure (ΔP) (MPa)	33.2
Radius a (km)	0.4
Mass increment (ΔM) (1MU $= 10^{12}$)	0.53 UM

(b)

Parameter	Source 1	Source 2
X	118° 54' 17"W	118° 55' 22"W
Y	37° 41'23.8"N	37° 37' 51"N
Depth c (km)	8.8	7.5
Increment magma pressure (ΔP) (MPa)	9.6	23.7
Radius a (km)	0.5	0.2
Mass increment (ΔM) (1MU $= 10^{12}$)	0.27 MU	0.2 MU

Then we have tried two spherical point sources to represent the deeper source beneath the resurgent dome and the south moat source (LANGBEIN *et al.*, 1995). Thus, we have doubled the number of unknowns. Table 2 (b) shows the predicted parameters of both sources. The free-air gravity changes computed using the elastic-gravitational model and the predicted parameters obtained from two sources GA inversion are shown in Figure 4. We can observe that the change in the pattern of the calculated residual gravity fits the data better than the one source inversion described above. The location of the maximum value is closer to the observed maximum residual gravity anomaly than the solution from the one source inversion. This value is the same order as the one source inversion (Figs. 2 and 3), although the fitness value of the single source inversion is better than the two sources inversion fitness value. Table 3 gives a summary of how well the single point source and two-source models fit the gravity data. The goodness of the fit (R_2), i.e., the model skill to explain the observed data, is slightly inferior for the two-source inversion model since it is more difficult to assess the parameter set for the second sphere. It seems like the second gravity signal originates from a larger source, and not from what is generally called the south-moat intrusion.

FIALKO *et al.* (2001) performed a joint inversion of the InSAR and two-color geodimeter data from the Long Valley Caldera for the period between 1996 and 1998

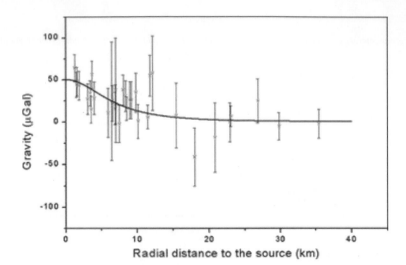

Figure 3
Comparison of observed (crosses with error bars) and synthetic free-air gravity change (solid line) (1982–1998) as a function of the radial distance from the point source for a source which parameters are described in Table 2.

that suggests that the deformation source has a shape of a steeply dipping prolate spheroid with a depth of ≈ 7 km and excess pressure of several MPa. The inferred location is in general agreement with a low velocity and high attenuation region beneath the resurgent dome revealed by seismic studies. Seismic evidence indicates that the main mass of the Long Valley magma chamber is about 10 km in diameter and that its roof is 8–10 km deep with smaller cupolas as shallow as 4–5 km (HILL *et al.*, 1985). BATTAGLIA *et al.* (2003) modeled the source of inflation combining geodetic and micro-gravity data. The inflation source, a vertical prolate ellipsoid, is located at 5.9 km deep beneath the resurgent dome. The results of the two-source inversion performed here are consistent with these results, but suggest that we cannot rule out that the gravity changes could be caused by a single, more spatially distributed source, replacing the magma chamber under Casa Diablo that is represented by two-point sources in the inverse problem.

TIAMPO *et al.* (2004) have performed a sensitivity analysis of the elastic-gravitational model against the GA inversion technique. The results establish that while the deformation measurements are very sensitive to the pressure and radius parameters, the gravity measurements are not. However, with increasing pressure the gravity measurements become more sensitive to both parameters. That is the reason we have performed another inversion with a wider range on the pressure limits for the first source, and on mass and radius limits for the second one. Although the fitness value is still less than the optimal value, the differences between pressure increments of both sources have diminished to 43.5 and

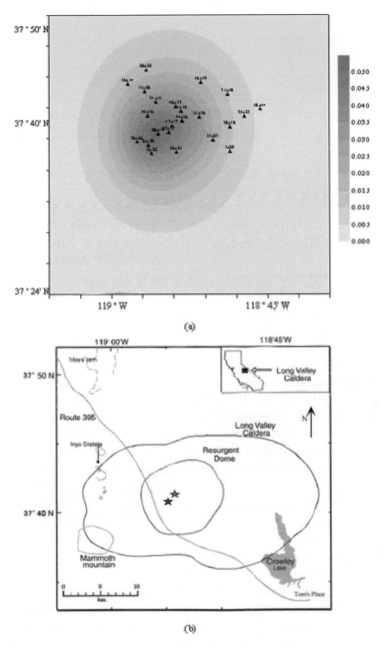

Figure 4
Same as Figure 2 but for two-sources inversion.

46.7 Mpa, respectively. This could indicate that free-air gravity changes can be interpreted by using a more spatially distributed source, and points out the necessity of a joint inversion of gravity and deformation measurements to

Table 3

Fit of the single- and two-source models to free-air gravity change. χ^2 is the chi-square value given by
expression (1). R^2 is the goodness of fit, if $R^2 = 1$, the model is able to explain the observed data and if
$R^2 = 0$ it is not able.

Number of Sources	χ^2	R^2
1	13.04	0.85
2	17.7	0.79

discriminate the role played by the pressure and mass variations in the volume change of the magma chamber.

The fitness values for both solutions (one- and two-source inversions) became stable after some iteration, although these values represent some FV less than optimal (Table 3). This fact implies the inversion results are preliminary and therefore additional or different geometry sources would need to be modeled for matching the residual gravity signal at the Long Valley Caldera.

The volume increase of the chamber is a combination of the volume of magma that enters the chamber in the case of inflation and other factors, including the compression of stored magma (e.g., JOHNSON et al., 2000). The volume of ground-surface uplift is composed of the combination of volume displayed by chamber expansion plus deformation-induced crustal dilation. Thus the chamber volume increase caused by overpressure in the magma chamber could not be equal to volume magma injection, i.e., overpressure could be produced by other factors. However, it is standard to assume the opposed hypothesis. Furthermore, while the deformation is mainly caused by source pressurization (e.g., FERNÁNDEZ et al., 1997), the gravity inversion is not very sensitive to the pressure and radius parameters (TIAMPO et al., 2004). Subsequently we can conclude that the pressure changes from the performed inversions would be unable to fit the observed deformation between 1982–1998. However, if we compute the volumetric expansion of the chamber caused by the intrusion mass, i.e., we assume the chamber volume increase is equal to the volume of magma that enters or leaves the cavity, and presumed a density of 3.3. 10^3 kg m^{-3} (BATTAGLIA et al., 1999) it yields deformation values inside the bounds estimated by BATTAGLIA et al. (1999). This predicted displacement fits the observed deformation data for this period of time. This result and previous theoretical works (FERNÁNDEZ et al., 1997, 2001a,b) point out that displacements and gravity data must be interpreted together whenever possible to discriminate the role played by overpressure and intrusion mass. A joint inversion would provide better constraints on intrusion volume changes and provide more insights into the physical process, since the elastic-gravitational model permits introduction of the pressurization of the magma chamber and the amount of magma recharge associated with the intrusion process (see e.g., FERNÁNDEZ et al., 2001a,b,c; TIAMPO et al., 2004).

6. Conclusions

We have investigated gravity data associated with currently active crustal magma bodies in the Long Valley Caldera, California, USA. Microgravity monitoring on active areas is a valuable tool for mapping the subsurface mass redistributions that are associated with volcanic activity (e.g., RYMER and WILLIAMS-JONES, 2000). Between July 1982 and July 1998 gravity within the Long Valley Caldera decreased substantially. The stations with the largest gravity decrease are all located on the resurgent dome which has experienced uplift during the past 20 years. The free-air gravity change manifests a prominent positive anomaly centered on the resurgent dome with a peak amplitude of 64 ± 16 μGal (BATTAGLIA et al., 1999). We have inverted free-air gravity change data by using a genetic algorithm technique. This technique is a valuable tool for geophysical problems in which data are contaminated by noise and are not continuously spatially distributed.

The results of the performed inversions (one or two spherical point-source inversion) fit the gravity anomaly centered under the resurgent dome. While one-source inversion matches the magnitude (within the rms limits), the two-source inversion matches both the magnitude and pattern of the observed residual gravity, although the single-source inversion fitness value is better than that of the two-source inversion (Table 3). However, given the large uncertainty in the gravity data, it is more difficult to assess the parameter set for the smaller, second sphere. The fitness value of both inversions indicates that the solutions represent a first approximation of the optimal source parameters. However the two-source inversion suggests that the gravity change could be caused by a more spatially distributed source under the resurgent dome. Therefore, additional or different geometry sources which may include the faulting mechanism should be included in the inversion procedure for fitting the pattern and magnitude of the free-air gravity change observed in the Long Valley Caldera. Finally, the results demonstrate the need to improve the inversion using more precise gravity data in Long Valley and to use them jointly with the displacement data.

Acknowledgments

The research by MC and JF has been funded under MCyT projects AMB99-1015-C02 and REN2002-03450. The research by KFT was conducted under NASA grants NAG5-3054. Research by JBR has been supported by US DOE grant DE-FG03-95ER14499. This research has been also partially supported with funds from a New Del Amo Program project. The authors are indebted to F. Beauducel and J. Langbein for helpful comments and suggestions.

REFERENCES

ABERS, G. (1985), *The Subsurface Structure of Long Valley Caldera, Mono County, California: A Preliminary Synthesis of Gravity, Seismic and Drilling Information*, J. Geophys. Res. *90*, B5, 3527–3636.

BAILEY, R. (1989), *Geologic Map of the Long Valley Caldera, Mono-Inyo Craters Volcanic Chain, and Vicinity, Eastern California*, Map I-1933, published by U.S. Geological Survey.

BAILEY, R., DALRYMPLE, G.B. and LANPHERE, M.A. (1976), *Volcanism, Structure and Geochronology of Long Valley Caldera, Mono County, California*, J. Geophys. Res. *81*, 725–744.

BATTAGLIA, M., ROBERTS, C. and SEGALL, P. (1999), *Magma Intrusion beneath Long Valley Caldera Confirmed by Temporal Changes in Gravity*, Science, *285*, 2119–2122.

BATTAGLIA, M., ROBERTS, C. and SEGALL, P. (2003), *The mechanics of Unrest at Long Valley Caldera, California. 2. Constraining the Nature of the Source Using Geodetic and Micro-Gravity Data*, J. Volcanol. Geotherm. Res. *127*, 219–245.

BEAUDUCEL, F., BRIOLE, P. and FROGER, J.L. (2000), *Volcano Wide Fringes in ERS SAR Interferograms of Etna: Deformation or Tropospheric Effect?*, J. Geophys. Res. *105*, B7, 16,391–16,402.

BEVINGTON, P.R. and ROBINSON, D.K., *Data Reduction and Error Analysis for the Physical Sciences* (McGraw-Hill, Inc., N.Y. 1992).

BILLINGS, S., KENNETT, B. and SAMBRIDGE, M. (1994), *Hypocenter Location: Genetic Algorithms Incorporating Problem Specific Information*, Geophys. J. Int. *118*, 693–706.

BOSCHETTI, F., DENTITH, M.C. and LIST, R.D. (1996), *Inversion of Seismic Refraction Data Using Genetic Algorithms*, Geophysics *61*, 1715–1727.

CASTLE, R.O., ESTREM, J.E. and SAVAGE, J.C. (1984), *Uplift across Long Valley Caldera, California*, J. Geophys. Res. *89*, 11,507–11,516.

DE NATALE, G. and PINGUE, F. (1993), *Ground Deformations in Collapsed Caldera Structures*, J. Volcanol. Geotherm. Res. *57*, 19–38.

DENLINGER, R.P. and RILEY, F. (1984), *Deformation of Long Valley Caldera, Mono County, California, from 1975–1982*, J. Geophys. Res. *89*, 8303–8314.

FERNÁNDEZ, J., RUNDLE, J.B., GRANELL, R.R. and YU, T.-T. (1997), *Programs to Compute Deformation due to a Magma Intrusion in Elastic-gravitational Layered Earth Models*, Comp. and Geosc. *23*, 231–249.

FERNÁNDEZ , J. and DÍEZ J.L. (1995), *Volcano Monitoring Design in Canary Islands by Deformation Model*, Cahiers Centre Eur. Géodyn. Séismol. *8*, 207–217.

FERNÁNDEZ, J., CHARCO, M., TIAMPO, K.F., JENTZSCH, G. and RUNDLE, J.B. (2001a), *Joint Interpretation of Displacement and Gravity Data in Volcanic Areas. A Test Example: Long Valley Caldera, California*, Geophys. Res. Lett. *28*, 1063–1066.

FERNÁNDEZ, J., TIAMPO, K.F., JENTZSCH, G., CHARCO, M. and RUNDLE, J.B. (2001b), *Inflation or Deflation? New Results for Mayon Volcano Applying Elastic-gravitational Modeling*, Geophys. Res. Lett. *28*, 12, 2349–2352.

FERNÁNDEZ, J., TIAMPO, K.F. and RUNDLE, J.B. (2001c), *Viscoelastic Displacement and Gravity Changes due to Point Magmatic Intrusions in a Gravitational Layered Solid Earth*, Geophys. J. Int. *146*, 155–170.

FIALKO, Y., KHAZAN, Y., and SIMONS, M. (2001), *Deformation due to Pressurized Horizontal Circular Crack in an Elastic Half-space, with Applications to Volcano Geodesy*, Geophys. J. Int. *146*, 181–190.

FOLCH, A., FERNÁNDEZ, J., RUNDLE, J.B. and MARTÍ, J. (2000), *Ground Deformation in a Viscoelastic Medium Composed of a Layer Overlying a Half-space: A Comparison between Point and Extended Sources*, Geophys. J. Int. *140*, 37–50.

GALLAGHER, K., SAMBRIDGE, M. and DRIJKONINGEN, G. (1991), *Genetic Algorithms: An Evolution From Monte-Carlo Methods for Strongly Nonlinear Geophysical Optimization Problems*, Geophys. Res. Lett. *18*, 2177–2180.

GRANT, F.S. and WEST, G.F., *Interpretation Theory in Applied Geophysics*, p. 583 (McGraw-Hill, New York 1965).

HASKELL, N.A. (1953), *The Dispersion of Surface Waves on Multilayered Media*, Bull. Seismol. Soc. Am. *43*, 421–440.

HILL, D.P. (1984), *Monitoring Unrest in a Large Silicic Caldera, the Long Valley-Inyo Craters Volcanic Complex in East-central California*, Bull. Volc. *47-2*, 371–395.

HILL, D.P., BAILEY, A. and RYALL, A.S. (1985), *Active Tectonic and Magmatic Process beneath Long Valley Caldera: An Overview*, J. Geophys. Res. *90*, B13, 11,111–11,120.

JACHENS, R.C. and ROBERTS, C.W. (1985), *Temporal and Areal Gravity Investigations at Long Valley Caldera, California*, J. Geophys. Res. *90*, 11,210–11,218.

JOHNSON, D.J., SIGMUNDSON F. and DELANEY, P.T. (2000), *Comment on "Volume of Magma Accumulation or Withdrawal Estimated from Surface Uplift or Subsidence, with Application to the 1960 Collapse of Kilauea Volcano" by P.T. Delaney and D.F. McTigue*, Bull. Volcanol. *61*, 491–493.

KISSLINGER, C. (1975), *Processes during the Matsushiro Japan Earthquake Swarm as Revealed by Levelling, Gravity and Spring Flow Observations*, Geology *3*, 2, 57–62.

LANGBEIN, J. (1989), *Deformation of Long Valley Caldera, Eastern California from Mid-1983 to Mid-1988: Measurements Using a two-color Geodimeter*, J. Geophys. Res. *94*, 3833–3849.

LANGBEIN, J., HILL, D.P., PARKER, T.N. and WILKINSON, S.K. (1993), *An Episode of Reinflation of the Long Valley Caldera, Eastern California: 1989–1991*, J. Geophys. Res. *98*, 15,851–15,870.

LANGBEIN, J., DZURISIN, D., MARSHALL, G., STEIN, R. and RUNDLE, J.B. (1995), *Shallow and Peripheral Volcanic Sources of Inflation Revealed by Modeling Two-color Geodimeter and Leveling Data from Long Valley Caldera, California, 1988–1992*, J. Geophys. Res. *100*, 12,487–12,495.

MOGI, K. (1958), *Relations between the Eruptions of Various Volcanoes and the Deformations of the Ground Surfaces around them*, Bull. Earthquake Res. Inst. Univ. Tokyo *36*, 99–134.

ROBERTS, C., JACHENS, R. and MORIN, R. (1988), *High-precision Stations for Monitoring Gravity Changes in Long Valley Caldera, California*, Open-File Report – U.S. Geological Survey OF 88–0050.

RYMER, H. (1994), *Microgravity Changes as a Precursor to Volcanic Activity*, J. Volcanol. Geotherm. Res. *61*, 311–328.

RYMER, H. and WILLIAMS-JONES, G. (2000), *Volcanic Eruption Prediction: Magma Chamber Physics from Gravity and Deformation Measurements*, Geophys. Res. Lett. *27*, 2389–2392.

RUNDLE, J.B. (1978), *Gravity Changes and the Palmdale Uplift*, Geophys. Res. Lett. *5*, 41–44.

RUNDLE, JB. (1980), *Static Elastic-gravitational Deformation of Layered Halfspace by Point Couple Sources*, J. Geophys. Res. *85*, 5355–5363.

RUNDLE, J.B. (1982), *Deformation, Gravity, and Potential Changes due to Volcanic Loading of the Crust*, J. Geophys. Res, *87*, 10,729–10,744 (Correction, J. Geophys. Res. *88*, 10,647–10,652, 1983).

STOFFA, P.L. and SEN, M.K. (1991), *Nonlinear Multiparameter Optimization Using Genetic Algorithms, Inversion of Plane-wave Seismograms*, Geophysics *56*, 1794–1810.

TAYLOR, J.R., *An Introduction to Error Analysis* (University Science Books, USA (1982), 225–230.

TIAMPO, K.F., RUNDLE, J.B., FERNÁNDEZ, J. and LANGBEIN, J. (2000), *Spherical and Ellipsoidal Volcanic Sources at Long Valley Caldera, California Using a Genetic Algorithm Inversion Technique*, J. Volcanol. Geother. Res. *102*, 189–206.

TIAMPO, K.F., FERNÁNDEZ, J., JENTZSCH, G., CHARCO, M. and RUNDLE, J.B. (2004), *Inverting for the Magmatic Source at Mayon, Philippines: Using a Genetic Algorithm to Define the Limits of a Volcanic Source* (in this issue).

WALSH, J.B. and RICE, J.R. (1979), *Local Changes in Gravity Resulting from Deformation*, J. Geophys. Res. *84* (B1), 165–170.

YU, T.T., FERNÁNDEZ, J. and RUNDLE, J.B. (1998), *Inverting the Parameters of an Earthquake-ruptured Fault with a Genetic Algorithm*, Comp. and Geo. *24*, 173–182.

(Received February 15, 2002, revised April 30, 2003, accepted May 9, 2003)

To access this journal online:
http://www.birkhauser.ch

Pure appl. geophys. 161 (2004) 1415–1431
0033–4553/04/071415–17
DOI 10.1007/s00024-004-2512-7

© Birkhäuser Verlag, Basel, 2004

❙ **Pure and Applied Geophysics**

Gravity Changes and Internal Processes: Some Results Obtained from Observations at Three Volcanoes

GERHARD JENTZSCH[1], ADELHEID WEISE[1],
CARLOS REY[2], and CARL GERSTENECKER[3]

Abstract—Temporal gravity changes provide information about mass and/or density variations within and below the volcano edifice. Three active volcanoes have been under investigation; each of them related to a plate boundary: Mayon/Luzon/Philippines, Merapi/Java/Indonesia, and Galeras/Colombia. The observed gravity changes are smaller than previously expected but significant. For the three volcanoes under investigation, and within the observation period, mainly the increase of gravity is observed, ranging from 1,000 nm^{-2} to 1,600 nms^{-2}. Unexpectedly, the gravity increase is confined to a rather small area with radii of 5 to 8 km around the summit.
At Mayon and Merapi the parallel GPS measurements yield no significant elevation changes. This is crucial for the interpretation, as the internal pressure variations do not lead to significant deformation at the surface. Thus the classical Mogi-model for a shallow extending magma reservoir cannot apply. To confine the possible models, the attraction due to changes of groundwater level or soil moisture is estimated along the slope of Merapi exemplarily by 2-D modelling. Mass redistribution or density changes were evaluated within the vent as well as deeper fluid processes to explain the gravity variations; the results are compared to the model incorporating the additional effect of elastic deformation.

Key words: Gravity changes at volcanoes, GPS, deformation, density changes, mass redistribution, modelling.

1. Introduction

For an improved risk assessment the understanding of the physical structure and mechanism of the volcano is important. Temporal gravity changes supply information about mass and/or density variations within the volcano system. Devesiculation processes are expressed by density changes, also combined with mass redistribution. The possible existence of a shallow magma reservoir is of interest as well. Additional information from other methods is needed, for example rock chemistry and gas

[1] Institut für Geowissenschaften, Friedrich-Schiller Universität, Burgweg 11, D-07749 Jena, Germany, E-mail: gerhard.jentzsch@uni-jena.de

[2] INGEOMINAS (Instituto de Investigationes en Geosciencias, Mineria y Quimica), Diagonal 53 No. 34-53, Santafe de Bogota, Colombia. E-mail: crey@ingeomin.gov.co

[3] Institut für Physikalische Geodäsie, Technische Universität Darmstadt, Petersstr. 13, D-64287 Darmstadt, Germany. E-mail: gerstenecker@geod.tu-darmstadt.de

analyses, to be integrated into the interpretation of the gravity changes (compare BROWN and RYMER, 1991).

Three active volcanoes have been under investigation: Mayon/Luzon/Philippines, Merapi/Java/Indonesia and Galeras/Colombia. All of them are located near a plate boundary, Merapi and Mayon are situated very near the coast (55 and 15 km from the summit).

The networks consist of 20 to 30 stations around the volcano with profiles to the summit, if possible. Usually 2 to 4 gravimeters, LaCoste and Romberg and Scintrex, are used in order to increase the number of observations and to decrease instrumental effects. The gravimeters are equipped with electronic feedback and electronic levels. Each gravity difference is observed at least three times with each instrument. Air-pressure variations as well as earth tides are corrected. Temporal changes of elevation are controlled by differential GPS measurements in parallel. The measurements are repeated several times with intervals of $\frac{1}{2}$, 1 or 2 years. Reference points in the vicinity of the volcanoes were chosen such that they are not affected by gravity and elevation changes. To allow the comparison and joint analysis of data of different gravimeters and campaigns, the calibration of the instruments is crucial.

Mayon/Philippines

The Philippines, located in the western Pacific Ocean, are part of an extensive island arc system. The geological and geophysical situation is mainly characterized by the collision between the Philippine and the Eurasian plates, yielding a complex tectonic state with main faults and lineaments extending through the entire crust down to the upper mantle (BISCHKE *et al.*, 1990; AURELIO, 1992). The active tectonics cause earthquakes and volcanic activities.

Mayon is the most famous volcano of the Bicol volcanic chain in the southeast of Luzon (13°15′ N, 123°41′ E, H: 2462 m). Its shape is a nearly perfect cone. Southwest of Mayon, rather close, the Legaspi lineament runs NW-SE across Legaspi City. According to AURELIO (1992) this could probably control the activity of the Mayon volcano (Fig. 1a). Ten activity periods were spread almost regularly over the 20th century. The last strong eruption was in 1984, the youngest one in 2001. Some smaller eruptions took place in February/March 1993, just between our first and second gravity campaigns. The decrease of Mayon's activities after the eruptions is usually rather slow, after one year crater glow still may be observable.

After the eruption of 1984, three seismological observatories were installed (PUNONGBAYAN *et al.*, 1990) and comprehensive research started. Even correlations with earth tides were examined. As the ocean loading tide around the Philippines varies strongly, it is proposed that there may be a triggering of volcano seismicity by the strain induced (JENTZSCH *et al.*, 1997). FE-modelling proves that during a critical phase the pressure increase in the vent due to cooling/crystallization is in the same order as the tidal stress (JENTZSCH *et al.*, 2001b).

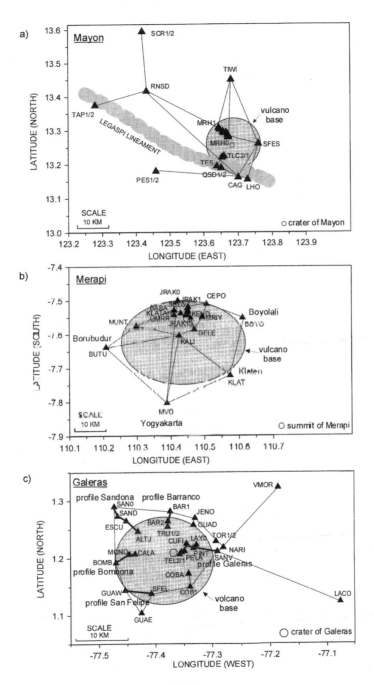

Figure 1

Sketches of the gravity and GPS networks at the three volcanoes (stations around and profiles): (a) Mayon, two profiles: MRH1 to MRH0 (Mayon Resthouse), QSD1 ... TLC2 (Tumpa Lahar Channel); (b) Merapi, one profile JRAK0 to JRAK15, (c) Galeras, 5 profiles.

The lava shows cyclical variation with basaltic and andesitic flows (NEWHALL, 1977, 1979). Actual geochemical analyses of samples of the volcanics from different lava flows exhibit typical arc-related trace element characters (JENTZSCH *et al.*, 2001a). The low degree of fractional crystallization of the lava samples argues against a shallow magma reservoir.

A network of 26 stations was installed around the volcano, including two profiles at the slope, towards the summit (maximum up to 850 m; JAHR *et al.*, 1995; see Fig. 1a). The gravity points are installed in areas of sand (lahars), pasture ground and basaltic rock. The total gravity range is 1800 μms^{-2} (1 mGal = 10^{-5} ms^{-2} = 10 μms^{-2}). During 4 years, 1992 until 1996, 5 campaigns of repeated gravity measurements were carried out with three gravimeters each (see Fig. 2a).

Merapi/Java/Indonesia

The Indonesian IDNDR decade volcano Merapi in Central Java is a high-risk stratovolcano (7° 32′ S, 110° 26′ E, H: 2985 m). Merapi is located about 55 km from the south coast of Java. The dominant magmatic mechanism of Merapi is andesitic and explosive. Large eruptions occur about every 30 years. The permanent activity is accompanied by small and intermediate eruptions. The lava of Merapi consists of basaltic andesite, the content of silicic acid is about 50 to 60 %. The viscosity of the lava leads to a continuously increasing lava dome. The state of instability is indicated by an increasing number of eruptions (block-and-ash flows, hot clouds and surges) that are triggered by a gravitational collapse of a part of the dome. This occurred about every five months during the last years. The mean cumulative ejected volume is 1 to $10 \cdot 10^6$ m^3/yr. Early seismological data show a zone with anomalously high attenuation of seismic waves 1 to 2 km below the summit, which was interpreted as a shallow magma reservoir (RATDOMOPURBO and POUPINET, 1995). But the seismic data from OHRNBERGER *et al.* (2000) do not show clear indications for an aseismic zone.

In 1997 a network of 18 points was installed in the form of three interconnected rings around Merapi in different levels, and attached a profile up to the summit (Fig. 1b; SNITIL *et al.* 2000), including some of the points of the French Group (JOUSSET *et al.*, 2000). The GPS-reference point of the Geodyssea-network near Borobudur (BUTU) is included within this network which covers a total gravity range of 7,510 μms^{-2}. The measurements started in 1997. Until 2000 five campaigns could be carried out, mainly in August/September (Fig. 2b). During this period an activity period occurred in July 1998, just before the third campaign.

Galeras/Colombia

Galeras volcano, South America's only IDNDR decade volcano, is one of the active volcanoes in the Northern Andes, at the convergence of West- and Central-Cordilleres, in South-West Colombia (1° 12′ N, 77° 21′ W, H: 4276 m). At least six

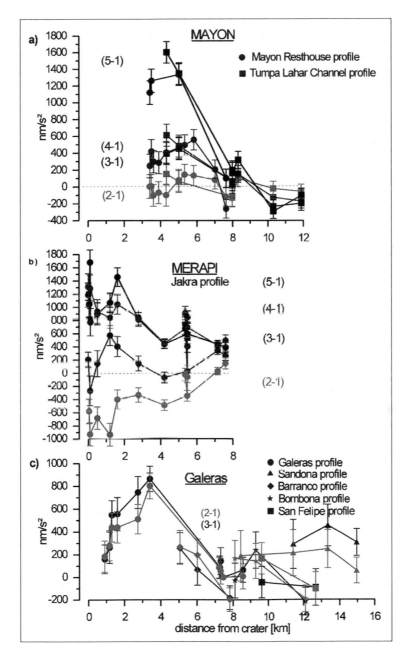

Figure 2

Observed gravity changes along profiles at the three volcanoes (relative to first campaign), error bars denote the standard deviation for a gravity difference: a) Mayon/Philippines: two profiles, campaigns: 1) Dec. '92, 2) April '93, 3) Dec. '93, 4) Dec. '94, 5) Dec. '96; volcanic activity: Feb./March '93. b) Merapi/Java: one profile, campaigns: 1) Aug. '97, 2) Feb. '98, 3) Aug. '98, 4) Aug. '99, 5) Aug. 2000; volcanic activity: July '98. c) Galeras/Colombia: main profile "Galeras" and four short profiles around the volcano, campaigns: 1) Sept. '98, 2) March '99, 3) March 2000; volcanic activity: Sept./Oct. '98.

major eruptions have been identified during the past 4,500 years. The high andesitic stratovolcano reactivated in 1988 after more than 40 years of rest. It is well known after the unexpected explosion of January 14, 1993, which killed several scientists and guides visiting the crater. Galeras is characterized by a varied pattern of seismic signals as well as by continuous activity with considerable variation of intensity. The most common events are pyroclastic flows. The main crater is surrounded by a rim of the ancient volcano, open to the west. The western slope of Galeras is not accessible after the slide of the west flank some 6000 years ago.

After August 1990, an inflationary tilt signal of about 400 μrad in radial direction and 60 μrad tangential was measured during a volcanic activity by two electronic tiltmeters installed below the rim of the caldera (ORDOÑEZ and REY, 1997; CALVACHE and WILLIAMS, 1997), referring to recent deformations.

The gravity network consists of 30 points, some of them originally established in the frame of the CASA GPS-geodynamic program (FREYMUELLER *et al.*, 1993). It includes two references with in a distance of more than 20 km on the eastern side of the volcano (Fig. 1c). The main profile leads along the road from Pasto (2,500 m) to the eastern rim of the caldera (highest point 4,005 m). Additional points are placed at the road around 7 to 12 km from the crater (to 1838 m deep), connected to four additional profiles leading upslope as far as the terrain allows. The connection of the profiles at the top is not possible (JENTZSCH *et al.*, 2000).

We completed three measuring campaigns (Fig. 2c). During the first campaign seismic activity of low level occurred.

2. Gravity Changes

From the least-squares adjustments of the observed gravity differences the standard deviations of the estimated gravity values of one campaign range from $\pm$ 60 to 150 nms^{-2} (1 σ, usually increasing with increasing elevation). Thus the temporal gravity changes have a standard deviation of $\pm$ 100 to 180 nms^{-2}. That means, generally for all three volcanoes: Only gravity changes of at least 300 to 400 nms^{-2} (2 σ) can be considered as significant.

The observed gravity changes are smaller than previously expected, nonetheless significant. For the three volcanoes under investigation, mainly an increase of gravity was observed to date. The ranges are 1,000 nms^{-2} to 1,600 nms^{-2}, spanning 2 to 4 years, at Galeras after half a year, providing that gravity is constant at the most distant points from the summit. The increase of gravity started after more or less strong volcanic activities.

At Mayon and Merapi the course of gravity changes as well as the areal distribution is similar. The gravity changes for the second and third campaigns, respectively, are hardly significant, however after four and five campaigns they are significantly part of a long-term trend of gravity increase. The distribution of the gravity changes over a rather short distance of 5 to 8 km from the summit is a special feature. In the more distant surroundings of the volcanoes generally no significant gravity changes are observed at the three volcanoes. The observed temporal gravity changes are presented for each volcano in Figure 2. In detail we state:

Mayon

At Mayon volcano along the two profiles the corresponding results are: The gravity increase started with the third campaign (after one year) with 400 nms^{-2}, just after an activity in February/March 1993, and continued until the fifth campaign (four years later) from 1,200 to 1,600 nms^{-2} (Fig. 2a). The temporal increase of gravity with 300 to 400 nms^{-2}/yr for the upper points is nearly linear. Gravity changes of $\leq$ 400 nms^{-2} for distant points are insignificant.

Merapi

At Merapi the second campaign is characterized by maximum gravity decrease of 1000 nms^{-2}. Then, after an activity in July 1998, just before the third campaign, gravity increased to nearly zero at the third campaign and finally to 1000 nms^{-2} for the fourth and fifth campaigns, after two and three years (Fig. 2b). The gravity increase vanishes at a distance of 5 to 7 km from the crater. As the second campaign took place during the rainy season (February/March 1998), the quality of the gravity data (the GPS data as well) of the second campaign especially along the summit profile, is problematic and thus the gravity decrease is actually insignificant.

The temporal gravity variations in the surroundings of Merapi are reduced to nearly zero at distances of 15 to 20 km from the summit, also reproducing the tendency to gravity decrease for the second campaign and gravity increase for the fourth and fifth campaigns, partly also for the third campaign. The most distant station (MVO at 25 km from the summit) is thought to be constant in gravity. The gravity increase at the top reaches 1,500 nms^{-2}, with strong lateral variations of approximately 500 nms^{-2} near the crater. This could be an effect of changing dome topography, estimated with up to 1,000 nms^{-2} for a point directly near the dome. However, the temporal variation of the dome topography is not yet observable.

Galeras

At Galeras the course of gravity changes is also positive although slightly different. At the foot around the volcano gravity is constant, here the gravity change is around zero within the error bars. Along the main profile leading to the top gravity increase reaches a maximum of 800 to 900 nms^{-2} at about 3 km from the summit, further decreasing to a gravity increase of about 200 nms^{-2} near the top (Fig. 2c) over a time span of half a year (second campaign). This is confirmed by the third campaign after one more year. The short profiles around Galeras show a similar tendency of gravity increase, including a decline with increasing distance from the summit (Fig. 2c). Although the gravity changes at these shorter profiles are merely significant, all profiles exhibit the same tendency, confirmed by the results of the third campaign.

Calibration of the Gravimeters

The calibration of the gravimeters generally is not better than $5 \cdot 10^{-5}$. As the gravity range at volcanoes is rather sizable (2 to $7 \cdot 10^3 \ \mu ms^{-2}$), due solely to the height differences, the calibration of all instruments is crucial. The maximum accuracy of 10^{-5} provides for a maximal gravity range of $7 \cdot 10^3 \ \mu ms^{-2}$, an uncertainty of $350 \ nms^{-2}$, being in the range of our signal. Consequently we need scale factors constant in time ranging $5 \cdot 10^{-5}$. The temporal stability of the scale factors is controlled by frequent calibration along calibration lines preferably close to the volcanoes, as it is critical to transfer the scale factors obtained for the gravity range of Germany to areas close to the equator. The problem of the calibration accuracy can only be reduced by the use of several gravimeters in parallel.

3. Elevation Control with GPS

The gravity variations must be reduced for elevation changes supplied by parallel differential GPS measurements (VÖLKSEN and SEEBER, 1995). At Galeras two receivers were available, three receivers at Mayon and Merapi (Trimble and Leica), in order to introduce full triangles into the adjustments. The elevation changes between two campaigns are obtained with an accuracy of about $\pm$ 3 to $5 \cdot 10^{-2}$ m at Mayon and Merapi, which is adequate to the achieved gravity changes (Fig. 3).

At Mayon and Merapi the parallel GPS measurements yield obviously no significant elevation changes (Figs. 3a,b). A gradient is indeterminable within the error bars. This is a very important aspect for the interpretation, as the internal pressure variations do not lead to significant deformation at the surface. Thus the classical Mogi-model for a shallow extending magma reservoir cannot apply in these two cases.

At Galeras no elevation changes could be derived as yet. Due to the use of only two receivers and disadvantages in the configuration of the GPS network, the quality of the data is not sufficient for all the observed points. More sophisticated analyses are still pending.

4. Influence of Ground Water and Soil Moisture

In general, we can distinguish different effects of ground water and/or soil moisture:

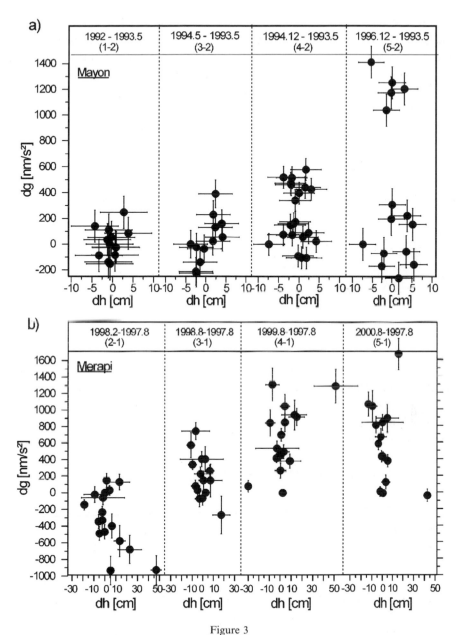

Figure 3
Elevation changes (GPS) versus gravity changes from five campaigns at a) Mayon/Philippines, gravity and elevation changes relative to second campaign, b) Merapi/Java/Indonesia, gravity and elevation changes relative to first campaign.

(1) Provided a homogeneous distribution of the available pore volume and water saturation, the gravity effect is eliminated because we measure gravity differences.

(2) In the case of laterally different pore volumes, density changes due to soil humidity occur and, using a Bouguer slab, we estimate an attraction effect of 419 nms^{-2} for a disk of 4 m thickness (porosity of 50%, change in the filling of the pore space of 50% leads to 250 kg/m^3). A lateral variation of the density change may reach a maximum effect of 100 nms^{-2}/m (400 nms^{-2}) (Fig. 4-left). As precipitation is considerable in tropical regions, this is a maximum estimation while the campaigns usually took place in the same season of the year, preferably during the dry season.

(3) The influence of humidity due to topography is roughly estimated with 2-D modelling. This indicates that the local topography 100 to 200 m around the point is effective. Due to the lack of more detailed information for a solid 3-D simulation this is suitable. The change of the gradient of the topography produces a very local and minor effect of $\leq$ 100 nms^{-2} (Fig. 4-center). The summit topography may cause a gravity effect due to humidity/ground water change of 200 nms^{-2} (Fig. 4-right). If the topography is smooth, there is no major effect in the expected gravity difference. A lateral "long-period" gravity effect increasing towards the summit can only reach about 100 nms^{-2}.

5. *Modelling of Gravity Changes*

At the three volcanoes under investigation, a gravity increase is observed after a more or less strong volcanic activity. At Mayon and Merapi the significant gravity

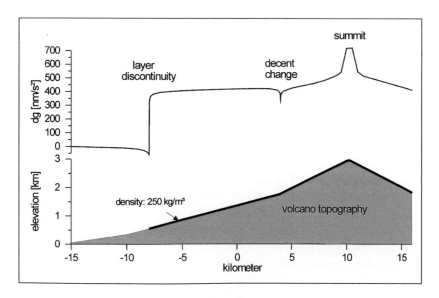

Figure 4

Exemplary 2-D modeling of the gravity effect due to density change in a 4-m thick surface layer, due to variation of ground water or soil moisture, in the cases of a semi-infinite layer, change of decent (topography), and summit topography of Merapi.

variations are confined to a radius less than 8 to 10 km from the summit, increasing systematically with decreasing distances. They are in the order of 1000 to 1500 nms^{-2}, significantly bigger than the estimated maximum effect of < 400 nms^{-2} due to ground water level changes.

In both cases the observed gravity changes are not accompanied by significant elevation changes. There is no significant coherence between changes of gravity and elevation. This is a very important factor for the interpretation, as the internal pressure variations do not lead to significant deformation at the surface. Thus the classical MOGI-model (1958) for a shallow extending magma reservoir cannot apply.

Generally we expect that due to an intrusion of magma, gravity is increased, and with decreasing activity gravity decreases. However in contrast, at Mayon and at Merapi similarly, gravity increase started after an activity phase, thus a gravity increase accompanies the decrease of volcanic activity. This can be interpreted by gravity changes due to mass redistribution or density changes for two different processes:
- for Mayon volcano a process within the vent and
- fluid processes for Merapi.

5.1 Vent Process vs. Elasto-gravitational Effects for the Mayon Volcano

Geochemical analyses of rock samples of Mayon indicate a rather undifferentiated magma (JENTZSCH et al., 2001a). Similar results are reported by BERRINO and CORRADO (1991) for Vesuvius and the Aeolian Islands/Italy, who explained gravity changes, not accompanied by vertical movements, by mass redistribution (see also EGGERS, 1987, for andesitic volcanoes).

The observed gravity increase at Mayon can be interpreted by a mass redistribution in the volcanic vent (density change or magma movement) which causes a mass change from above the gravity points to below (highest elevations of the gravity points are about 850 m and 560 m at the two profiles). The model is quite basic (JENTZSCH et al., 2001a): It consists of a vertical cylinder in the place of the volcanic vent. The mass removed downwards from the vent is distributed in a layer below the volcano edifice (radius: 2.5 km; JAHR et al., 1998). Besides the density change, the controlling variable is the radius of the vent. In accordance with RYMER (1991), who assumed at Poás volcano a diameter of several hundred meters for the vent area, for Mayon radii between 200 m and 400 m are tested (Table 1). The density changes are estimated with 400 kg/m^3 and 600 kg/m^3. The observed gravity changes (last campaign) are compared in Table 1 to the computed gravity changes for the points closest to the summit, for different combinations of parameters. In Figure 5a the model curves are compared to the trend of the observed gravity change along both profiles at Mayon.

With this model we can explain not only the observed increase in gravity, but also the range and restriction of the changes to the area of the volcano (radius 8 to 10 km).

Table 1

Modelled maximum gravity changes at Mayon along two profiles in [nms^{-2}] for the model parameters of the vent: radius and density difference

Model version	Radius [m]	Density diff. [kg/m^3]	Profile Tumpa Lahar Channel	Profile Mayon Resthouse
1	200	400	155	297
2	200	600	233	445
3	300	400	349	668
4	300	600	524	1002
5	400	400	621	1188
6	400	600	931	1782
max. observ.			1579	1339

On the other hand, FERNÁNDEZ *et al.* (2001) and TIAMPO *et al.* (2004) showed, that by elastic-gravitational modelling of gravity changes and displacements together, the curve fitting is considerably better. Here, the assumption is inflation rather than deflation. The boundary condition "no elevation changes" is used to constrain elastic deformation. It is also possible that both effects are present; in the vent model no deflation of the area is assumed. The occurrence of the eruption in 2001 may support this idea.

5.2 Deep Fluid Process for Merapi

Beneath Merapi MÜLLER *et al.* (2000) found a conductive layer nearly 1000 m below the surface, with a thickness of roundly 1000 m. Mean resistivity for the conductor is 20 Ωm. While this layer continues below Java extending to the coast (RITTER *et al.*, 1998), they suggest a water-saturated layer. With a realistic porosity reaching 15% in volcanic sediments, the fluid is expected to be very saline (MÜLLER *et al.*, 2000). The origin could be meteoric water which is thought to be transported thermally upwards into the summit area.

Temporal density variations could occur within this conductive layer by varying water content. Already very small density changes of 2 to 4 kg/m^3 within a 1000-m thick layer can produce attraction effects in the order of 1000 nms^{-2}. Figure 5b provides a sketch of the layer beneath the volcano for certain radial-symmetric cases (lateral constant density change or varying with the distance). The upper part of Figure 5b shows the attraction effect over the distance from the summit. Compared to the observation this suggests eventual variations of water content varying laterally,

▶

Figure 5

Modelling of the observed gravity changes: (a) Mayon/Philippines: modelled vent process (- - - - disc model), compared with observed gravity changes along two profiles (—●—, —■—); (b) Merapi/Java: modelled gravity effect (above) of a semi infinite water saturated layer beneath the volcano in a depth of ca. 1000 m (below).

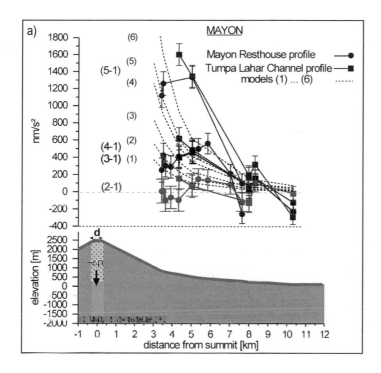

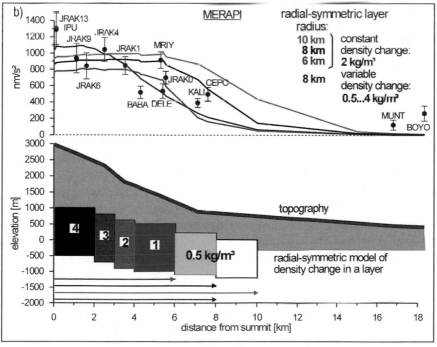

maximal near the center of the volcano. The distribution of the observed gravity points give rise to a slight asymmetry. Temporal variations of the water saturation within this layer could be connected with the hydrothermal system of Merapi.

Chromatographic gas measurements at the top of Merapi yield a very high water content of $>90\%$ with relatively strong short-periodic variations (5 hours) while the long periodic behavior is yet unknown (ZIMMER *et al.*, 2000). However as the water present in the fumarole is saturated with dissolved atmospheric nitrogen and noble gases, it is concluded to be mainly rain water, circulating within the volcano edifice.

Possible causes of the local gravity variations reaching 500 nms^{-2} on top of Merapi are:

• variation of the dome topography (no GPS measurements at the top of the dome; not corresponding to lacking elevation changes of the area),
• density variation within the vent system in the upper 1000 m, and
• variation of ground water/soil moisture.

Although this is not significant, at Merapi there is a slight trend observable to a gradient between gravity and elevation change among some upper points (Fig. 3b). In addition, a gravity effect from varying density in the vent is possible, which can produce a local effect. Thus, a combination of both models 5.1 and 5.2 is also possible.

Despite lacking boundary conditions, considerations regarding the effect of elastic deformation are taken into account, gravity and elevation changes are interpreted together in one model applying elasto-gravitational modelling for Mayon (FERNÁNDEZ *et al.*, 2001).

6. Conclusions

The aim is a better understanding of the internal physics of the volcanoes under investigation. The relatively small but significant gravity changes at the three volcanoes approach the threshold of being traceable due to the extreme condition and the large total gravity range at volcanoes, point to the importance of the calibration of the gravimeters. However the tendency of temporal gravity changes and above all the comparison between several volcanoes can provide further knowledge leading to ideas of a more realistic interpretation.

Mayon and Merapi are similar concerning the observed increase of gravity during decreasing volcanic activity not accompanied by elevation changes. The restriction of the changes to the fairly small area of the volcanoes is also an observed phenomenon (at Galeras as well). The observed gravity variations at Mayon, Merapi and Galeras which extend over several kilometers from the summit cannot be caused solely by ground water or soil moisture variation, because these variations near the surface may not surmount 400 nms^{-2}, and they will only occur at single points. Effects related to the topography can appear very locally with 100 nms^{-2}, at the summit

200 nms^{-2} are possible. There are no indications of a shallow magma reservoir below the volcanoes or even within the volcano edifice.

Nonetheless, although the general position of Mayon and Merapi is similar (subduction, located close to the coast), at both volcanoes the interpretation leads to different models:

(1) For Mayon both inflation and deflation (mass redistribution in the vent system) are possible, which can even be combined. The amount of the observed gravity change can be explained as well as the areal distribution. There is no hint to ground water/soil moisture effects and there is no further data to constrain the modelling.

(2) For Merapi additional information is available suggesting a density and/or mass variation within a deep layer: e.g., fluid (water) can explain the temporal gravity changes regarding the amount as well as the distribution around the volcano edifice. This effect could act coupled with the activity of the volcano by the pressure variation inside the edifice.

The trend of the gravity changes along the top-profile at Merapi generally is rather similar to that observed at Mayon volcano.

With regard to the monitoring strategy (early warning) microgravity can be of use for long- and mid-term prediction once or twice a year. Of course this cannot compete with seismic monitoring for short-term prediction as it is done at Merapi (RATDOMOPURBO and POUPINET, 1995; OHRNBERGER et al., 2000), and Galeras (SEIDL et al., 1998a,b), and is being improved at Mayon volcano.

7. Acknowledgements

This research was partly supported by Deutsche Forschungsgemeinschaft (DFG, Mayon, Merapi), by Gesellschaft für Technische Zusammenarbeit (GTZ, Mayon), and by Bundesanstalt für Geowissenschaften und Rohstoffe (BGR, Hannover, Galeras). Under the research contract of the European Commission (ENV4-CT96-0259) detailed studies were possible (Mayon). We thank D. Lelgeman, Technical University Berlin, and the Bundesamt für Kartographie und Geodäsie in Frankfurt for making available L&R gravimeters. For the least-squares adjustments of the gravity observations the program system GRAV by late H.-G. Wenzel was used.

The project at Mayon volcano was a cooperation with G. Seeber, TU Hannover (GPS-measurements), U. Schreiber, University of Essen (Geology and Geochemistry) and R.S. Punongbayan, PHIVOLCS (Philippine Institute of Volcanology and Seismology). The activity at Merapi is part of the project MERAPI (Interdisciplinary Research at a High-Risk Volcano), supported by the Volcanological Survey of Indonesia (VSI) and the Gadjah Mada University, Dep. of Physics, Yogyakarta (I. Suyanto). The measurements at Galeras are carried out in cooperation with INGEOMINAS (Instituto de Investigationes en Geociencias, Mineria y Quimica,

GPS), M. Ordoñez carried out the GPS measurements and A. Bermudez and Gloria Moncayo-Gamez participated in the gravity measurements. Marta Calvache, the head of the Volcanological Observatory at Pasto, provided for the local logistics. The Nariño University in Pasto also supported our work (late A. Cabrera).

Two anonymous reviewers provided valuable comments and suggestions to improve the manuscript. All this is gratefully acknowledged.

REFERENCES

AURELIO, M. A. (1992), *Tectonique du segment central de la faille Philippine*, Academie de Paris, Ph.D. Thesis, Université Pierre et Marie Curie.

BISCHKE, R. E., SUPPE, J., and PILAR, R. D. (1990), *A New Branch of the Philippine Fault System as Observed from Aeromegnetic and Seismic Data*, Tectonophysics *183*, 243–264.

BERRINO, G. and CORRADO, G. (1991), *Gravity Changes and Volcanic Dynamics*, Cahier du Centre Europ. Geod. et de Seismol., Vol. 4, Proc. of the Workshop: *Geodynamical Instrumentation Applied to Volcanic Areas*, Oct. 1–3, 1990, Walferdange, Luxemburg, 305–323.

BROWN, G. and RYMER, H. (1991), *Microgravity Monitoring at Active Volcanoes: A Review of Theory and Practice*, Cahier Centre Europ. Geod. et de Seismol., Vol. 4, Proc. of the Workshop: *Geodynamical Instrumentation Applied to Volcanic Areas*, 1990, Walferdange, Luxembourg, 279–304.

CALVACHE, M. L. and WILLIAMS, S. N. (1997), *Emplacement and Petrological Evolution of the Andesitic Dome of Galeras*, J. Volcanol. and Geotherm. Res. 77, 57–9.

EGGERS, A. A. (1987), *Residual Gravity Changes and Eruption Magnitudes*. J. Volcanol. and Geotherm. Res. *33*, 210–216.

FERNÁNDEZ, J., TIAMPO, K. F., JENTZSCH, G., CHARCO, M., and RUNDLE, J. B. (2001), *Inflation or Deflation? New Results for Mayon Volcano Applying Elastic-gravitational Modeling*, Geophys. Res. Lett. *28*, 2349–2352.

FREYMUELLER, J., KELLOGG, J., and VEGA, V. (1993), *Plate Motions in the North Andean Region*, J. Geophys. Res. *98*, 21,853–21,863.

JAHR, T., JENTZSCH, G., and DIAO, E., (1995), *Microgravity Measurements at Mayon Volcano, Luzon, Philippines*, Cahier du Centre Europ. de Geod. et de Seismol., Vol. 8, Proc. of the Workshop: *New Challenges for Geodesy in Volcanoes Monitoring*, 1993, Walferdange, Luxembourg, 307–317.

JAHR, T., JENTZSCH, G., PUNONGAYAN, R. S., SCHREIBER, U., SEEBER, G., VÖLKSEN, C., and WEISE, A. (1998), *Mayon Volcano, Philippines: Improvement of Hazard Assessment by Microgravity and GPS-measurements?* Proceedings Int. Symp. on *Current Crystal Movement and Hazard Assessment* (IUGG, IAG), Wuhan, Nov. 1997, Seismological Press, Beijing, pp. 599–608.

JENTZSCH, G., HAASE, O., KRONER, C., WINTER, U., and PUNONGBAYAN, R. S. (1997), *Tidal Triggering at Mayon Volcano?* Cahier Centre Europ. de Geod. et de Seismol., Vol. 14, Proc. of the Workshop: *Short Term Thermal and Hydrological Signatures Related to Tectonic Activities*, 1995, Walferdange, Luxembourg, pp. 95–104.

JENTZSCH, G., CALVACHE, M., BERMUDEZ, A., ORDONEZ, M., WEISE, A., and MONCAYO, G. (2000), *Microgravity and GPS at Galeras/Colombia: The New Network and First Results*, In (Buttkus, B., Greinwald, S., Ostwald, J., eds.), 2. Merapi-Galeras Workshop, 1999, Deut. Geophys. Ges., IV/2000, pp. 65–68.

JENTZSCH, G., PUNONGBAYAN, R. S., SCHREIBER, U., SEEBER, G., VÖLKSEN, C., and WEISE, A. (2001a), *Mayon volcano, Philippines: Change of Monitoring Strategy after Microgravity and GPS Measurements*, J. Volcanol. and Geotherm. Res. *109*, 219–234.

JENTZSCH, G., HAASE, O., KRONER, C., and WINTER, U. (2001b), *Mayon Volcano, Philippines: Some Insights into Stress Balance*, J. Volcanol. and Geotherm. Res. *109*, 205–217.

JOUSSET, Ph., DWIPA, S., BAUDUCEL, F., DUQUESNOY, T., and DIAMANT, M. (2000), *Temporal Gravity at Merapi during the 1993–1995 Crisis: An insight into the Dynamical Behaviour of Volcanes*, J. Volcanol. and Geotherm. Res. *100*, 289–320.

MOGI, K. (1958), *Relations between the Eruptions of Various Volcanoes and the Deformation of the Ground Surfaces Around them*, Bull. Earthquake Res. Inst. Tokyo *36*, 99–134.

MÜLLER, M., HÖRDT, A., and NEUBAUER, F. M. (2000), *A LOTEM Survey on Mt. Merapi 1998—first Insights into 3-D Resistivity Structure*. In (Buttkus, B., Greinwald, S., Ostwald, J., Eds.,: 2. Merapi-Galeras Workshop, 1999, Deut. Geophys. Ges., IV/ 2000, pp. 43–47.

NEWHALL, C. G. (1977), *Geology and Petrology of Mayon Volcano, Southeastern Luzon, Philippines*, Master Thesis Geology, Univ. California (Davis), 292 pp.

NEWHALL, C. G. (1979), *Temporal Variation in the Lavas of Mayon Volcano, Philippines*, J. Volcanol. and Geotherm. Res. *6*, 61–83.

OHRNBERGER, M., WASSERMANN, J., BUDI, E. N., and GOSSLER, J., (2000), *Continuous Automatic Monitoring of Mt. Merapi's Seismicity*. In (Buttkus, B., Greinwald, S., Ostwald, J., Eds.), 2. Merapi-Galeras Workshop, 1999, Deut. Geophys. Ges., IV/2000, pp. 103–108.

ORDOÑEZ, M. and REY, C. A. (1997), *Deformation Associated with the Extrusion of a Dome at Galeras Volcano, Colombia, 1990–1991*, J. Volcanol. and Geotherm. Res. *77*, 115–120.

PUNONGBAYAN, R. S., PENA, O. D., RAMOS, E. G., UMBAL, J. V., RUELO, H. B., BAUTISTA, L. P., and TAYAG, J. C. (1990), *Operation Mayon*, Philippine Institute of Volcanology and Seismology (unpublished).

RATDOMOPURBO, M. A. and G. POUPINET, (1995), *Monitoring Temporal Change of Seismic Velocity in a Volcano: Application to the 1992 Eruption of Mt. Merapi (Indonesia)*, Geophys. Res. Lett. *22* (7), 775–778.

RITTER, O., HOFFMANN-ROTHE, A., MÜLLER, A., DWIPA, S., ARSADI, E. M., MAHFI, A., NURNUSANTO, I., BYRDINA, S., ECHNACHT, F., and HAAK, V. (1998), *A Magnetotelluric Profile Across Central Java, Indonesia* Geophys. Res. Lett. *25* (23), 4265 4268.

RYMER, H. (1991), *The Use of Microgravity for Monitoring and Predicting Volcanic Activity: Poás volcano, Costa Rica*, Cahier Centre Europ. de Geod. et de Seismol., Vol. 4, Proc. of the Workshop: *Geodynamical Instrumentation Applied to Volcanic Areas*, 1990, Walferdange, Luxembourg, pp. 325 331.

SEIDL, D., CALVACH, M., BANNERT, D., BUTTKUS, B., FABER, E., GREINWALD, S., HELLWEG, M., and RADEMACHER, R. (1998a), *The Galeras Multiparameter-Station*, Deut. Geophys. Ges., Mitt. III, 9–11.

SEIDL, D., HELLWEG, M., BUTTKUS, B., GOMEZ, D., and TORRES, R. (1998b), *Seismic Signals at Galeras Volcano*, Deut. Geophys. Ges., Mitt. III, 9–11.

SNITIL, B., GERSTENECKER, C., JENTZSCH, G., LÄUFER, G., SETIAWAN, A., and WEISE, A. (2000), *Gravity and GPS-measurements at Mt. Merapi: Results of 4 Campaigns*, In (Buttkus, B., Greinwald, S., Ostwald, J., eds.), 2. Merapi-Galeras Workshop, 1999, Deut. Geophys. Ges., IV/2000, pp. 65–68.

TIAMPO, K. F., FERNÁNDEZ, J., JENTZSCH, G., CHARCO, M., and YU, T.-T. (2004), *A Two-source Inversion for Modeling Deformation and Gravity Changes Produced by a Magmatic Intrusion Considering Existing Ground Topography at the Mayon Volcano, Philippines*, Pure Appl. Geophys. *161*, 7.

VÖLKSEN, C. and SEEBER, G. (1995), *Establishment of a GPS Based Control Network at Mayon Volcano*, Cahier Centre Europ. Geod. et de Seismol., Vol. 8, Proc. of the Workshop: *New Challenges for Geodesy in Volcanoes Monitoring*, 1993, Walferdange, Luxembourg, pp. 99–113.

ZIMMER, M., ERZINGER, J., and SULISTIYO, Y. (2000), *Continuous Chromatographic Gas Measurements on Merapi Volcano, Indonesia*. In (Buttkus, B., Greinwald, S., Ostwald, J., eds.), 2. Merapi-Galeras Workshop, 1999, Deut. Geophys. Ges., IV/2000, 87–91.

(Received February 11, 2002, revised May 6, 2003, accepted May 13, 2003)

 To access this journal online:
http://www.birkhauser.ch

Pure appl. geophys. 161 (2004) 1433–1452
0033–4553/04/071433–20
DOI 10.1007/s00024-004-2513-6

❙ Pure and Applied Geophysics

New Results at Mayon, Philippines, from a Joint Inversion of Gravity and Deformation Measurements

K. F. Tiampo[1], J. Fernández[2], G. Jentzsch[3],
M. Charco[2], and J. B. Rundle[4]

Abstract—In this paper, we detail the combination of the genetic algorithm (GA) inversion technique with the elastic-gravitational model originally developed by Rundle and subsequently refined by Fernández and others. A sensitivity analysis is performed for the joint inversion of deformation and gravity to each of the model parameters, illustrating the importance of proper identification of both the strengths and limitations of any source model inversion, and this technique in particular. There is a practical comparison of the theoretical results with the inversion of geodetic data observed at the Mayon volcano in the Philippines, where there are gravity changes without significant deformation, after the 1993 eruption.

Key words: Genetic algorithm, elastic-gravitational layered earth model, volcanic source, gravity, displacement.

1. Introduction

Volcanic activity almost inevitably produces geodetic effects such as deformation and gravity changes. As a result, the most promising ways of studying volcano- and earthquake-related phenomena today include the use of high-precision crustal deformation data. On the basis of this fact and the high levels of precision attainable, different geodetic techniques have proven to be a powerful tool in the monitoring of volcanic activity, making it possible to detect ground motion and gravity changes that reflect magma rising from depth, at times months or weeks before the magma flow leads to earthquakes or other eruption precursors

[1] Cooperative Institute for Research in Environmental Sciences, University of Colorado, Boulder, CO 80309, USA and Dept. of Earth Sciences, University of Western Ontario, London, Ontario, Canada. E-mail: ktiampo@uwo.ca.
[2] Instituto de Astronomía y Geodesia (CSIC UCM), Fac. C. Matemáticas, Ciudad Universitaria, 28040–Madrid, Spain.
[3] Institute for Geosciences, FSU Jena, Burgweg 11, D-07749 Jena, Germany.
[4] Center for Computational Science and Engineering, University of California, Davis, CA 95616, USA.

(e.g., DELANEY and McTIGUE, 1994; RYMER, 1996; DVORAK and DZURISIN, 1997; MASSONET and FEIGL, 1998; DZURISIN et al., 1999; FERNÁNDEZ et al., 1999; DZURISIN, 2000; RYMER and WILLIAMS-JONES, 2000). However, applying these long-term geodetic monitoring techniques to volcanically active zones inevitably involves the subsequent interpretation of the deformations and gravity changes using a variety of modeling techniques.

Mathematical deformation models are essential tools for the latter task, and include spherical and ellipsoidal point sources, finite sources, magma migration, collapse structures and fluid migration (e.g., MOGI, 1958; RUNDLE, 1982; DAVIS, 1986; BONAFEDE, 1991; FERNÁNDEZ and RUNDLE, 1994; LANGBEIN et al., 1995; DE NATALE and PINGUE, 1996; DVORAK and DZURISIN, 1997; TIAMPO et al., 2000). Such models are also useful for analyzing precursory phenomena and optimizing the development of observation networks. Normally, simple deformation models are used because of the ease of calculating the inverse problem. However, sometimes it is better to use models that are as realistic as possible, particularly when it is desirable to learn about the physics of the source process.

Obviously, quantitative models cannot purport to cover all different physical and chemical aspects of volcanoes. It is important to select and focus on key phenomena. Present knowledge of the critical stages prior to volcanic eruption in large part is based on elastostatic views; by studying volcanic unrest in terms of mechanical models involving overpressure in magma chambers and conduits. MOGI (1958) applied a center of dilatation (point pressure source) in an elastic half-space to interpret the ground deformation produced in volcanic areas. The Mogi model has been extensively applied in modeling ground deformations in volcanic areas and has been successful primarily in explaining the vertical component of the ground deformations. However, this model often poses difficulties in simultaneously modeling observed displacements and gravity changes (KISSLINGER, 1975; RYMER et al., 1993; RYMER, 1996), and there is a large body of evidence for ground deformations and seismicity at calderas and other volcanic areas that cannot be modeled by these purely elastic effects (e.g., EGGERS, 1987; BERRINO and CORRADO, 1991; BONAFEDE, 1991; RYMER et al., 1993; DE NATALE et al., 1997; GAETA et al., 1998; JAHR et al., 1998; WATANABE et al., 1998; JENTZSCH et al., 2001b; PRITCHARD and SIMONS, 2002).

In this paper, we detail the combination of the genetic algorithm (GA) inversion technique with the elastic-gravitational model originally developed by RUNDLE (1980, 1982) and subsequently refined by FERNÁNDEZ and RUNDLE (1994) and FERNÁNDEZ et al. (1997), including a sensitivity analysis of both deformation and gravity to each of the model parameters. There is a practical comparison of the theoretical results with the modeling of geodetic data observed at the Mayon volcano in the Philippines, where there are gravity changes without significant deformation and where the more complicated inversion provides new insights into the underlying source mechanisms.

2. Inversion Methodology

2.1. The Genetic Algorithm

Many geophysical optimization problems are nonlinear and result in objective functions with a rough fitness landscape and several local minima. Consequently, local optimization techniques such as linearized matrix inversion and steepest descent can converge prematurely to a local minimum. In addition, the success in obtaining an optimum solution can depend strongly on the choice of the starting model. GAs have proven themselves an attractive global search tool suitable for the irregular, multimodal fitness functions typically observed in nonlinear optimization problems in the physical sciences. Because of their initial random and progressively more deterministic sampling of the parameter space, these algorithms offer the possibility of efficiently and relatively rapidly locating the most promising regions of the solution space. The GA is different from other optimization and search procedures in four ways: (1) the GA works with a *coding* of the parameter set, not the parameters themselves; (2) the GA searches a population of points, not a single point; (3) the GA uses the objective function information in a forward model, *not* derivatives, in its auxiliary knowledge; and (4) the GA makes use of probabilistic transition rules, not deterministic ones (GOLDBERG, 1989).

In general, geophysical inverse problems involve employing large quantities of measured data, in conjunction with an efficient computational algorithm that explores the model space to find the global minimum associated with the optimal model parameters. In the GA, the parameters to be inverted for are coded as genes, and a large population of potential solutions for these genes is searched for the optimal solution. Several recent studies have employed GAs to invert for seismic structure (STOFFA and SEN, 1991; BOSCHETTI et al., 1996), hypocenter relocation (BILLINGS et al., 1994), seismic phase alignment (WINCHESTER et al., 1993), mantle velocity structure (CURTIS et al., 1995; NEVES et al., 1996), crustal velocity structure (JIN and MADARIAGA, 1993; BATTACHARYYA et al., 1998), mantle viscosity (KIDO et al., 1998), fault zone geometry (YU et al., 1998), and a variety of volcanic sources (TIAMPO et al., 2000; FERNÁNDEZ et al., 2001a).

The basic structure of the GA code used here is modified from MICHALEWICZ (1992, appendix). The process begins by representing the model to be optimized as a real-value string. Starting with an initial range of models, these algorithms progressively modify the solution by incorporating the evolutionary behavior of biological systems. The fitness of each solution is measured by a quantitative, objective function, the fitness function, FV. Next, the fittest members of each population are combined using probabilistic transition rules to form a new offspring population. Copying strings according to their fitness values means that strings with a better value of fitness have a higher probability of contributing one or more offspring in the next generation. This procedure is repeated through a large number of generations until the best solution is obtained, based on the fitness measure (MICHALEWICZ, 1992). It has been demonstrated that those members of the population with a fitness value greater than the average fitness of the population itself will increase in number exponentially, effectively accelerating the convergence of the inversion process (HOLLAND, 1975; GOLDBERG, 1989).

Our program, shown schematically in Figure 1a, employs a random number generator to produce an initial set of 100 potential values for each of the model parameters, which are coded as genes. For example, if the model to be fit is the spherical Mogi source, then there are four parameters to be optimized—x and y location, depth and expansion volume. One gene for each model parameter is assigned to a particular member of that initial population, creating 100 potential solutions to the inversion problem. These members are ranked, from best to worst, according to an external fitness function. Our fitness function is obtained from the chi-square fit to the difference between the measurements (gravity and or deformation) and the appropriate values calculated for each potential solution. The members with the lowest chi-square value are the fittest and are selected to contribute to the next generation, where the genetic operations of crossover and mutation take place.

Crossover and mutation alter the new, fitter population in a process of controlled yet random information exchange, as shown in Figure 1(b,c) (WRIGHT, 1991). The crossover, or mating, process is repeated for a large portion of the entire population, while mutation, the random alteration of a small number of genes, is introduced to ensure against the occasional loss of valuable genetic material (MICHALEWICZ, 1992). After completion of both crossover and mutation, the population is reevaluated as above, and the process is repeated, exploiting information in past generations to search the parameter space with improved performance. Specific features of the this GA include real-valued genes, an elitist function, and a windowing fitness function, in order to ensure higher precision, continuous incorporation of the fittest parameter into the gene pool, and to prevent search stagnation.

This process modifies the initial population over subsequent generations, where each member's fitness is calculated in the evaluate subroutine (Fig. 1a). In this work, we inverted for six parameters—the x and y location of the volcanic source, the depth to the source, and its radius, mass, and internal pressure change.

2.2 The Fitness Function—The Forward Problem

The algorithm seeks the "fittest" model, i.e., the model that produces solutions to the given problem that are closest to the observed measurements. In the case of a volcanic source or sources, it is the model whose parameters, when input into the forward source model, produce surface deformations and gravity changes that best match the observed measurements. Given the expected range of values for the model parameters, the algorithm randomly selects a set of models and proceeds to evolve them to produce better, fitter models. The fitness value, *FV*, for any one model, is

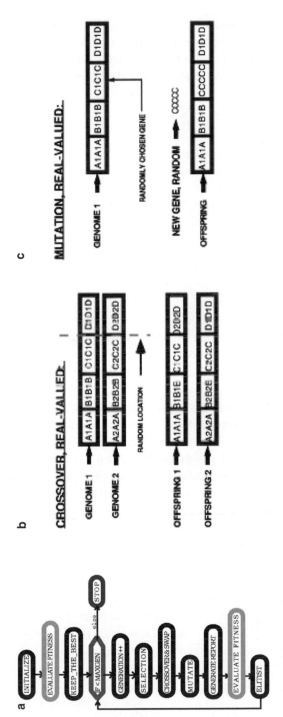

Figure 1

Schematic of (a) genetic algorithm (GA) program; (b) crossover operation; and (c) mutation operation.

calculated in the fitness subroutine, which contains the geophysical information specific to this problem. The gravity or deformation pattern for the expected volcanic source(s) is calculated in this subroutine and then the value of chi-square is calculated for that deformation pattern as it relates to the measurement pattern.

$$\chi^2 = \sum_k \frac{(C_K - E_k)^2}{\sigma_k}, \tag{1}$$

where E_k are the expected or measured data, C_k are the calculated deformations or gravity changes, σ_k is measurement standard deviation, and k is total number of measurements (TAYLOR, 1982; BEVINGTON and ROBINSON, 1992). The GA picks the fittest members of the population based upon the maximum fitness value, yet the value of chi-square decreases as the model approaches the correct solution. We therefore chose the fitness value to be $FV = \frac{1}{\chi^2}$, effectively converting the fitness to a continuously increasing function.

2.3 The Elastic-Gravitational Layered Earth Model

In this problem, the forward model, or fitness function, used by the GA contains the geophysical information relating the gravity changes and surface deformations for the particular volcanic source model.

While in the past the simple Mogi model has been used for the computation of surface deformations due to volcanic intrusion, it often poses difficulties in simultaneously modeling observed displacements and gravity changes (JAHR *et al.*, 1998; RYMER *et al.*, 1993; RYMER, 1996). In addition, gravity calculations must be modeled separately, requiring an iterative inversion process. More complicated models have been proposed for the modeling of deformation (DAVIS, 1986; YANG *et al.*, 1988; BONAFEDE, 1991; LANGBEIN *et al.*, 1995; DE NATALE *et al.*, 1997; FOLCH *et al.*, 2000; TIAMPO *et al.*, 2000; FERNÁNDEZ *et al.*, 2001b), but the joint modeling of gravity and deformation has remained elusive, despite the obvious benefits (BONAFEDE and MAZZANTI, 1998; JENTZSCH *et al.*, 2001b; BATTAGLIA and SEGALL, 2004).

Here we use a model in which the Earth is represented as a stack of plane layers having both elastic properties and mass (RUNDLE, 1980, 1982); hence both elasticity and self-gravitation are potentially important effects. The numerical formulation required to obtain the integration kernels for a layer overlying a half-space is described in RUNDLE (1981, 1982), and FERNÁNDEZ *et al.* (1994a,b). FERNÁNDEZ *et al.* (1997) presented an elastic-gravitational model consisting of a stratified half-space of homogeneous layers that takes into account the interaction between the mass of the intrusion, the ambient gravity field, and the effect caused by the change of pressure in the magmatic system chamber. The magmatic source can be located in any of the layers or the half-space. The GRAVW program set is formed by the GRAVW1, GRAVW2, GRAVW3, and GRAVW4, where the number included in the name of the program indicates the number of elastic layers overlying the homogeneous elastic half-space. The program used here is as described for the GRAVW1 program, one layer overlying a half-space (FERNÁNDEZ, 1997).

The required input data include the characteristics of the crustal model (layer thicknesses, Lame constants and densities), the x and y coordinates of the points where we want to compute the effects, and the magma intrusion characteristics (radius, mass, pressure and depth). The units of data input are km for distances, 10^3 kg m^{-3} for densities, 10^{12} Pa for the elastic parameters, 10^{12} kg (1 MU) for mass and 0.1 MPa for pressure values. The output of GRAVW1 includes various effects computed for each measurement location, including vertical and radial displacement, tilt, vertical strain, surface gravity change, free air gravity change, Bouguer gravity change, and sea-level change (FERNÁNDEZ et al., 1997).

Figure 2 shows the results for a two-layer model overlying a half-space (GRAVW2). The half-space is considered to be homogeneous, isotropic and elastic. Shown are vertical deformation, tilt, radial displacements, vertical strain, and surface gravity change for a both a center of expansion and a point mass intrusion located at 6 km in depth, with a pressure increment of 10^4 bars, a radius of 1 km, and a mass,

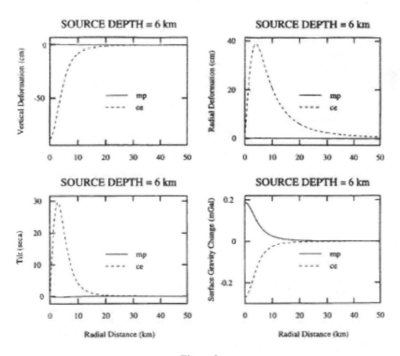

Figure 2
Vertical and horizontal deformation, tilt, and surface gravity change due to a center of expansion (ce) with a radius of 1 km and a pressure of 10^4 bars and a point mass (mp) of 1 MU located at 6 km in depth, using GRAVW2 (FERNÁNDEZ et al., 1997).

where applicable, of 1 MU. The model parameters are as given in FERNÁNDEZ *et al.* (1997). Note that while the gravitational contribution of the point mass to the deformation is negligible compared to that of the pressure, this is not the case for the surface gravity change, illustrating why both effects must be taken into account in the modeling of deformation and gravity data.

2.4 Sensitivity Analyses

Sensitivity tests were conducted for two separate elastic-gravitational sources. Each was located at 2.0 km in depth, centered at origin, and both with a mass of 1 MU. However, they are different in intensity size—the larger one, shown in Figure 3, has a pressure of 250 bars and a radius of 1.5 km, while the smaller, shown in Figure 4, has a pressure of 80 bars and a radius of 0.75 km. In each case, in order to test the sensitivity of each parameter, one parameter was varied while the other five remained fixed. In addition, the sensitivity of deformation measurements was tested versus the sensitivity of the associated gravity measurements. Results are

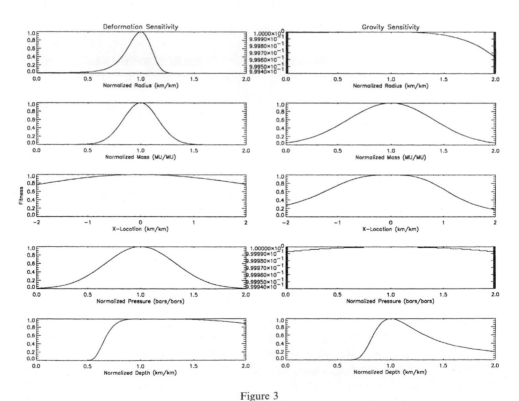

Figure 3
Sensitivity of the inversion fitness to the various elastic-gravitational model parameters, relative to both deformation, on the left, and gravity on the right. Smaller source, with a pressure of 80 bars and a radius of 0.75 km (centered at (0,0), depth of 2 km, mass of 1 MU).

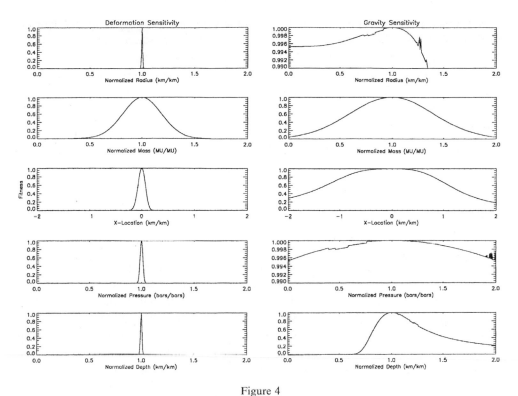

Figure 4
Sensitivity of the inversion fitness to the various elastic-gravitational model parameters, relative to both deformation, on the left, and gravity on the right. Larger source, with a pressure of 250 bars and a radius of 1.5 km (centered at (0,0), depth of 2 km, mass of 1 MU).

shown in Figures 3 and 4 (note the variation in the y-axis ranges). The fitness is plotted against the normalized variable in question.

Several features are immediately obvious. Not surprisingly, the x (or y) distance is a steeply varying parameter, with a single, easily defined maximum, for both gravity and deformation measurements. This corresponds with the GAs ability to invariably locate the source in the x-y plane, as a result of the distinct, location-dependent pattern created by a magmatic source. However, for the smaller source, it is the *gravity* measurements that better define the x-location, while in the case of a larger source, the *deformation* measurements are more sensitive to the x-location. This same reversal of gravity and deformation roles also occurs in the case of the depth parameter. This illustrates the importance of incorporating both gravity and deformation measurements into the volcanic modeling process.

In the case of both deformation and gravity measurements, the sensitivity to the mass is also a well-defined maximum. Not surprisingly, while the deformation measurements are very sensitive to the pressure and radius parameters, the gravity measurements are not. However, with increasing pressure, the gravity measurements become more sensitive to both parameters. Also, notice the steep decrease in fitness for the radius as the source radius increases past the correct value, while the fitness remains flat for the smaller values. This implies that *any* inversion will tend to solve for values of the radius that are too small, particularly for low pressures. This analysis demonstrates the importance of thoroughly understanding of the source model is used in any inversion and the impact of the various measurement techniques on the determination of the individual model parameters.

3. Application to Mayon, Philippines

3.1. Geology

Mayon volcano is part of the Bicol volcanic chain southeast of the island of Luzon, Philippines (see Fig. 5), part of the Legaspi Lineament of the central Philippine fault system, which runs NW-SE across Legaspi City. The Bicol chain, ten volcanic peaks, of which Mayon and Bolusan are still active, is made up of volcanic rocks, accompanied by Tertiary and Pleistocene sediments (JENTZSCH *et al.*, 2001a). Mayon itself is a nearly perfect cone with an altitude of 2462 m. Almost one million people live in the vicinity of Mayon volcano, and Legaspi City is situated only 12 km away from the crater. Mayon lava varies from basaltic through basaltic andesites,

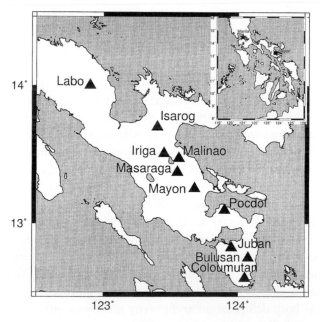

Figure 5
Location of Mayon volcano, Philippines.

and into the andesitic region, with a pattern of three basaltic flows followed by six to ten andesitic flows (NEWHALL, 1979). Its activity is quasiperiodic; in the last century it erupted nearly every 10 years, with the last eruption in 2001. In early 1993, just after the start of gravity measurements in December 1992, an eruption occurred, with ash fall and a lava flow of approximately 10 million m^3 (VÖLKSEN and SEEBER, 1995; JAHR et al., 1998; JENTZSCH et al., 2001b). In addition to the data discussed here, three seismological stations exist, but the coverage prior to 1993 was not homogeneous, and not adequate to determine precursory seismicity (JENTZSCH et al., 2001b).

After the 1993 eruption, a second campaign was carried out in May 1993 to monitor the decrease in activity. Height control was provided by parallel global positional system (GPS) measurements. Five microgravity and differential GPS campaigns were carried out within four years. The error on the gravity differences between the points derived from a least-squares adjustment of the data of each campaign is approximately ± 12 μGal (JENTZSCH et al., 2001b). The gravity and GPS measurements covered two profiles up towards the summit (one of them up to an elevation of 850 m) in a network of 26 points, and connected to a local and a regional network around the volcano with an extension of 40 km by 50 km, via two points at the opposite side of the Legaspi lineament (see Fig. 6a), although the closest gravity measurements are along the Mayon-Resthouse profile, still more than 3 kms from the crater (Figs. 6 and 7).

There was no significant gravity change between the first and the second campaign (December 1992, May 1993), despite the activity in February/March, 1993.

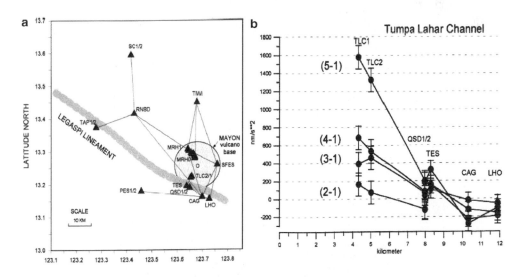

Figure 6
a) Gravity and GPS network around Mayon crater and b) microgravity measurements over the horizontal distance from the crater, for the Tumpa-Lahar-Channel profile, epochs 2-1, 3–1, 4–1, and 5–1 (JENTZSCH et al., 2001b).

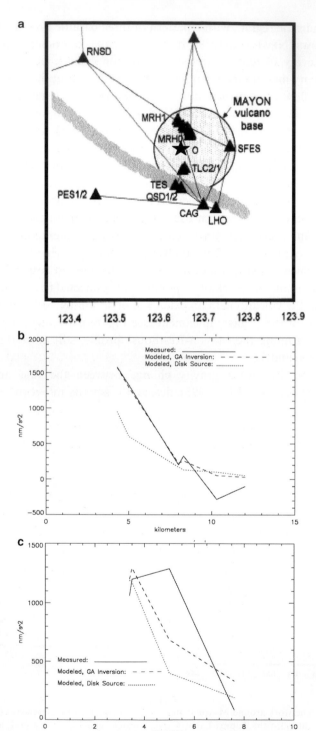

The gravity increase between May, 1993 and December, 1996 reached almost 150 µGal ($\pm$ 14 µGal), increasing with elevation and decreasing with distance from the crater. Note that a detailed discussion of the effect of groundwater on the gravity measurements can be found in JENTZSCH et al., (2001b), but the resulting conclusion, based on the lack of seasonal variation in the data, the location of the preponderance of measurements on consolidated rock, and the steep slopes of the upper reaches of the network, result in the conclusion that the estimated maximum effect due to water level changes are on the order of 50 µGal on the lower slopes of the volcano (JENTZSCH et al., 2001b). There remains, therefore, a significant residual gravity signal to be explained near the volcanic crater.

The microgravity measurements for the Tumpa-Lahar-Channel profile are shown in Figure 6b. The gravity changes are restricted to a radius of 8 km around the volcano summit, a variation of around 30 µGal per year if a constant process is assumed. Yet, the differential GPS measurements carried out at the same locations with three parallel receivers revealed *no significant changes of elevation* during this time, within their accuracy of $\pm$ 3 to 4 cm.

The connection between the variation of gravity and elevation due to mass and density changes is usually interpreted in relation to two gradients. If gravity follows the free air gradient (FAG), no subsurface change in the mass has occurred, while data following the gradient after the standard Bouguer correction (BCFAG) implies mass changes (e.g., BROWN and RYMER, 1991; RYMER et al., 1993; RYMER, 1996). Departures from these gradients are used to model volcanic processes. The results at Mayon cannot be related to the gravity gradients FAG and BCFAG because the gravity changes drawn against height differences from one campaign to the next show no gradient, while it is unusual for gravity to increase with decreasing activity (JENTZSCH et al., 2001b).

As a result, the observed gravity increase, unaccompanied by significant changes in elevation, cannot be explained with the classical Mogi model. Originally, JAHR et al. (1998) and JENTZSCH et al. (2001b) explained the gravity changes as density changes within a vent system. This model is based on a redistribution of the mass via the migration of magma down the vent system, and explains about 50% of the observed signal as well as the pattern of observed gravity changes (see the computed results, dotted lines in Fig. 7b). However, the inability of this model to account for a large part of the observed gravity change data signal prompted further analysis using a more complicated elastic-gravitational model. The elastic-gravitational model has several advantages for this particular inversion. First, the numerical formulation allows for the joint inversion of deformation and gravity. Second, the incorporation

◄

Figure 7

a) Location of the inverted single source relative to the crater and the monitoring stations (star); b) inversion results for the Tumpa-Lahar-Channel profile, epoch 5-1, compared to the measure values and the vent motion model; and c) the same results for the Mayon-Resthouse profile.

of the elastic-gravitational effect introduces a long-wavelength effect to the gravity and deformation field that may prove important in an area with significant topography. Finally, with the correct combination of mass, pressure, and radius, inside a volcano that likely contains a complicated, preexisting vent system, it may be possible to model the unusual measurements recorded at Mayon.

3.2 Inversion Results

The recent eruptions of Mayon volcano gives rise to the assumption that a process of injection might have taken place, instead of deflation. We test this hypothesis by modeling the observed data, including both the measured gravity changes and the GPS deformation measurements, using the GA inversion technique combined with elastic-gravitational Earth models detailed above. Originally, the data was inverted based upon the assumption of a single source, located some distance below the crater. The elastic-gravitational model provided a much better solution than either the simple Mogi source or the vent magma model, as can be seen in Figure 7, for both the Tumpa-Lahar-Channel profile (Fig. 7a) and the Mayon-Resthouse profile (Fig. 7b). The particular characteristics obtained for this intrusion are a depth of 1.975 km below the base of the volcano, 94 bars pressure, a radius of 1.48 km, and 0.68 MU for mass (1 MU = 1 Mass Unit = 10^{12} kg) for epoch 5-1. The results for this elastic-gravitational inversion (dashed line) are compared with the gravity measurements for (solid line), as well as that for the model that redistributes magma down the volcanic vent (dotted line). However, while the results for the Mayon-Resthouse profile for the same epoch reproduces the total magnitude and character of the measured signal, they do not extend as far from the vent, nor does the maximum for the modeled source coincide with the maximum seen in the data. In addition, the need to match gravity measurements far from the crater results in a source radius that is unreasonably large, and a resulting magma density that is too low. As a result, a GA inversion for a two-source model was implemented, in order to study the possible asymmetries in the source mechanism.

Results for the two-source inversion to the epoch 5–1 data are shown in Figure 8. While the fit to the Tumpa-Lahar-Channel profile (Fig. 8a) data is similar to the single-source inversion, the fit to the data along the Mayon-Resthouse profile has increased significantly, with the maximum calculated gravity now matching the location of the data maximum. Interestingly, the two sources are quite different. This source, located at a depth of 1.02 km below the base of the volcano, at 24 bars, a

▶

Figure 8
a) Location of the two sources relative to the crater and the monitoring stations (grey star is the lighter, shallower source, black star is the deeper magmatic source); b) inversion results for the Tumpa-Lahar-Channel profile, epoch 5-1, compared to the measure values and the vent motion model; and c) the same results for the Mayon-Resthouse profile.

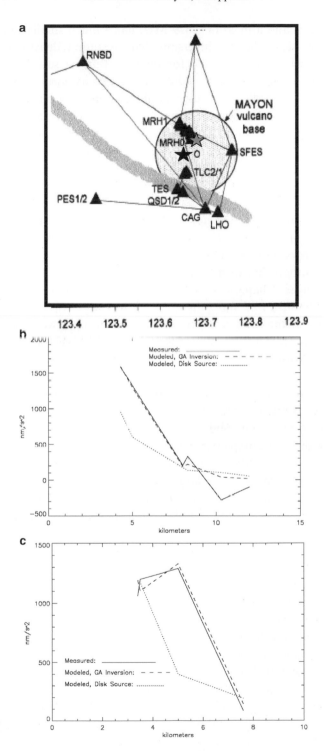

radius of 0.312 km, and a mass of 0.629 MU, has a much smaller mass and radius than that of the single source inversion. The resulting density of intrusion is approximately 4950 kg/m^3, too high for magma. But, as seen in the analysis of Figure 3, the lack of sensitivity to radius of the gravity measurements will cause the inversion to underpredict the radius, easily accounting for the 25% error in the radius necessary to overestimate the density by this much. In addition, despite the incorporation of the deformation data in the inversion, the high variance relative to the measurements themselves means that, as again can be seen in the sensitivity analysis (Section 2.4), it is difficult in any inversion to accurately specify the pressure or radius. The deformation calculated from this joint inversion, in all cases, is well within the variance of the measured data.

The second source is located north of the caldera and is much shallower and broader than the first source, at a depth of only 0.22 km below the base of the volcano, with 75 bars pressure, a radius of 1.54 km, and a mass of 0.378 MU. With a density much less than that of magma, this suggests that it results from significant groundwater changes. Changes in well water and outflow have been noticed by the local inhabitants both prior and subsequent to Mayon eruptions (JENTZSCH *et al.*, 2001). In particular, the local residents had to deepen their wells prior to the 1993 eruption, but measurements of the wells are not available for after the eruption (ALBANO, 2001). However, these changes were difficult to correlate with the existing gravity measurements (JENTZSCH *et al.*, 2001b). In addition, while the steep slopes of the Mayon-Resthouse profile would seem to preclude a significant water table (JENTZSCH *et al.*, 2001b), there are recorded instances where dilatation, due to changes in the local strain field, results in the storage of large quantities of groundwater, with an accompanying decrease in gravity (KISSLINGER, 1975). This would account for the lower well levels prior to 1993. Following the eruption, crack closure results in the expulsion of large amounts of water and an *increase* in the relative gravity accompanied by no increase, or even a decrease, in surface deformation (KISSLINGER, 1975). Similar models could account for this second source along the volcanic slope, where the gravity measurements recorded not a change in the groundwater table per se, but a temporary storage of water as a result of underlying tectonic and volcanic stress changes.

Interpretation of these results suggests that the primary gravity changes at Mayon are better explained by reinjection of magma below the volcano caldera following the 1993 eruption rather than mass redistribution in the volcanic vent. Note that these results do not preclude some combination of mass redistribution down the vent, as proposed by JENTZSCH *et al.* (2001b), in conjunction with additional mass injection below the vent. Concurrent with this intrusion, localized changes in the strain field caused the expulsion of water stored at some point during and/or after the previous eruption. Additional work should explore the potential of a model that uses the genetic algorithm to optimize the combination of these features.

4. Discussion and Conclusions

This paper deals with integration of a GA inversion technique with implementation of a numerical technique for the calculation of ground deformation and gravity changes caused by a point source magma intrusion into an elastic-gravitational multilayered earth model. Unlike other techniques, the GA neither totally relies on prior constraints (manually provided by the programmer) nor runs a trial-and-error scheme. An additional advantage is that the selection process rejects automatically those data with a large bias. The combination of the GA flexibility of the inversion and the more complicated and realistic GRAVW1 program has provided a new interpretation of gravity and deformation data from the Mayon volcano. Here, we have detailed the results of a sensitivity analysis for each parameter and type of measurement, using a synthetic testing procedure, and have shown that the joint inversion for gravity and deformation measurements takes into account the varying importance of each parameter over a range of sizes.

In order to properly interpret cases where gravity changes exist without deformation, and *vice versa*, analysis should include adequate mass and pressure combinations for the intrusion. While simpler models are unable to explain gravity changes without displacements or *vice versa*, the elastic-gravitational model, which accounts for the interaction of the mass of the intrusion with the ambient gravity field and redistribution of densities inside the crust, can explain more unusual volcanic regions.

The reasons for this become clear when one studies both Figure 2 and the sensitivity analysis shown in Figures 3 and 4. If, for example, the effects on the intrusion are not due primarily to pressure changes in the intrusion, but to a considerable amount of magma recharge, elastic-gravitational models should be used. This approach accounts for the effect of opposite signs for each type of source. In addition, the depth and location are better determined by either deformation or gravity, depending on the size of the source. Note that, even in the case of Mayon, as no joint measurements were possible at the crater itself, the internal mechanism must remain open to final interpretation. Displacements and gravity changes must be interpreted together wherever possible, in order to discriminate between the roles played by the mass and pressure changes, which otherwise are difficult to interpret correctly. Correct interpretation of the observed geodetic signals has implications in the development of monitoring and alert systems for the mitigation of volcanic hazards.

Note that there are a number of important effects not considered here. For example, we do not consider topographic effects for two reasons, although these can be important under certain conditions (CHARCO et al., 2004). First, we wished to compare with earlier results that also did not consider topography. Second, as the gravity and GPS stations are not located at the source, or the top of the cone, the topographic effect will be less important. In future work, such effects as source geometry, topography, and multiple layer rheologies will be modeled as well for an even more realistic interpretation.

Acknowledgements

We are grateful to M. Bonafede and an anonymous reviewer for their critical review of an earlier version of this work, and C. Kisslinger for useful suggestions. The research by KFT and JBR was conducted under NASA grants NAG5-3054 and NGT5-30025, respectively. The research by JF and MC has been funded under MCyT projects AMB99-1015-C02 and REN2002-03450. This research has been also partially supported with funds from a New Del Amo Program project. KFT would also like to acknowledge the Instituto de Astronomía y Geodesia, Universidad Complutense de Madrid, for its hospitality during the completion of this work.

REFERENCES

ALBANO, S. E., SANDOVAL, T., and TOLEDO, R. (2001), *Groundwater at Mayon volcano*, EOS Trans., AGU, *82*, 47.

BERRINO, B. and CORRADO, G. (1991), *Gravity changes and volcanic dynamics*, Cahier du Centre Europeén de Geodynamique et de Seismologie *4*, Proc. Workshop: *Geodynamical Instrumentation Applied to Volcanic Areas*, Oct. 1–3, 1990, Walferdange, Luxembourg, 305–323.

BATTAGLIA, M., ROBERTS, C., and SEGALL, P. (1999), *Magma Intrusion beneath Long Valley Caldera Confirmed by Temporal Changes in Gravity*, Science *285*, 2119–2122.

BATTAGLIA, M. and SEGALL, P. (2004), *The interpretation of Gravity Changes and Crustal Deformation in Active Volcanic Areas*, Pure Appl. Geophys. *161*, 1453–1467.

BEVINGTON, P. R. and ROBINSON, D. K., *Data Reduction and Error Analysis for the Physical Sciences* (McGraw-Hill, Inc., N.Y. 1992).

BHATTACHARYYA, J., SHEEHAN, A. F., TIAMPO, K. F., and RUNDLE, J. B. (1998), *Using Genetic Algorithms to Model Regional Waveforms for Crustal Structure in the Western United States*, BSSA *89*, 202–214.

BILLINGS, S., KENNETT, B., and SAMBRIDGE, M. (1994), *Hypocenter Location: Genetic Algorithms Incorporating Problem Specific Information*, Geophys. J. Int. *118*, 693–706.

BONAFEDE, M. (1991), *Hot Fluid Migration, an Efficient Source of Ground Deformation: Application to the 1982–1985 Crisis at Campi Flegrei-Italy*, J. Volcanol. Geoth. Res. *48*, 187–198.

BONAFEDE, M. and MAZZANTI, M. (1998), *Modeling Gravity Variations Consistent with Ground Deformation in the Campi Flegrei Caldera (Italy)*, J. Volcanol. Geoth. Res. *81*, 137–157.

BOSCHETTI, F., DENTITH, M. C., and LIST, R. D. (1996), *Inversion of Seismic Refraction Data Using Genetic Algorithms*, Geophys. *61*, 1715–1727.

BROWN, G. C. and RYMER, H. (1991), *Microgravity Monitoring at Active Volcanoes, A Review of Theory and Practice*, Cahiers du Centre Européen de Geodynamique et Seismologie *4*, 279–304.

Charco et al. (2004) *New Results at Mayon, Philippines, from a Joint Inversion of Gravity and Deformation Measurements*, Pure. Appl. Geophys. *161*, 1433–1452.

CURTIS, A., DOST, B., TRAMPERT, J., and SNIEDER, R. (1995), *Shear-wave Velocity Structure Beneath Eurasia from Surface Wave Group and Phase Velocities in an Inverse Problem Reconditioned Using the Genetic Algorithm*, EOS Trans., AGU *76*, 386.

DAVIS, P.M. (1986), *Surface Deformation due to Inflation of an Arbitrarily Oriented Triaxial Ellipsoidal Cavity in an Elastic Half-space, with Reference to Kilauea Volcano, Hawaii*, J. Geophys. Res. *91*, 7429–7438.

DE NATALE, G., PETRAZZUOLI, S. M., and PINGUE, F. (1997), *The Effect of Collapse Structure on Ground Deformations in Calderas*, Geophys. Res. Lett. *24*, 1555–1558.

DELANEY, P. T. and McTIGUE, D. F. (1994), *Volume of Magma Accumulation or Withdrawal Estimated from Surface Uplift or Subsidence, with Application to the 1960 Collapse of Kilauea Volcano*, Bull. Volcanol. *56*, 417–424.

DVORAK, J. J. and DZURISIN, D. (1997), *Volcano Geodesy, the Search for Magma Reservoirs and the Formation of Eruptive Vents*, Rev. Geophys. *35*, 343–384.

DZURISIN, D. (2000), *Volcano Geodesy: Challenges and Opportunities for the 21st century*, Phil. Trans. Roy. Soc. Lond. A *358*, 1547–1566.

DZURISIN, D., WICKS, C. Jr., and THATCHER, W. (1999), *Renewed Uplift at the Yellowstone Caldera Measured by Leveling Surveys and Satellite Radar Interferometry*, Bull. Volcanol. *61*, 349–355.

EGGERS, A. A. (1987), *Residual Gravity Changes and Eruption Magnitudes*, J. Volcanol. Geoth. Res. *33*, 201–216.

FERNÁNDEZ, J., CARRASCO, J. M., RUNDLE, J. B., and ARAÑA, P. (1999), *Geodetic Methods for Detecting Volcanic Unrest, A Theoretical Approach*, Bull. Volcanol. *60*, 534–544.

FERNÁNDEZ, J., CHARCO, M., TIAMPO, K. F., JENTZSCH, G., and RUNDLE, J. B. (2001a), *Joint Interpretation of Displacement and Gravity Data in Volcanic Areas. A Test Example: Long Valley Caldera, California* , Geophys. Res. Lett. *28*, 1063–1066.

FERNÁNDEZ, J. and RUNDLE, J. B. (1994a), *Gravity Changes and Deformation due to a Magmatic Intrusion in a Two-layered Crustal Model*, J. Geophys. Res. *99*, 2737–2746.

FERNÁNDEZ, J., RUNDLE, J. B., GRANELL, R., and YU, T.-T. (1997), *Programs to Compute Deformation due to a Magma Intrusion in Elastic-gravitational Layered Earth Models*, Comp. and Geosci. *23*, 231–249.

FERNÁNDEZ, J., TIAMPO, K. F., JENTZSCH, G., CHARCO, M., and RUNDLE, J. B. (2001b), *Inflation or Deflation? New Results for Mayon Volcano Applying Elastic-gravitational Modeling*, Geophys. Res. Lett. *28*, 2349–2352.

FERNÁNDEZ, J. and RUNDLE, J. B. (1994b), *FORTRAN Program to Compute Displacement, Potential and Gravity Changes Resulting from a Magma Intrusion in a Multilayered Earth Model*, Computers and Geosciences, *20*, 461–510.

FOLCH, A., FERNÁNDEZ, J., RUNDLE, J. B., and MARTI, J. (2000), *Ground Deformation in a Viscoelastic Medium Composed of a Layer Overlying a Half-space: A Comparison between Point and Extended Sources*, Geophys. J. Int. *140*, 37–50.

GAETA, F. S., DE NATALE, G., and ROSSANO, S. (1998), *Genesis and Evolution of Unrest Episodes at Campi Flegrei Caldera, the Role of the Thermal Fluid-Dynamical Processes in the Geothermal System*, J. Geophys. Res. *103*, 20,921–20,933.

GOLDBERG, D. E., *Genetic Algorithms in Search, Optimization, and Machine Learning* (Addison Wesley, Reading, MA 1989).

HOLLAND, J. H., *Adaptation in Natural and Artificial Systems* (MIT Press, Cambridge, MA 1975).

JAHR, T., JENTZSCH, G., PUNONGBAYAN, R. S., SCHREIBER, U., SEEBER, G., VÖLKSEN, C., and WEISE, A. (1998), *Mayon Volcano, Philippines: Improvement of hazard assessment by microgravity and GPS?*, Proc. Int. Symp. *On Current Crustal Movement and Hazard Assessment* (IUGG, IAG), Wuhan. Seismological Press, Beijing, 599–608.

JENTZSCH, G., HAASE, O., KRONER, C., SEEBER, G., and WINTER, U. (2001a), *Mayon Volcano, Philippines: Some Insights into Stress Balance*, J. Volcanol. Geotherm. Res. *109*, 205–217.

JENTZSCH, G., PUNONGBAYAN, R. S., SCHREIBER, U., SEEBER, G., VÖLKSEN, C., and WEISE, A. (2001b), *Mayon Volcano, Philippines: Change of Monitoring Strategy after Microgravity and GPS Measurements*, J. Volcanol. Geotherm. Res. *109*, 219–234.

JIN, S. and MADARIAGA, R. (1993), *Background Velocity Inversion with a Genetic Algorithm*, Geophys. Res. Lett. *20*, 93–96.

KIDO, M., YUEN, D. A., ČADEK, O., and NAKAKUKI, T. (1998), *Mantle Viscosity Derived by Genetic Algorithm Using Oceanic Geoid and Seismic Tomography for Whole-mantle versus Blocked-flow Situations*, Phys. Earth Plan. Int. *107*, 307–326.

KISSLINGER, C. (1975), *Processes during the Matsushiro, Japan Earthquake Swarm as Revealed by Leveling, Gravity, and Spring-flow Observations*, Geology *3*, 57–62.

LANGBEIN, J., DZURISIN, D., MARSHALL, G., STEIN, R., and RUNDLE, J. (1995), *Shallow and Peripheral Volcanic Sources of Inflation Revealed by Modeling Two-color Geodimeter and Leveling Data from Long Valley Caldera, California, 1988–1992*, J. Geophys. Res. *100*, 12,487–12,495.

MASSONET, D. and FEIGL, K. L. (1998), *Radar Interferometry and its Application to Changes in the Earths Surface*, Rev. Geophys. *36*, 441–500.

MICHALEWICZ, Z., *Genetic Algorithms + Data Structures = Evolution Programs* (Springer-Verlag, New York, NY 1992).

MOGI, K. (1958), *Relations between the Eruptions of Various Volcanoes and the Deformations of the Ground Surfaces around them*, Bull. Earth. Res. Inst. Tokyo *36*, 99–134.

NEVES, F. A., SINGH, S. C., and PRIESTLEY, K. F. (1996), *Velocity Structure of Upper-mantle Transition Zones beneath Central Eurasia from Seismic Inversion Using Genetic Algorithms*, Geophys. *22*, 523–552.

NEWHALL, C.G (1979), *Temporal Variation in the Lavas of Mayon Volcano, Philippines*, J. Volanol. Geoth. Res. *6*, 61–83.

PRITCHARD, M. E. and SIMONS, M. (2002), *A Satellite Geodetic Survey of Large-scale Deformation of Volcanic Centers in the Central Andes*, Nature *418*, 167–171.

RUNDLE, J. B. (1980), *Static Elastic-gravitational Deformation of a Layered Half-space by Point Couple Sources*, J. Geophys. Res. *85*, 5355–5363.

RUNDLE, J. B. (1982), *Deformation, Gravity and Potential Changes due to Volcanic Loading of the Crust*, J. Geophys. Res. *87*, 10,729–10,744.

RYMER, H., *Microgravity Monitoring*, In *Monitoring and Mitigation of Volcano Hazards* (Springer 1996) pp. 169–197.

RYMER, H., MURRAY, J. B., BROWN, G. C., FERRUCI, F., and McGUIRE, W. J. (1993), *Mechanism of Magma Eruption and Emplacement at Mt. Etna between 1989 and 1992*, Nature *361*, 439–441.

RYMER, H. and WILLIAMS-JONES, G. (2000), *Volcanic Eruption Prediction, Magma Chamber Physics from gravity and Deformation Measurements*, Geophys. Res. Lett. *27*, 2389–2392.

STOFFA, P. L. and SEN, M. K. (1991), *Nonlinear Multiparameter Optimization Using Genetic Algorithms, Inversion of Plane-wave Seismograms*, Geophys. *56*, 1794–1810.

TAYLOR, J. R., *An Introduction to Error Analysis* (University Science Books, USA 1982) pp. 225–230.

TIAMPO, K. F., RUNDLE, J. B., FERNÁNDEZ, J., and LANGBEIN, J. (2000), *Spherical and Ellipsoidal Volcanic Sources at Long Valley Caldera, California Using a Genetic Algorithm Inversion Technique*, J. Volcanol. Geotherm. Res. *102*, 189–206.

VÖLKSEN, C. and SEEBER, G. (1995), *Establishment of a GPS-based Control Network at Mayon Volcano*, Cahiers du Centre Européen de Geodynamique et de Seismologie *8*, 99–113.

WATANABE, H., OKUBO, S., SAKASHITA, S., and MAEKAWA, T. (1998), *Drain Back Process of Basaltic Magma in the Summit Conduit Detected by Microgravity Observation at Izu-Oshima Volcano, Japan*. Geophys. Res. Lett. *25*, 2865–2868.

WINCHESTER, J. P., CREAGER, K. C., and McSWEENEY, T. J. (1993), *Better Alignment through Better Breeding: Phase Alignment Using Genetic Algorithms and Cross-correlation Techniques*, EOS, AGU *74*, 394.

WRIGHT, A. H., *Genetic algorithms for real parameter optimization*. In *Foundations of Genetic Algorithms* (Morgan Kaufmann Publishers, San Mateo, CA 1991).

YANG, X. M., DAVIS, P. M., and DIETERICH, J. H. (1988), *Deformation from Inflation of a Dipping Finite Prolate Spheroid in an Elastic Half-space as a Model for Volcanic Stressing*, J. Geophys. Res. *93*, 4249–4257.

YU, T. T., FERNÀNDEZ, J., and RUNDLE, J. B. (1998), *Inverting the Parameters of an Earthquake-ruptured Fault with a Genetic Algorithm*, Comp. and Geo. *24*, 173–182.

(Received February 3, 2002, revised February 26, 2003, accepted April 25, 2003)

 To access this journal online:
http://www.birkhauser.ch

Pure appl. geophys. 161 (2004) 1453–1467
0033–4553/04/071453–15
DOI 10.1007/s00024-004-2514-5

© Birkhäuser Verlag, Basel, 2004

▌Pure and Applied Geophysics

The Interpretation of Gravity Changes and Crustal Deformation in Active Volcanic Areas

M. BATTAGLIA[1,2], and P. SEGALL[1]

Abstract — Simple models, like the well-known point source of dilation (Mogi's source) in an elastic, homogeneous and isotropic half-space, are widely used to interpret geodetic and gravity data in active volcanic areas. This approach appears at odds with the real geology of volcanic regions, since the crust is not a homogeneous medium and magma chambers are not spheres. In this paper, we evaluate several more realistic source models that take into account the influence of self-gravitation effects, vertical discontinuities in the Earth's density and elastic parameters, and non-spherical source geometries. Our results indicate that self-gravitation effects are second order over the distance and time scales normally associated with volcano monitoring. For an elastic model appropriate to Long Valley caldera, we find only minor differences between modeling the 1982–1999 caldera unrest using a point source in elastic, homogeneous half-spaces, or in elasto-gravitational, layered half-spaces. A simple experiment of matching deformation and gravity data from an ellipsoidal source using a spherical source shows that the standard approach of fitting a center of dilation to gravity and uplift data only, excluding the horizontal displacements, may yield estimates of the source parameters that are not reliable. The spherical source successfully fits the uplift and gravity changes, overestimating the depth and density of the intrusion, but is not able to fit the radial displacements.

Key words: Volcano geodesy, gravity, crustal deformation, calderas, models.

1. Introduction

Quite simple models are widely used to interpret geodetic and gravity data in active volcanic areas. An example is the well-known point source of dilation (Mogi's source), used to approximate the behavior of a pressurized spherical magma chamber, embedded in an elastic, homogeneous, and isotropic half-space (EGGERS, 1987; DVORAK and DZURISIN, 1997). Mogi's source models successfully reproduce displacement and gravity changes at many volcanoes during either uplift or subsidence (MCKEE *et al.*, 1989; BERRINO, 1994; BATTAGLIA *et al.*, 1999). In addition, some authors use the linear gravity/height correlation from the point source model to study the physics of magma chambers (RYMER and WILLIAM-JONES, 2000) or investigate the likelihood of volcanic eruptions (BERRINO *et al.*, 1992; RYMER,

[1] Department of Geophysics, Stanford University, Stanford CA 94305-2215, U.S.A.
[2] Corresponding author. Now at: Seismo Lab, UC Berkeley, 215 McCone Hall, Berkeley CA 94720-4760, U.S.A. E-mail: battag@seismo.berkeley.edu

1994). This approach appears at odds with the complex geology of volcanic regions, since the crust is not a homogeneous medium and magma chambers are not spheres. But what should a model include to obtain a better insight into the physics of volcanoes? Some authors claim that elastic-gravitational models can be a far more appropriate approximation to problems of volcanic load in the crust than the more commonly used purely elastic models (e.g., RUNDLE, 1982; FERNÁNDEZ et al., 2001a, 2001b). Vertical discontinuities in the Earth's density and elastic parameters can play an important role when modeling gravity changes induced by deformation (FERNÁNDEZ et al, 1997; BONAFEDE and MAZZANTI, 1998). We apply models including one (or more) of the above features to the 1982–1999 period of unrest at Long Valley caldera to evaluate: (a) if elasto-gravitational models are a more appropriate approximation to problems of volcanic load in the crust than the purely elastic models at the space and time scales associated with volcano monitoring; (b) the importance of vertical discontinuities in the Earth's density and elastic parameters when modeling displacement and gravity changes induced by a point source of dilation; (c) the bias introduced using a point source of dilation model to reproduce geodetic and gravity data if the magma intrusion does not posses a spherical symmetry.

Our results show that (a) self-gravitation effects due to coupling between elasticity and gravity potential are second order over the distance and time scales normally associated with volcano monitoring. (b) For an elastic model appropriate to Long Valley caldera, we find only minor differences between modeling the intrusion using a point source in a homogeneous or layered medium. (c) A simple experiment of matching deformation and gravity data from an ellipsoidal source (YANG et al., 1988; CLARK et al., 1986) using a spherical source suggests that the standard approach of fitting a center of dilation to gravity and uplift data only, excluding the horizontal displacements, can yield estimates of the source parameters that are not reliable. In our experiment, the spherical source successfully fits the uplift and gravity changes, inferring a deeper location (8.5 km instead of 6 km) and a larger density (4500 kg/m^3 instead of the actual 2500 kg/m^3) for the intrusion, but is not able to fit the radial displacements.

2. Coupling between Elastic and Gravitational Effects

The complete solution for gravity and deformation changes in volcanic regions should include the coupled effects of gravity and displacement changes (FERNÁNDEZ et al., 1997). To determine if these effects are important on space and time scales associated with volcano monitoring, we perform a dimensional analysis of the fully coupled elasticity and potential equations (see POLLITZ, 1997 for similar discussion in the earthquake context). We approximate the Earth with an isotropic, elastic sphere. The origin of the coordinate system is taken at the center of this sphere. Following the approach by LOVE (1911, p. 89), the density ρ_0, the pressure p_0 and the potential V_0 (with the corresponding gravitational acceleration $-g_0\mathbf{e}_r = \nabla V_0$ define the initial state of equilibrium

$$\rho_0 \nabla V_0 = \nabla p_0. \tag{1}$$

After the intrusion of mass in a spherical magma body of initial radius a (approximated by the superposition of a point source of dilation and a mass point source at $\mathbf{x} = \mathbf{x}_m$, see Fig. 1), the Earth's surface deforms by $\mathbf{u}$ ($|\mathbf{u}| \ll a$). Perturbations in the density ρ, pressure p and potential V are defined by

$$\begin{aligned} \rho &= \rho_0 - \nabla \cdot (\rho_0 \mathbf{u}) \\ p &= p_0 - \mathbf{u} \cdot \nabla p_0 \quad , \\ V &= V_0 + V_p + V_m \end{aligned} \tag{2}$$

where V_p is the change in the Earth's potential due to inflation of a massless cavity and V_m is the potential due to the mass $\Delta M = \rho_m \Delta V$ of the intrusion (ρ_m and ΔV are respectively the density and volume of the intrusion). The equilibrium equation is

$$\nabla \cdot \mathbf{s} - \nabla p + \rho \nabla V + \mathbf{F}_p + \mathbf{F}_m = 0, \tag{3}$$

where $\mathbf{s}$ is the elastic stress tensor, $\mathbf{F}_p$ and $\mathbf{F}_m$ are the body force density corresponding to a point source of dilation and a mass point source. Substituting

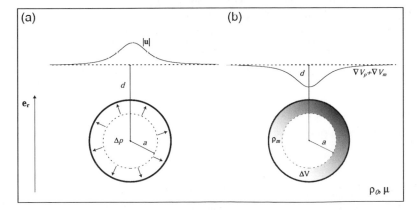

Figure 1
Coordinate system and parameters for the scaling problem. The effect of a magma body intrusion is approximated by two contributions: (a) a pressurized mass-less cavity; (b) a spherical mass intrusion with no pressure change. Scaling parameters (Table 1): u — vertical displacement; ρ_0 — crust density; μ — shear modulus; d — depth of the magma chamber; a — radius of the magma chamber; ρ_m — magma intrusion density; ΔP — pressure change; ΔV — volume change; $\nabla V_p + \nabla V_m$ — changes in gravity, see (2).

(2) in (3), we get the following equations describing the potentials V_p and V_m, and the elastic deformation (RUNDLE, 1982)

$$\nabla^2 V_p = 4\pi G \nabla \cdot (\rho_0 \mathbf{u}), \quad \nabla^2 V_m = -4\pi G \rho_m \delta(\mathbf{x} - \mathbf{x}_m) \tag{4}$$

$$\nabla \cdot \mathbf{s} - g_0[\nabla(\rho_0 \mathbf{u} \cdot \mathbf{e}_r) - \mathbf{e}_r \nabla \cdot (\rho_0 \mathbf{u})] + \rho_0[\nabla V_p + \nabla V_m] + \mathbf{F}_p + \mathbf{F}_m = 0. \tag{5}$$

The second and third term on the left-hand side of (5) depend on g_0. The fourth and fifth term on the left-hand side of (5) depend on G. The g_0 dependent part scales as $g_0 \rho_0 u / d$, where u and d are characteristic distance scales set respectively by the vertical displacement and the depth of the magma chamber. From (4), ∇V_p scales as $G\rho_0 u$. The scaling of ∇V_m is given by the expression for the gravity change associate to a mass point source, $G\rho_m \Delta V / d^2$ (EGGERS, 1987). The elastic stress s scales as $\mu u / d$. The shear modulus μ is most commonly estimated from seismic wave speeds. However, the dynamic modulus (μ_d) may exceed the quasi-static shear modulus (μ_s). The ratio μ_d / μ_s depends on several factors including the porosity and applied pressure p. For granite and tuff (CHENG and JOHNSTON, 1981), $\mu_d / \mu_s \approx 0.1$ at low pressures ($p \leq 0.1$ GPa, depth ≤ 3.5 km) and $\mu_d / \mu_s \approx 1$ when $p \geq 0.2$ GPa (depth ≥ 7.0 km).

Using typical parameter values (see Table 1), we can show that the potentials V_p and V_m have the same order of magnitude, but are negligible compared to the elastic term

$$\frac{\nabla V_p}{\nabla V_m} \sim \frac{ud^2}{\Delta V} \sim 1, \quad \frac{\rho_0 \nabla V_p}{\nabla \cdot \mathbf{s}} \sim \frac{G\rho_0^2 d^2}{\mu} \sim 10^{-5}. \tag{6}$$

A similar scaling analysis for the relative importance of g_0 and elastic terms gives

$$\frac{g_0[\nabla(\rho_0 \mathbf{u} \cdot \mathbf{e}_r) - \mathbf{e}_r \nabla \cdot (\rho_0 \mathbf{u})]}{\nabla \cdot \mathbf{s}} \sim \frac{\rho_0 g_0 d}{\mu} \sim 10^{-2} \tag{7}$$

that indicates that g_0 terms are negligible as well. $\mathbf{F}_p$ scales as

$$F_p \sim \frac{|\mathbf{M}_p \nabla \delta(\mathbf{x} - \mathbf{x}_m)|}{\Delta V} \sim \frac{\lambda + 2\mu}{\mu} \frac{\pi a^3 \Delta P}{a} \frac{1}{\Delta V} \sim \frac{\mu}{a}, \tag{8}$$

where $\mathbf{M}_p \nabla \delta(\mathbf{x} - \mathbf{x}_m)$ is the body-force equivalent to a point source of dilation (AKI and RICHARDS, 1980, p. 61), λ and μ are the elastic moduli, ΔP the pressure change of

Table 1

Value of scaling parameters. The volume change ΔV corresponding to a vertical displacement u is estimated using a point source model (EGGERS, 1987)

d m	ρ_0, ρ_m kg/m^3	u m	μ Pa	ΔV m^3	a M
10^3	$3 \cdot 10^3$	1	$3 \cdot 10^9$	10^6	10^3
10^4	$3 \cdot 10^3$	1	$3 \cdot 10^{10}$	10^8	10^3

the point source of dilation and δ the Dirac's delta function. $\mathbf{F}_m$ scales as (ZHONG and ZUBER, 2000)

$$\mathbf{F}_m = -\rho_m g_0 \delta(\mathbf{x} - \mathbf{x}_m)\mathbf{e}_r \sim \rho_m g_0. \tag{9}$$

The ratio between the two body-forces gives

$$\frac{\mathbf{F}_m}{\mathbf{F}_p} \sim \frac{\rho_0 g_0 a}{\mu} \sim 10^{-2}. \tag{10}$$

$\mathbf{F}_p$ is larger than $\mathbf{F}_m$ and the equilibrium equation (5) reduces to

$$\nabla \cdot \mathbf{s} + \mathbf{F}_p = 0. \tag{11}$$

In summary, we can see from (6), (7) and (10) that the coupling between gravity and elasticity is negligible in the space scale associated with volcano monitoring. Changes in the spherical magma body pressure push the deformation, while the potential V_p (due to coupling between gravity and deformation) is of the same order of magnitude as the potential V_m (due to mass intrusion). It is worth noting that in the special case of a spherically symmetric source in a homogeneous medium, RUNDLE (1978) and WALSH and RICE (1979) show that the change in gravity actually results only from the mass of the intrusion. That is, the change in gravity due to coupling between gravity and deformation cancel ($|\nabla V_p| = 0$). We will further investigate in the next section the coupling between elasticity and gravity.

3. Layered Earth Model

In the second step of our investigation into the interpretation of gravity and deformation changes, we will compare estimates of the parameters (depth, volume, mass, density) of the deformation source both for a homogeneous and layered half-space model of Long Valley caldera, California (Fig. 2 and Fig. 3). The goal is to evaluate the various deformation models including elasto-gravitational effects and vertical discontinuities in the Earth's density and elastic parameters (e.g., FERNÁNDEZ et al., 2001a). We also check the results of the dimensional analysis performed in Section 2.

Over the past two decades, Long Valley caldera has shown persistent unrest with recurring earthquake swarms, uplift of the resurgent dome by over 80 cm and the onset of diffuse magmatic carbon dioxide emissions around the flanks of Mammoth Mountain on the southwest margin of the caldera (BAILEY and HILL, 1990, SOREY et al., 1993; LANGBEIN et al., 1995). Several sources of deformation have been identified in Long Valley caldera, although their geometry, depth and volume are not yet well constrained. Surveys of two-color EDM and leveling networks indicate that the principal sources of deformation are the intrusion of a magma body beneath the resurgent dome, and right lateral strike-slip within the south moat of the caldera (LANGBEIN *et al.*, 1995). In addition, there is evidence for dike intrusion beneath the south moat (SAVAGE and COCKERHAM, 1984) and Mammoth Mt. (HILL *et al.*, 1990). The intrusion beneath the resurgent dome has been confirmed by gravity measurements (BATTAGLIA *et al.*, 1999). For the purpose of this work, we will use uplift and residual gravity data collected in Long Valley caldera between 1982 and 1999.

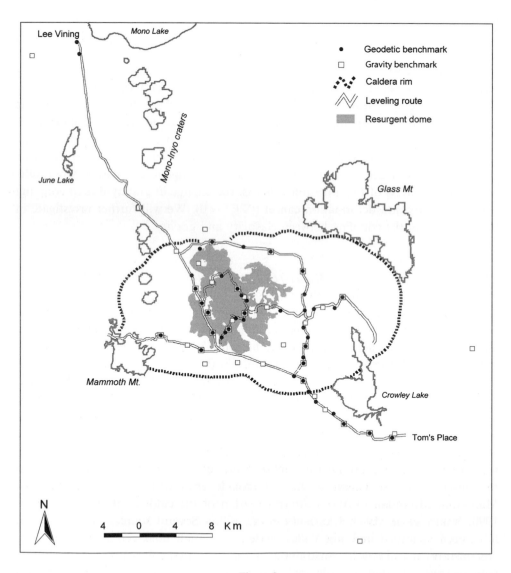

Figure 2
Map of Long Valley showing the location of the leveling and gravity benchmarks.

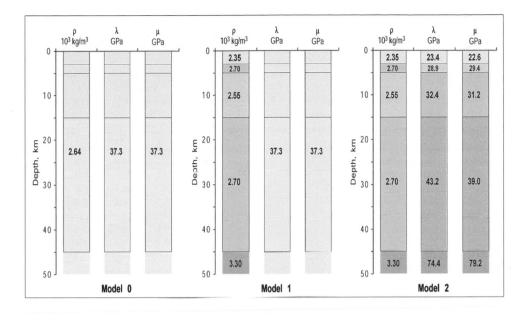

Figure 3

The layered model proposed to study gravity and deformation changes in Long Valley caldera (see Table 2). Model 0 represents a homogeneous medium with average values of the parameters; *Model 1* is elastically homogeneous, but with density stratification; *Model 2* is elastically inhomogeneous.

The Earth model proposed for Long Valley caldera includes 5 layers, the fifth being the mantle (see Table 2, Fig. 3). The thickness, density and seismic velocities assigned to the four layers representing the crust have been estimated from published works on regional gravity, seismic tomography and geology of Long Valley (KISSLING *et al.*, 1984; CARLE, 1988; DAWSON *et al.*, 1990; PONKO and SANDERS, 1994; SACKETT *et al.*, 1999). We use the numerical code developed by FERNÁNDEZ *et al.* (1997) to compute gravity changes and deformation due to an isotropic point source of dilation in a layered half-space. This code solves the fully coupled system of elastic-gravitational equations. Gravity changes and deformation due to a spherical intrusion in a homogeneous half-space are computed using the analytical point source approximation (EGGERS, 1987).

The numerical experiments are carried out considering separately (i) the effects on gravity changes and uplift due to a pressurized magma chamber cavity with no mass change and (ii) the effects of mass intrusion only with no magma chamber overpressure. Note that individually (i) and (ii) do not possess a geologic equivalent, because the geologically meaningful solution is given by the superposition of (i) and (ii). Furthermore, to study the effect of the density and elastic moduli stratification, we consider three different cases (see Fig. 3): (a) homogeneous medium (*Model 0*); (b) density stratification in an elastically homogeneous medium (*Model 1*); (c) an elastically inhomogeneous medium (*Model 2*). The results of the numerical

Table 2

Layered model of Long Valley caldera (see Fig 3b). Crosssection modified after SACKETT et al., (1999), density model after CARLE (1988), velocity model after KISSLING et al., (1984), depth of the Moho after DAWSON et al., (1990)

Layers	Crosssection	Thickness km	Depth Km	Density Model 10^3 kg/m^3	Velocity model 10^3 m/s		λ_d GPa	μ_d GPa
					V_S	V_P		
1	Caldera fill bishop tuff	3	3	2.35	3.1	5.4	23.4	22.6
2	Basement	2	5	2.70	3.3	5.7	28.9	29.4
3	Solidified magma chamber	10	15	2.55	3.5	6.1	32.4	31.2
4	Basement	30	45	2.70	3.8	6.7	43.2	39.0
5	Mantle			3.30	4.9	8.4	74.4	79.2
0	Homogeneous model	45		2.64			37.3	

computation for the homogeneous medium are compared with the analytical results. For every one of the seven (six numerical and one analytical) experiments, we find the solution that best fits the gravity and deformation data from Long Valley caldera using a least squares algorithm. To compare the results, we use two quantitative indicators:

$$\chi^2 = r^T \Sigma^{-1} r, \quad R^2 = 1 - \frac{r^T \Sigma^{-1} r}{u^T \Sigma^{-1} u}, \tag{12}$$

where r is the difference between the observed and predicted displacements, and Σ the data covariance matrix, u is the observed displacements. χ^2 is a measure of the error in fitting the experimental data with a model (the smaller χ^2 the better the fit), while R^2 is a measure of the ability of the model to explain the data. If $R^2 = 1$, the model is able to explain all variations in the observed data, if $R^2 = 0$, the model cannot explain the observed data.

For a pressurized mass-less cavity, the fit of the computed uplift to the Long Valley data differs significantly for the three different structural models (Fig. 4a, Table 3). The only noticeable difference is the slightly deeper source (9.4 km instead of 8.8 km, a 7% increase) obtained for the elastically inhomogeneous medium (*Model 2*). This can be explained by the greater compliance of the shallower layers above the source, compared with the homogeneous case (*Model 0*). The numerical results for the uplift in the elastically homogeneous medium (*Model 0* and *Model 1*) are identical to the analytical point source solution. This confirms the conclusion derived from the dimensional analysis that the effect of the elasto-gravitational coupling on the displacement is negligible. The gravity change results (Fig. 4b) indicate that the contribution from the potential ϕ_p is negligible for all practical

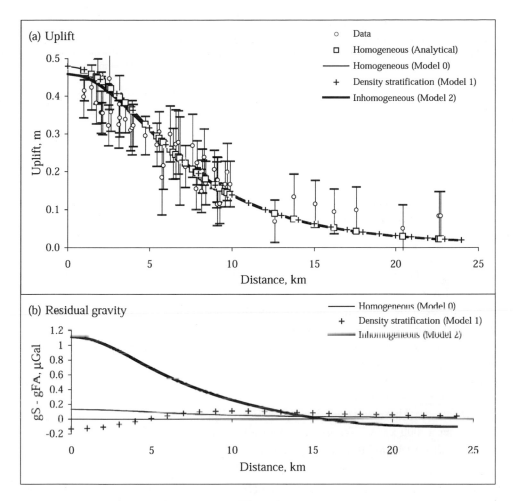

Figure 4

Pressurized massless cavity. (a) Match between experimental and modeled uplift. The plot shows no major differences between modeling the intrusion using a point source in a homogeneous or layered medium (see Table 3). (b) The plot shows the difference (residual gravity) between the surface gravity change (g_s) and the free-air effect (g_{FA}). The residual gravity corresponds to contributions to gravity changes from the elastic-gravitational coupling and density stratification. The maximum change (1 μGal) is about 3% of the residual gravity due to mass intrusion, well below the typical errors of 10 μGal for a relative gravity survey.

purposes. The numerical results for the elastically homogeneous media (*Model 0* and *Model 1*) show a maximum difference between the surface gravity and the free-air effect of less than 0.2 μGal (Fig. 4b). This is consistent with the numerical results of RUNDLE (1978) and the analytical results of WALSH and RICE (1979) that the change in gravity observed for a massless, spherically symmetrical, dilatational source in a homogeneous medium is equal to the free-air effect only, or $|\nabla V_p| = 0$. The contribution $|\nabla V_p|$ is practically negligible (about 1 μGal) in the inhomogeneous

Table 3

Results from numerical experiment

	Pressurized cavity				Mass intrusion			
	Depth km	Volume km^3	χ^2	R^2	Depth Km	Mass 10^{12} kg	χ^2	R^2
Point source (analytical)	8.8	0.16	57	0.99	8.8	0.45	37	0.66
Model 0 (homogeneous)	8.8	0.16	57	0.99	8.8	0.45	37	0.66
Model 1	8.8	0.16	57	0.99	8.8	0.45	37	0.66
Model 2 (inhomogeneous)	9.4	0.16	31	1.00	9.2	0.48	37	0.65

medium (*Model 3*, Fig. 4b) as well. Note that for a massless intrusion all the gravity changes (except the free-air effect) are well below the typical errors of 10 μGal (e.g., BATTAGLIA et al., 1999).

The results for the case study of a point mass intrusion with no pressure change are very similar. The maximum displacement induced by the mass intrusion is around 3 mm (Fig. 5a), or about 1% of the uplift due to cavity pressurization, indicating that the contribution to the uplift from ∇V_p is practically negligible. The fit of the computed gravity changes to the Long Valley data (Fig. 5b, Table 3) do not show a significant difference between the three cases proposed. The estimated mass in the inhomogeneous medium (*Model 2*) is slightly larger (0.48 × 10^{12} kg instead of 0.45 × 10^{12} kg, a 7% increase) and deeper (9.2 km instead of 8.8 km, a 5% increase) than that inferred for the elastically homogeneous medium (*Model 0* and *Model 1*). The estimated density in the inhomogeneous medium (*Model 2*) is 7% higher (3000 kg/m^3 versus 2800 kg/m^3) than that for the elastically homogeneous medium (*Model 0* and *Model 1*).

4. Source Geometry

A very common approach to infer the deformation source parameters (depth, volume, mass, density) is to match gravity and uplift data to the predictions of an isotropic center of dilation (e.g., MOGI, 1958; EGGERS, 1987; MCKEE et al., 1989; BERRINO, 1994; BATTAGLIA et al., 1999; FERNÁNDEZ et al., 2000; RYMER and WILLIAM-JONES, 2000). A major shortcoming of this technique is that we may fit the wrong model to the data, because different source models produce very similar vertical deformations (DIETERICH and DECKER, 1975). This may yield estimates of the density and depth of the intrusion that are not reliable. Consider, for example, modeling data generated from an ellipsoidal source (YANG et al., 1988; CLARK et al., 1986), assuming incorrectly a spherical symmetry for the source. The parameters of the actual ellipsoidal source are depth = 6 km, volume change = 0.2 km^3, mass change = 0.5 × 10^{12} kg and density = 2500 kg/m^3. The spherical source fits the uplift data well (see Fig. 6a) but predicts a deeper location for the intrusion (8.5 km

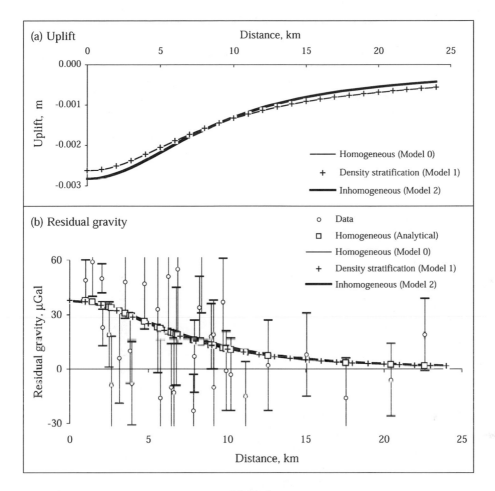

Figure 5
Spherical mass intrusion with no pressure change. (a) Uplift due to the mass intrusion. The maximum uplift (3 mm) is about 1% of the uplift due to cavity pressurization. (b) Match between experimental and modeled residual gravity. Again, the plot shows no major differences between modeling the intrusion using a point source in a homogeneous or layered medium (see Table 3).

instead of 6 km). The spherical model also requires a larger mass (0.9 instead of 0.5×10^{12} kg) to obtain the same gravity signal (Fig. 6b). The estimated volume increase is close to the correct value (0.2 km^3), because the spherical source is more efficient in causing vertical deformation. The spherical model fit appears reasonable and is able to explain about 99% of the uplift and gravity data. It is only when we compare the actual and computed radial displacements (Fig. 6c) that we realize that the spherical model is not appropriate. Using the parameters of the spherical source to estimate the intrusion density, we infer a value of 4500 kg/m^3 instead of the actual 2500 kg/m^3. Note that a straight line can successfully fit the ellipsoidal data gravity/ height correlation as well (Fig. 6d). The linear correlation between gravity and height

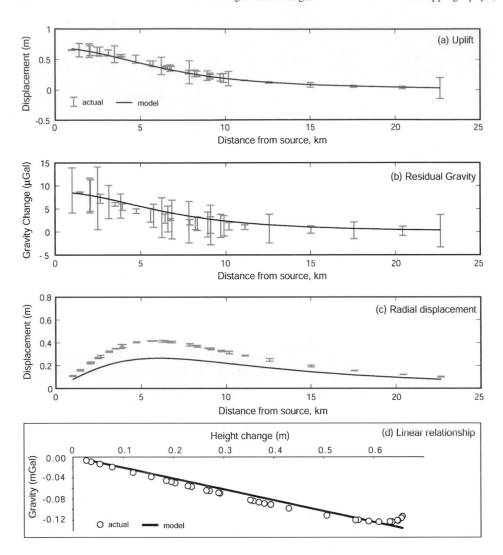

Figure 6

Bias due to incorrectly assessing the source shape. Actual ellipsoidal source (YANG et al., 1988; CLARK et al., 1986): depth = 6 km, volume = 0.2 km^3, mass = 0.5 × 10^{12} kg, density = 2500 kg/m^3. (a) (b) and (c) Fitting a spherical source (solid line) to a data set created using an ellipsoidal source (error bars). Uncertainties for the synthetic data set are 6 cm for uplift, 20 μGal for the residual gravity and 6 mm for the radial displacement. Inferred spherical source: depth = 8.5 km, volume = 0.2 km^3, mass = 0.9 × 10^{12} kg, density = 4500 kg/m^3. (d) A straight line fits the ellipsoidal data gravity/height correlation ($R^2 = 0.94$). The linear correlation between gravity and height changes is considered to be a special characteristic of spherically symmetric magma bodies (EGGERS, 1987). In this case, the inferred density is 3670 kg/m^3 and $\Delta g / \Delta u = 206 \pm 4$ μGal/m.

changes is considered to be a special characteristic of spherically symmetric magma bodies (EGGERS, 1987). In this case, the inferred density is 3670 kg/m^3 instead of the actual 2500 kg/m^3.

5. Summary and Conclusions

Combined geodesy and gravity measurements allow us to infer the density of intrusive bodies, and better constrain deformation sources in volcanic areas. In this work, we investigate three factors that can help in obtaining a more realistic picture of the intrusive body: (a) coupling between elastic and gravitational effects, (b) a layered Earth model, with one or more layers with differing densities and elastic properties and (c) non-spherical source geometries.

The first two factors investigated in this work do not affect the estimate of the source parameters significantly. Coupling between elastic and gravitational effects (self-gravitation) is second order over the time and distance scales normally associated with volcano monitoring. For a Maxwell material and at times long compared to the relaxation time, the stresses will relax to the point where the source terms are no longer balanced by the divergence of the stresses in the equilibrium equations. In this limit the self-gravitational effects cannot be ignored. We find no major differences between modeling the intrusion using a point source in a homogeneous or layered medium for an elastic model appropriate to Long Valley caldera.

Our results indicate that the critical step in the interpretation of the field data is the choice of the source model used to inverting geodetic and gravity data to infer the actual deformation source parameters. A simple experiment of matching deformation and gravity data from an ellipsoidal source using a spherical source shows that the standard approach of fitting a Mogi's source to gravity and uplift data only, excluding the horizontal displacements, can yield estimates of the source parameters that are not reliable. In our experiment, the spherical source fits the uplift and gravity data well (Figs. 6a and 6b), estimates correctly the volume increase (0.2 km^3), but predicts a deeper location (8.5 km instead of 6 km), a larger mass (0.9 instead of 0.5×10^{12} kg) and a larger density (4500 kg/m^3 instead of the actual 2500 kg/m^3) for the intrusion. Only by comparing the actual and the modeled radial displacements (Fig. 6c), we can demonstrate that the spherical model is not appropriate. It is important to note that a center of dilation can bias the results, overestimating the depth, mass and density of the intrusion. To obtain a reliable estimate of the depth and density of the intrusion, inversion of geodetic and gravity data should (a) not assume that the source possesses a spherical symmetry and (b) include not only the uplift and residual gravity, but the horizontal deformation as well (DIETERICH and DECKER, 1975). The ellipsoidal model can be used to invert geodetic and gravity data without assuming any particular orientation or symmetry, but the modeling requirements can be substantially more complicated than those for the Mogi's source (e.g., TIAMPO et al., 2000).

Acknowledgements

We would like to thank J. Fernández for providing the numerical code. Reviews by F.H. Cornet and K.F. Tiampo greatly helped to improve the manuscript. Support for this research was provided by DoE, Office of Basic Energy Science, grant DE-FG03-99ER14962.

REFERENCES

AKI, K. and RICHARDS, P. G., *Quantitative Seismology: Theory and Methods* (W. H. Freeman and Co., San Francisco 1980).

BAILEY, R. A. and HILL, D. P. (1990), *Magmatic Unrest at Long Valley Caldera, California, 1980–1990,* Geoscience Canada *17*, 175–179.

BATTAGLIA, M., ROBERTS, C., and Segall, P. (1999), *Magma Intrusion beneath Long Valley Caldera Confirmed by Temporal Changes in Gravity,* Science *285*, 2119–2122.

BERRINO, G., RYMER, H., BROWN, G. C., and CORRADO, G. (1992), *Gravity-height Correlations for Unrest at Calderas,* J. Volcanol. Geotherm. Res. *53*, 11–26.

BERRINO, G. (1994), *Gravity Changes Induced by Height-mass Variations at the Campi Flegrei Caldera,* J. Volcanol Geotherm. Res *61*, 293–309.

BONAFEDE, M. and MAZZANTI, M. (1998), *Modelling Gravity Variations Consistent with Ground Deformation in the Campi Flegrei Caldera (Italy),* J. Volcanol Geotherm. Res. *81*, 137–157.

CARLE, S. (1988), *Three-dimensional Gravity Modeling of the Geologic Structure of Long Valley Caldera,* J. Geophys. Res. *93*, 13,237–13,250.

CHENG, C. H. and JOHNSTON, D. H. (1981), *Dynamic and Static Moduli,* Geophys. Res. Lett. *8*, 39–42.

CLARK, D. A., SAUL, S. J., and EMERSON, D. W. (1986), *Magnetic and Gravity Anomalies of a Triaxial Ellipsoid,* Exploration Geophysics *17*, 189–200.

DAWSON, P. B., EVANS, J. R., and IYER, H. M. (1990), *Teleseismic Tomography of the Compressional Wave Velocity Structure beneath the Long Valley Region, California,* J. Geophys. Res. *95*, 11,021–11,050.

DIETERICH, J. H. and DECKER, R. (1975), *Finite Element Modeling of Surface Deformation Associated with Volcanism,* J. Geophys. Res *80*, 4094–4102.

DVORAK, J. J. and DZURISIN, D. (1997), *Volcano Geodesy: The Search for Magma Reservoirs and the Formation of Eruptive Vents,* Rev. Geophys. *35*, 343–384.

EGGERS, A. (1987), *Residual Gravity Changes and Eruption Magnitudes,* J. Volcanol. Geotherm. Res. *33*, 201–216.

FERNÁNDEZ, J., RUNDLE, J., GRANELL, R., and YU, T. (1997), *Programs to Compute Deformation due to a Magma Intrusion in Elastic-gravitational Layered Earth Model,* Comput. and Geosci. *23*, 231–249.

FERNÁNDEZ, J., CHARCO, M., TIAMPO, K. F., JENTZSCH, G., and RUNDLE, J. B. (2001a), *Joint Interpretation of Displacement and Gravity Data in Volcanic Areas. A Test Example: Long Valley Caldera, California,* Geophysic. Res. Lett. *28*, 1063–1066.

FERNÁNDEZ, J., TIAMPO, K. F., JENTZSCH, G., CHARCO, M., and RUNDLE, J. B. (2001b), *Inflation or Deflation? New Results for Mayon Volcano Applying Elastic-gravitational Modeling,* Geophysic. Res. Lett. *28*, 2349–2352.

HILL, D. P., ELLSWORTH, W. L., JOHNSTON, M. S., LANGBEIN, J. O., OPPENHEIMER, D. H., PITT, A. M., REASENBERG, P. A., SOREY, M. L., and McNUTT, S. R. (1990), *The 1989 Earthquake Swarm beneath Mammoth Mountain, California; an Initial Look at the 4 May through 30 September Activity,* Bull. Seismol. Soc. Am. *80*, 325–339.

KISSLING, E., ELLSWORTH, W., and COCKERHAM, R. (1984), *Three-dimensional Structure of the Long Valley Caldera Region, California, by Geotomography,* Open-File Report - U.S. Geological Survey, OF *84–0939*, 188–220.

LANGBEIN, J., DZURISIN, D., MARSHALL, G., STEIN, R., and RUNDLE, J. (1995), *Shallow and Peripheral Volcanic Sources of Inflation Revealed by Modeling Two-color Geodimeter and Leveling Data from Long Valley Caldera, California, 1988–1992*, J. Geophys. Res. *100*, 12487.

LOVE, E. H., *Some Problems in Geodynamics* (Cambridge University Press, New York, 1911).

MCKEE, C., MORI, J., and TALAI, B., *Microgravity Changes and Ground Deformation at Rabaul Caldera, 1973–1985*. In IAVCEI Proc. in *Volcanology* (ed., Latter, J.) (Springer-Verlag, Berlin 1989), pp. 399–431.

MOGI, K. (1958), *Relations of the Eruptions of Various Volcanoes and the Deformation of Ground Surfaces around them*, Bull. Earthq. Res. Inst. Tokio Univ. *36*, 94–134.

POLLITZ, F. (1997), *Gravitational Viscoelastic Postseismic Relaxation on a Layered Spherical Earth*, J. Geophys. Res. *102*, 17,921–17,941.

PONKO, S. and SANDERS, C. (1994), *Inversion for P- and S-wave Attenuation Structure, Long Valley Caldera, California*, J. Geophys. Res. *99*, 2619–2635.

RUNDLE, J. (1978), *Gravity Changes and the Palmdale Uplift*, Geophys. Res. Lett. *5*, 41–44.

RUNDLE, J. (1982), *The Deformation, Gravity, and Potential Changes due to Volcanic Loading of the Crust*, J. Geophys. Res *87*, 10,729–10,744.

RYMER, H. (1994), *Microgravity Change as a Precursor to Volcanic Activity*, J. Volcanol. Geotherm. Res *61*, 311–328.

RYMER, H. and WILLIAMS-JONES, G. (2000), *Volcanic Eruption Precursors: Magma Chamber Physics from Gravity and Deformation Measurements*, Geophys. Res. Lett. *27*, 2389–2392.

SACKETT, P. C., MCCONNELL, V. S., ROACH, A. L., PRIEST, S. S., and SASS, J. H. (1999), *Long Valley coring Project, 1998—Preliminary Stratigraphy and Images of Recovered Core*, U.S. Geol. Survey Open-File Report 99–158.

SAVAGE, J. and COCKERHAM, R. (1984), *Earthquake Swarm in Long Valley Caldera, California, January 1983; Evidence for Dike Inflation*, J. Geophys. Res. *89*, 8315–8324.

SOREY, M. L., KENNEDY, B. M., EVANS, C. W., FARRAR, C. D., and SUEMNICHT, G. A. (1993), *Helium Isotope and Gas Discharge Variations Associated with Crustal Unrest in Long Valley Caldera, California, 1989–1992*, J. Geophys. Res *98*, 15,871–15,889.

TIAMPO K. F., RUNDLE, J. B., FERNÁNDEZ, J., and LANGBEIN J. (2000), *Spherical and Ellipsoidal Volcanic Sources at Long Valley Caldera, California, Using a Genetic Algorithm Inversion Technique*, J. Volcanol Geotherm. Res. *102*, 189–206.

YANG, X., DAVIS, P.M, and DIETERICH, J. H. (1988), *Deformation from Inflation of a Dipping Finite Prolate Spheroid in an Elastic Half-space as a Model for Volcanic Stressing*, J. Geophys. Res. *93*, 4249–4257.

WALSH, J., and RICE, J. (1979), *Local Changes in Gravity Resulting from Deformation*, J. Geophys. Res. *84*, 165–170.

ZHONG, S. and ZUBER, M. T. (2000), *Long-wavelength Topographic Relaxation for Self-gravitating Planets and its Implications for the Time-dependent Compensation of Surface Topography*, J. Geophys. Res. *105*, 4153–4164.

(Received November 12, 2002, revised January 31, 2003 accepted February 3, 2003)

 To access this journal online:
http://www.birkhauser.ch

Pure appl. geophys. 161 (2004) 1469–1487
0033–4553/04/071469–19
DOI 10.1007/s00024-004-2515-4

© Birkhäuser Verlag, Basel, 2004

❚ Pure and Applied Geophysics

Intrusive Mechanisms at Mt. Etna Forerunning the July-August 2001 Eruption from Seismic and Ground Deformation Data

A. Bonaccorso[1], S. D'Amico[1], M. Mattia[1], and D. Patanè[1]

Abstract—In this work we present seismological and ground deformation evidence for the phase preparing the July 18 to August 9, 2001 flank eruption at Etna. The analysis performed, through data from the permanent seismic and ground deformation networks, highlighted a strong relationship between seismic strain release at depth and surface deformation. This joint analysis provided strong constraints on the magma rising mechanisms. We show that in the last ten years, after the 1991–1993 eruption, an overall accumulation of tension has affected the volcano. Then we investigate the months preceding the 2001 eruption. In particular, we analyse the strong seismic swarm on April 20–24, 2001, comprising more than 200 events (M_{max} = 3.6) with prevalent dextral shear fault mechanisms in the western flank. The swarm showed a ca. NE-SW earthquake alignment which, in agreement with previous cases, can be interpreted as the response of the medium to an intrusive process along the approximately NNW-SSE volcano-genetic trend. These mechanisms, leading to the July 18 to August 9, 2001 flank eruption, are analogous to ones observed some months before the 1991–1993 flank eruption and, more recently, in January 1998 before the February-November 1999 summit eruption.

Key words: Ground deformation, volcano seismology, Mt. Etna Volcano, intrusive mechanism.

1. Introduction

In the last twenty years, geophysical investigations have played an increasingly important role in studies of Mt. Etna eruptive processes. In particular, seismological and ground deformation studies provided the best information during recent years, in which several important lateral eruptions have occurred. It is well known that the study of seismic activity is a very powerful tool for understanding the inner structure of a volcano. Unfortunately, at Mt. Etna, the permanent seismic network has been of low density for a long time, limiting the accuracy of hypocentral locations, and consequently our expertise on Mt. Etna magma dynamics in the shallow 5 kilometres of crust (e.g., Gresta *et al.*, 1998; Barberi *et al.*, 2000; Patanè and Privitera, 2001). Only since the 90s have seismic data been available in digital form by a sufficient number of stations (increased in time) equipped also with three-component sensors (e.g., Patanè *et al.*, 1999; Patanè and Privitera, 2001; Patanè *et al.*,

[1] Istituto Nazionale di Geofisica e Vulcanologia – Sezione di Catania, Piazza Roma, 2, 95123 Catania, Italy. E-mail: bonaccorso@ct.ingv.it

2003a). This allows us to put high-quality constraints on seismic activity occurring at most depths and to better study the recent eruptive activity.

For the different recent lateral eruptions, ground deformation data analysis and modelling has indicated preparation times which lasted from months to several years. The modelling showed that the final uprising system, i.e., the tensile mechanisms inside the volcano edifice, involved the same structure oriented approximately NNW-SSE. A resumé of the ground deformation modelling of three principal recent lateral eruptions (1981, 1989, 1991–1993) and the implications in terms of associated precursors is discussed in BONACCORSO (2001). The NNW-SSE trending fault zone, older than Mt. Etna, borders on the eastern margin of Sicily crossing the volcano and characterises the main eruptive trend (e.g., FRAZZETTA and VILLARI, 1981; LO GIUDICE and RASÀ, 1986). The relationship of this regional trend with the regional tectonic setting, volcanological features and geophysical constraints is widely discussed in BONACCORSO *et al.* (1996).

In a wider and integrated context, a multidisciplinary investigation focussed on the eruption precursors should be implemented in order to better understand the possible modification of the volcanic activity. Recently, interesting results have been obtained through the integrated analysis of seismic and ground deformation data. Short-term precursors were observed by seismic foci migration with stress field modification (PATANÈ *et al.*, 1994; BONACCORSO *et al.*, 1996; Cocina et al., 1998) and transient variations on tilt during the two months preceding the 1991-1993 eruption (BONACCORSO and GAMBINO, 1997). Especially, the occurrence of strong seismic swarms, which occurred in the southwestern sector of the volcano along an approximately NE-SW structure, has been observed in relationship with magma movement through the upper crust and/or inside the volcanic edifice. This seismicity forerunning the 1991-1993 eruption has been interpreted as a response of the medium to the intrusive episode occurring across the volcano-genetic NNW-SSE structural trend (Fig. 1). Analogous conclusions have been reached for the swarm which occurred on January 1998 again in the southwestern flank (BONACCORSO and PATANÈ, 2001). Also in this case the seismic swarm was associated with an intrusion along the NNW-SSE trend which then led to the February-November 1999 summit eruption (Fig. 1).

In this work first we furnish wide evidence that after the 1991–1993 eruption, a tension accumulation characterised the volcano. Then we focus our study on investigating the final intrusive mechanisms which occurred during the period before the last lateral eruption occurring on July-August, 2001 (Research Staff of INGV-CT, 2001) as inferred by seismic strain release and surface 3-D deformation provided by the GPS permanent network. In particular, we analyse the strong seismic swarm of April 20–24, 2001. We show that this event is associated with the transition between a phase of continuous accumulation of tension below the volcano and the beginning of an intrusion process. Furthermore, we point out that this kind of seismicity can be, once again, considered a confirmation of the south western flank response to the intrusion processes occurring across the volcano-genetic, approximately NNW-SSE structural trend. We discuss the implications of this stress readjustment, which is to be considered a valid precursor.

2. Seismic and Ground Deformation Permanent Networks

The Mt. Etna permanent seismic network, run by the Catania Section of the National Institute of Geophysics and Volcanology (hereafter, INGV-CT), is the

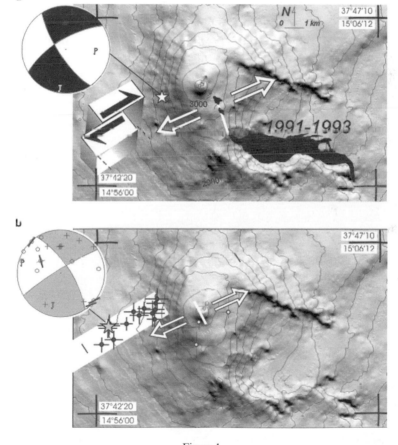

Figure 1
(a) Sketch model of the rupture mechanism associated with the 1991–1993 flank eruption (BONACCORSO *et al.*, 1996); (b) epicentral map of the January 9–14, 1998 seismic swarm. The position of the surface projection of the modelled shallow dike is also drawn (BONACCORSO and PATANÈ, 2001). The stars represent the main shocks of the swarms ($M_d = 4.5$ and $M_d = 3.7$, respectively). The focal solutions are also reported. The swarms were interpreted as shear response to the intrusion process along the NNW-SSE structural trend.

integration of two seismic networks running separately until 2000 by Istituto Internazionale di Vulcanologia (IIV) and Sistema Poseidon (Fig. 2). The former network was composed of 14 seismic stations, 10 of which were one-component analog stations, equipped with short-period sensors (1 s), and the other four three-component stations, two of which were equipped with broadband sensors (30 s). The Sistema Poseidon network was composed of 56 stations, spread out over eastern Sicily, 39 of which were deployed on Mt. Etna and equipped with short-period sensors (1 sec). Also in this case, excluding four three-component stations, the entire network was composed of one-component analog stations. Signals from remote sites are transmitted by radio or cable to Catania where they are continuously recorded on drum recorders and digitalized at a sampling rate varying from 100 Hz in continuous mode to 200 Hz in trigger mode.

The permanent network for the continuous tilt monitoring is composed of eight bi-axial electronic borehole tiltmeters positioned at about 3–5 m depth (Fig. 2). The complete configuration has been operating since 1990. The stations are equipped with Applied Geomechanics Inc. tiltmeters model 722 (CDV, MSC, MEG, SPC stations) with 0.1 μrad precision or model 510 (DAM, MDZ, MMT stations) with

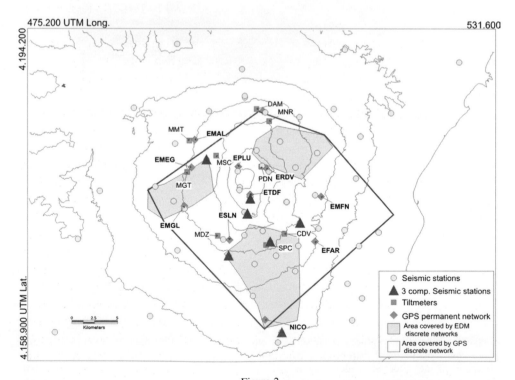

Figure 2
Mt. Etna seismic and ground deformation networks run by the Catania Section of the National Institute of Geophysics and Volcanology (INGV-CT).

0.01 μrad precision. The radial tilt, i.e., the component directed toward the crater, has positive signal variation that means crater up. The second component, the tangential one, is orthogonal to the radial and a positive signal variation means uplift in the anticlockwise direction. The control datalogger is programmed for 48 data/day sampling (1 sample every 30 minutes) and includes acquisition of the two tilt components, air and ground temperatures, and instrumental control parameters, such as power supply and dc/dc converter voltage. The data are transmitted to INGV-CT via radio-link. Temperature noise is present in the borehole types, whose electrolytic bubble sensors can be affected by diurnal and seasonal temperature variations (AGI, 1993; BONACCORSO et al., 1999). The tilt data are filtered from the seasonal temperature noise using the linear correlation with the ground temperatures. In 1997 a high precision long-base fluid tiltmeter was installed at the Pizzi Deneri observatory (PDN.OBS) in the high (2850 m a.s.l.) northeastern volcano flank. The instrumentation is positioned along two underground tunnels, where two 80-m long orthogonal tubes are filled with mercury, whose vertical changes at the extremities are measured by laser sensors giving a real precision of 0.01 μrad (BONACCORSO et al., 1998). It is stable and not affected by temperature noise because both the tilt measurement is obtained as the difference between vertical values of the mercury level at the extremities of the tubes and the sensors (optical laser) are very stable for temperature variations.

Recently the ground deformation continuous monitoring was upgraded with the installation of the GPS permanent network, which began to operate in November 2000. Only six stations were working until March 2001, while a configuration of 13 stations was reached in the late spring-summer of 2001 (Fig. 2). A PC-based master station, on which a software program (called Genesis Master) calls the remote stations (once or more per day), downloads GPS data through a Remote Access Server. Each remote station is based on an industrial PC motherboard and the transmission of data is performed by a GSM cellular modem. The gathered data are processed with software called EOLO (AMORE et al., 2002), that one or more times per day processes the data and stores the processed baselines in a database. The same software displays data in terms of length variations between the benchmarks, and an automatic subroutine calculates the strain parameters and displays the areal dilatation, the shear components, the components of the strain tensor, etc. In off-line mode, a more accurate analysis can be performed with the Trimble Geomatics Office software. The network is adjusted using a same local reference point (NICO - Nicolosi), which can be considered fixed with respect to the summit area. Due to the small extension of the measured area, the tilt or bias introduced by differences in the satellite ephemeris reference systems are negligible. Therefore, the minimal constraint approach using one of the network benchmarks as a fixed point ensures the stability of the reference frame during different days. The error ellipse calculated at the confidence interval of 95% is of the order of a few millimeters. In Table 1 the main features of data processing are reported.

Table 1

Method of GPS observations and analysis

Receiver	Trimble 4700
Daily session	6–24 h
Data sampling	30 sec.
Elevation cutoff	15°
Software	Trimble TGO
Orbit	precise eph. (NGS)
Tropospheric model	Saastamoinen's model with standard atmospheric conditions
Reference point	NICO (Nicolosi)

3. Stress Accumulation after the 1991-1993 Lateral Eruption

During the 1991-1993 eruption, the most significant lateral eruption in the last three centuries both in terms of duration (476 days) and lava erupted (ca. 250 $\cdot$ 10^6 m^3), the ground deformation showed a deflation of the volcano edifice, analysis of which constrained an intermediate storage at a depth of 3–4 km b.s.l. (BONACCORSO, 1996). A renewal of tension accumulation was observed in the whole volcanic edifice after the end of this eruption. The ground deformation showed an inflation (PUGLISI *et al.*, 2001) which uniformly increased during the following years. The areal deformation, accumulated by EDM and GPS networks from 1994 to 2001, indicated an overall continuous accumulation of tension inside the volcano (Fig. 3). It is noteworthy that this accumulation did not appear equilibrated by the energy discharge that occurred through the summit eruption (February-November, 1999), fed by a small dyke emplacement (BONACCORSO and PATANÈ, 2001), the January-May 2001 summit eruption, and the several tens of strong explosive events at the summit craters during the 1998–2001 (GVN, 1998; GVN, 2000; LA DELFA *et al.*, 2001; RESEARCH STAFF OF INGV-CT, 2001).

After the 1991-1993 deflation, the tilt signals evidenced a modest but progressive positive trend during the following years. The radial components, filtered from the seasonal thermic noise, are shown in Figure 4. The modest accumulation of tilt changes could be explained by a near uniform lifting of the volcano edifice. Also this pattern could be compatible with a tension accumulation inside the volcano. It is interesting to observe that in the last four years data of the summit long-base fluid tiltmeter has shown a near continuous deflation with about 14 microradians accumulated (Fig. 5) rather than inflation seen in the geodetic data. This interesting aspect could be an effect of the summit topography with average slope larger than 20° where a deep pressuring source can produce radial tilt opposite in sign to what would be expected for a flat topography (CAYOL and CORNET, 1998; WILLIAM and WADGE, 1998).

An overview of the seismicity occurring at Mt. Etna since 1990 is given by the rate of the earthquakes ($M_d \geq 2.5$) expressed as a number of events per day (inset in

Fig. 6). In Figure 6 the cumulative strain release is also reported. In detail, it is possible to observe that the 1991-1993 eruption was accompanied by a period of more or less total absence of earthquakes. The reappearance of such events was observed only since May 1993. The level of seismicity, however, was low until June 1996. Thereafter a significant increase in the seismic strain release gradient was observed and remained practically constant throughout 1997. In this last period seismicity affected most of the volcanic edifice (PATANÈ *et al.*, 1999) and was located at depths ranging from sea level to ca. 25 km. A strong increase in activity was recorded during the swarm occurring on January, 9–14, 1998 in the western flank of the volcano. Also this swarm was associated with an intrusive process

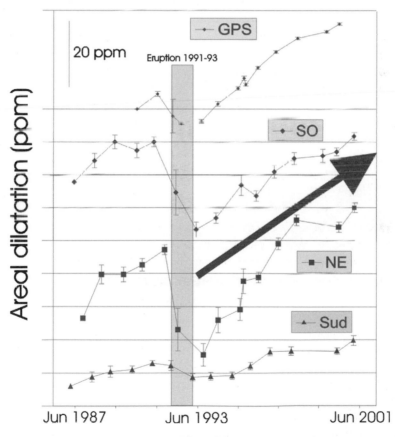

Figure 3
Planar areal dilatation calculated from the line length changes in the different monitored sectors through the EDM and GPS networks during the last 15 years. The areal dilatation represents mean values of the deformation in these areas. It is clearly evident how the 1991–1993 lateral eruption was accompanied by an areal contraction phase (deflation). After this eruption a near-continuous expansion accompanied the volcano edifice until 2001. A partial attenuation of the areal dilatation positive trend is recorded astride the February-November 1999 eruption.

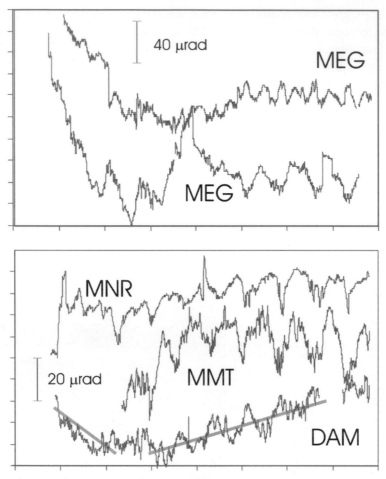

15/08/90 28/12/91 11/05/93 23/09/94 05/02/96 19/06/97 01/11/98 15/03/00 28/07/01

Figure 4

Radial tilt recorded at the shallow borehole stations during the period 1990–2001 before the July 2001 eruption onset. The data have been filtered from thermic seasonal noise using a linear correlation with the ground temperature. After the 1991–1993 deflation the signals evidence a modest but progressive inflation. The signals of the stations of MDZ, CDV, and SPC are not shown due to several interruptions which resulted from the electronic and instrumental problems which often caused the removing and re-installation of the instrumentation.

which provoked a shallow dike uprising inside the volcano edifice (BONACCORSO and PATANÈ, 2001). This dike was interpreted as the emplacement leading to the January-February 1999 eruption, which allowed a partial tension release. This interpretation is in agreement with the partial attenuation of the areal dilatation trend (see Fig. 3) and the low seismic level mainly recorded in the second half of 1999.

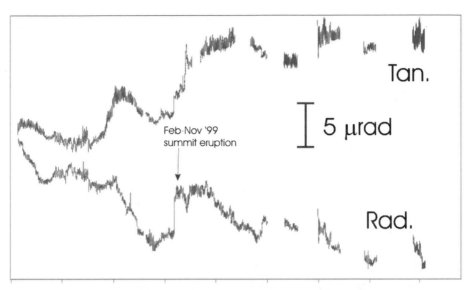

Figure 5

Radial tilt recorded at the long-base mercury tiltmeter during 1997–2001 before the July 2001 eruption onset (raw data). The station is located in the northeastern summit area (2850 m a.s.l.).

After several months of low seismic activity, a significant variation in the strain release gradient was observed from November 5, 2000, when a seismic swarm occurred in the upper southern flank of the volcano (115 earthquakes; $M_{max} = 3.6$) at a depth of ca. 8–10 km. This swarm marked the beginning of a relevant positive variation in the cumulative strain release curve (Fig. 6). In the following months other relevant swarms occurred and the average number of "background" events also increased together with the mean magnitude.

4. January – April 2001 Seismicity and Ground Deformation

Focusing our attention now on the spatial distribution of seismicity during the period 1 January–19 April 2001, in Figure 7 the map of epicentral locations and the related west-east cross section are reported. Earthquakes were located using HYPOELLIPSE code (LAHR, 1989) so as to allow for the difference in altitude of the seismic stations and to correctly detect the hypocentre spatial patterns in the volcanic cone. The 1-D velocity model used was derived by HIRN et al. (1991), with sea level assumed as the reference (zero) elevation. We selected 232 events using the following thresholds: number of observations (N) $\geq$ 10, the horizontal 68% confidence limit in the least well-constrained direction (ERH) $\leq$ 1.0 km, the 68% confidence limit for depth (ERZ) $\leq$ 1.5 km, root-mean-square residual (RMS) $\leq$ 0.3 s,

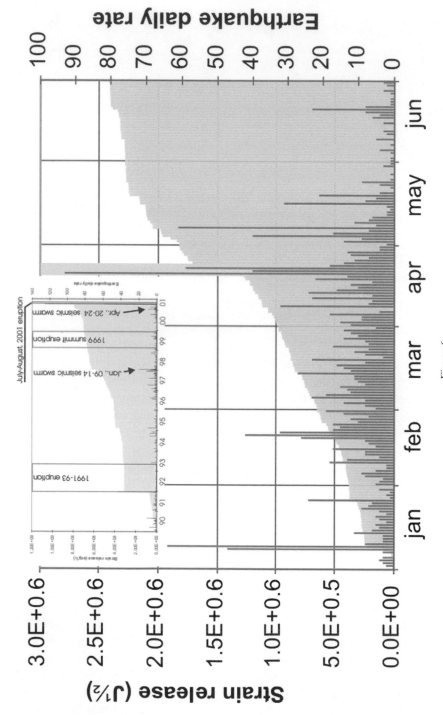

Figure 6

Number of earthquakes ($M_d \geq 1.0$) and related strain release vs. time recorded in the Etnean area during January-July 2001 time period. In the inset, the number of earthquakes ($M_d \geq 2.5$) and related strain release vs. time recorded during 1990-July 2001 is also shown.

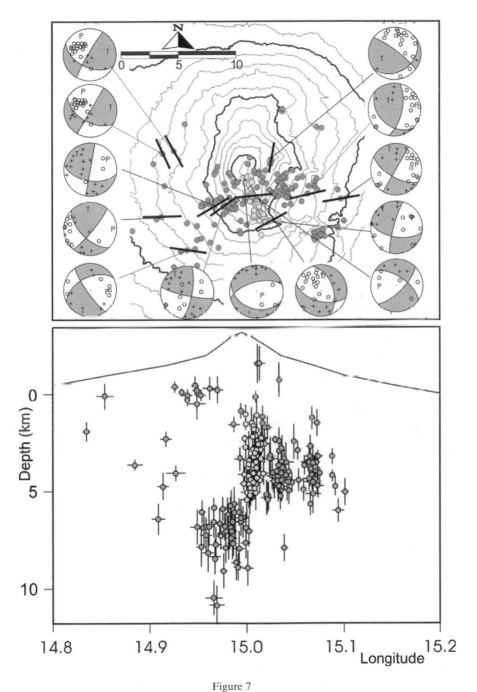

Figure 7
Map of epicentral locations and related W-E cross section of the best constrained earthquakes recorded between January and 19, April, 2001. FPSs of some stronger earthquakes and the related azimuth distribution of the horizontal projection of *P* axes are also shown.

largest azimuthal separation in degrees between stations as seen from the epicenter (GAP) ≤ 180°. Earthquake epicenters spread in a wide area covering most of the southern and eastern part of the volcano (Fig. 7). A peculiar feature of the earthquakes spatial distribution is its depth pattern. In fact, in the W-E cross section we can observe that the seismicity of the eastern flank is confined to the depth range 2–6 km b.s.l., while, in the central-western flank shocks are confined in two main different volumes, under ca. 6 km and above 1 km b.s.l., respectively. Moreover, since the second half of February, an increase in the seismicity has also been observed in the central upper part of the volcano. Earthquake locations show a ca. NNW-SSE alignment, and focal depths range between 1 and 5 km b.s.l. (light grey circles in Fig. 7). In Figure 7 thirteen Fault Plane Solutions (FPS) of the stronger earthquakes which occurred in this period and the related spatial *P*-axes distribution are also reported. These FPSs have been selected on the basis of rigorous criteria (number of polarities ≥15, number of polarity discrepancies ≤2, focal planes uncertainty < 20°, unique and unambiguous solution). FPSs show prevalent strike-slip rupture mechanisms and, in several cases, these strike-slip faults present a remarkable normal component. Normal dip-slip mechanisms were also observed along the alignment oriented NNW-SSE where seismicity was mainly clustered. Southwards of the central craters the *P*-axes orientation indicate a ca. E-W direction, coherently with the ground deformation results which indicated a similar orientation of the displacement vectors for this sector of the volcano. Conversely, at greater depth (> 15 km) earthquakes located in the western and northwestern sector of the volcano displays a different orientation of *P*- and *T*- axis with respect to the shallower ones. In fact, we usually observe that *P*-axes are ca. NNW-trending below 15-km depth, suggesting a closer relation with the regional stress at greater depth (PATANÈ and PRIVITERA, 2001; PATANÈ *et al.*, 2003). It is noteworthy, excepting the deep earthquakes recorded in the southwestern sector (Jan.-April, 2001), that this seismicity shows similar features to that observed during the July, 12–18 seismic swarm, which led to the lateral eruption of July 17 to August 9 (RESEARCH STAFF OF INGV-CT, 2001), with prevalent dip-slip rupture mechanism (PATANÈ *et al.*, 2003a). The main difference is that before the eruption the foci were located at shallow depth.

A powerful contribution to continuous ground deformation monitoring was provided by the GPS permanent network. Coherently with the renewal of seismic activity observed since November 2000, during the early months of 2001 the daily length variations between the stations of the GPS permanent network followed the general trend measured once per year since the end of the 1991-1993 eruption, and confirmed that the inflation of the volcano continued (Fig. 8). Moreover, the height changes showed a clear positive trend during the period between January and the 20th of April 2001 in agreement with the continuation of the inflation phase. The distribution of the displacement vectors together with the vertical changes calculated for the period January 12 to April 20 is shown in Figure 9.

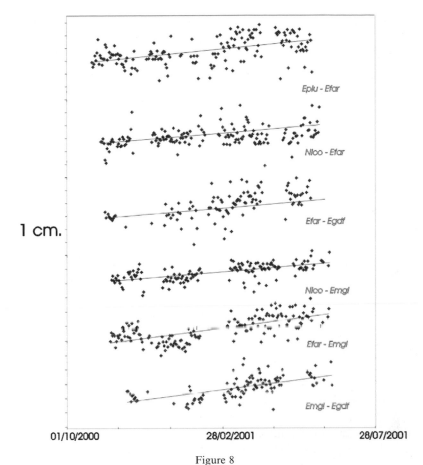

Figure 8

Example of line length changes observed by the daily sessions of the GPS permanent network during January-June 2001. Continuing the previous eight-year trend, a near continuous expansion is still evident. This expansion was to be interrupted by the July 2001 eruption.

5. The Swarm of April 20-24, 2001

On April 20, 2001 a strong seismic swarm started and more than 200 events ($M_{max} = 3.6$) were recorded in about four days. We selected 41 events using the following thresholds: number of observations $\geq$ 10, ERH $\leq$ 1.0 km, ERZ $\leq$ 1.0 km, RMS $\leq$ 0.30 s, GAP $\leq$ 180°. The epicenters distributed in the southwestern upper part of the edifice clearly define a NE-SW alignment located between ca. 5 and 10 km depth (Fig. 10). The epicentral distribution is in agreement with the rupture mechanism obtained by the focal plane solution of the main shock (star in Fig. 10) and then constrains the fault plane to be along a NE-SW direction. In Figure 8 we report the FPSs of ten shocks selected on the basis of rigorous criteria (number of polarities $\geq$ 15, number of polarity discrepancies $\leq$ 2, focal planes uncertainty < 20°,

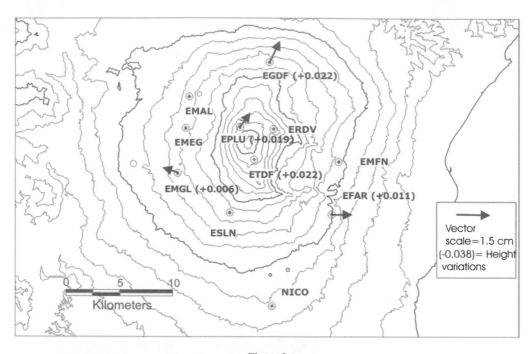

Figure 9
Displacement vectors and vertical changes (ellipsoidal) calculated for the time interval January 01, 2001 to April 26, 2001 The reference point is NICO. The error ellipse calculated at the confidence interval of 95% is of the order of a few millimetres.

unique and unambiguous solution). The analysis was performed using the FPFIT algorithm (REASENBERG and OPPENHEIMER, 1985) to plot first motion data and to evaluate nodal planes and the orientation of main strain axes. The good azimuth coverage of the investigated area well constrained the single focal solutions for most of the events. Coherently with the rupture mechanism of the main shock (star in Fig. 10), which is of dextral strike-slip, the majority of the other earthquakes present comparable rupture mechanisms. In Figure 10 the spatial distribution of the horizontal projection of the *P*-axes for these mechanisms is also reported. *P*-axes directions are mostly distributed almost horizontally, showing a prevalence of compression in the N90°–110°E direction.

Important information pertaining to the deformative behaviour of the volcano during the seismic crisis is provided by the GPS permanent network. The positive areal dilatation pattern, which was evidenced by EDM and GPS discrete networks since 1994 and was also confirmed during the early months by daily GPS measurements, appears interrupted during the week of the swarm (Fig. 9). The main aspect is represented by the negative vertical changes which indicate a lowering at the entire volcano scale (Fig. 11). This behaviour is compatible with magma movements from depth into the shallow (3–5 km) reservoir (PATANÈ *et al.*,

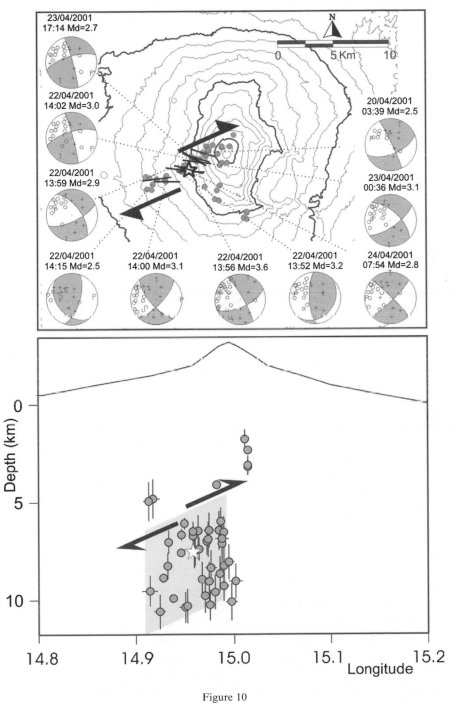

Figure 10

Epicentral map and hypocentral W-E cross section of the best constrained earthquakes of the swarm recorded between, April 20–24, 2001. Some examples of FPSs and the related azimuth distribution of the horizontal projection of *P* axes are also reported.

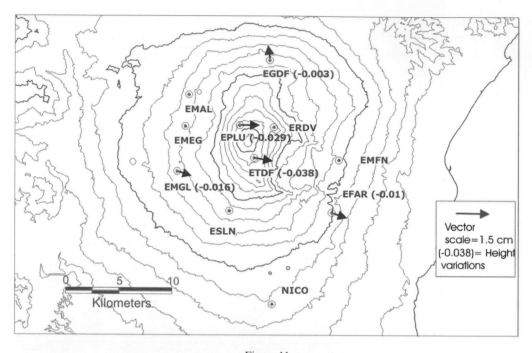

Figure 11
Displacement vectors and vertical changes calculated for the time interval 20–26 April, astride the seismic swarm. The reference point is NICO. The error ellipse calculated at the confidence interval of 95%. is of the order of a few millimetres.

2003b) which causes a temporary decrease of the pressure at depth and the edifice deflation.

6. *Concluding Remarks*

In this work we present an overview of seismic activity and ground deformation observed at Mt. Etna in recent years, focusing our attention on the last few months before the July-August 2001 eruption. After the 1991–1993 eruption, the areal dilatation highlights that the volcano is recharged with a near continuous constant rate, whereas during the same period the seismicity mainly exhibits two different behaviours: it is practically negligible (except during some swarms which occurred from late 1994 until early 1995) until June 1996 and then gradually increases from July 1996 until the beginning of 1999, when a ten-month summit eruption took place. During this time interval (1996–1999) the spread and deep seismicity and the overall and marked inflation of the entire edifice suggest that the pressure centre could be several kilometres deep (5–10) below the sea level. In this period the tension accumulation was partially released through an intense summit activity (lava fountains, overflows from central craters, and summit eruptions). Thereafter, the seismicity remained at low levels until November 2000, when a strong seismic swarm marked a new significant modification in the seismic strain release. In the ensuing months the average number of "background" events increased together with the mean magnitude, and several swarms also occurred. A marked change in the state of the volcano was represented by the April 20–24, 2001 seismic swarm, which occurred in the southwestern flank. Following this swarm, a further increase in the strain release gradient was observed until May 20, slightly before the beginning of a cycle of Strombolian eruptions at the southeast Crater. This strong seismic swarm represents the response of the medium to a deep magma intrusive episode that started in the preceding months and uprose at a shallower depth in the following months, and was a forerunner of the July 17 to August 10, 2001 lateral eruption (RESEARCH STAFF OF INGV-CT, 2001). This intrusion seems to be testified by the seismicity clustered along the NNW-SSE direction, which occurred in the period January-April, 2001 and by a clear pattern evidenced by geodetic data. Particularly, the polarization of the displacement vectors in the east direction can be considered as the result of magma intrusion along the same structures activated seismically.

Gravity measurements before the 2001 eruption show a progressive negative variation which reversed its trend between July 2001 and the beginning of August 2001 when the eruption occurred. This has been interpreted as the effect of a 2–3 * 10^{11} kg mass decrease within a source 2–3 km b.s.l. that could have supplied the 2001 eruption (RESEARCH STAFF OF INGV-CT, 2001). However, in our opinion, the interpretation of gravity changes alone may not be sufficient to describe an intrusion or magma recharge processes (FERNÁNDEZ et al., 2001a,b). In fact, gravity measurements on Mt. Etna are performed periodically along a profile of 19 stations crossing east-west the uppermost southern flank of the volcano and no continuous recording was still available.

From the integration of seismological and geodetic data, in this work we have evidence that before the July-August 2001 eruption an intrusive mechanism took place along the volcano-genetic NNW-SSE structural trend, where dike emplacement caused several recent eruptions (e.g., BONACCORSO, 2001). In particular, similar mechanisms were observed before the 1991–1993 eruption (e.g., BONACCORSO et al., 1996) and in January 1998 (BONACCORSO and PATANÈ, 2001). As regards the April 2001 swarm, an interesting contribution is provided by the new GPS permanent network which during the period of the swarm (20–24 April) showed a general lowering of the volcano edifice. This behaviour could be explained by the intrusion process which is characterised by a mass transfer from a deeper storage zone (5–10 km b.s.l.) to the upper part of the volcano.

The uprising of melt from the overpressured reservoir causes a depressuring of the same reservoir. The lowering of the edifice recorded during the magma ascent and the associated swarm would be the consequent deformation effect recorded on the surface. This interesting aspect would confirm the possible existence of magma storages inside Mt. Etna and provide the basis for future investigations.

In conclusion, the integration of seimicity, i.e., the fast strain release along preferential structures, together with the ground deformation, i.e., slow movement of the overall edifice, proves to be a powerful tool to further knowledge of the volcano and to understand its dynamics.

Acknowledgments

We acknowledge the helpful comments and suggestions of P. Davis and J Fernandez regarding the manuscript. This work was supported by grants from the Gruppo Nazionale per la Vulcanologia of INGV.

REFERENCES

AMORE, M., BONACCORSO, A., FERRARI, F., and MATTIA, M. (2002), *EOLO: Software for the Automatic On-line Treatment and Analysis of GPS Data for Environmental Monitoring*, Computer and Geosciences *28*, 2, 271–280

APPLIED GEOMECHANICS INCORPORATED (1993), *Notes on Temperature Coefficients and Long-term Drift of Applied Geomechanics Tiltmeters*, Report no. A-93-1001, AGI, Santa Cruz, California.

BARBERI, G., COCINA, O., NERI, G., PRIVITERA, E., and SPAMPINATO, S. (2000), *Volcanological Inferences From Seismic Strain Tensor Computation at Mt. Etna Volcano, Sicily*, Bull. Volcanol. *62*, 318–330.

BONACCORSO, A., FALZONE, G., RAIA, G., and VILLARI, L. (1998), *Application for New Technology for Ground Deformation Continuous Detection*, Acta Vulcanol. *10 (1)*, 7–12.

BONACCORSO, A., FALZONE, G., and GAMBINO S. (1999), *An Investigation into Shallow Borehole Tiltmeters*, Geophys. Res. Lett. *26*, 1637–1640.

BONACCORSO, A., FERRUCCI, F., PATANÈ, D., and VILLARI, L. (1996), *Fast Deformation Processes and Eruptive Activity at Mt. Etna (Italy)*, J. Geophys. Res. *101*, 17,467—17,480.

BONACCORSO, A. (1996), *A Dynamic Inversion for Modelling Volcanic Sources through Ground Deformation Data (Etna 1991–1992)*, Geophys. Res. Lett. *23*, 451–454.

BONACCORSO, A. and GAMBINO, S. (1997), *Impulsive Tilt Variations at Mount Etna Volcano (1990–1993)*, Tectonophysics *270*, 115–125

BONACCORSO, A. (2001*), Mt Etna Volcano: Modelling of Ground Deformation Patterns of Recent Eruptions and Considerations on the Associated Precursors*, J. Volcanol. Geotherm. Res. *109*, 99–108.

BONACCORSO, A. and PATANÈ, D. (2001), *Shear Response to an Intrusive Episode at Mt. Etna Volcano (January 1998) Inferred through Seismic and Tilt Data*, Tectonophysics, *334*, 61–75.

CAYOL, V. and CORNET F. H. (1998), *Effects of Topography on the Interpretation of the Deformation Field of Prominent Volcanoes – Application to Etna*, Geophys. Res. Lett. *23*, 1979–1982.

COCINA, O., NERI, G., PRIVITERA, E., and SPAMPINATO, S. (1998), *Seismogenic Stress Filed beneath Mt. Etna (South Italy) and Possible Relationships with Volcano-tectonic Features*, J. Volcanol. Geotherm. Res. *83*, 335–348.

FERNÁNDEZ, J., CHARCO, M., TIAMPO, K. F., JENTZSCH, G., and RUNDLE, J. B. (2001a), *Joint Interpretation of Displacement and Gravity in Volcanic Areas. A Test Example: Long Valley Caldera, California*, Geophys. Res. Lett. *28*, 1063–1066.

FERNÁNDEZ, J., TIAMPO, K. F., JENTZSCH, G, CHARCO, M., and RUNDLE, J. B. (2001b), *Inflation or Deflation?: New Results for Mayon Volcano Applying Elastic-gravitational Modeling*, Geophys. Res. Lett. *28*, 2349–2352.

FRAZZETTA, G. and VILLARI, L. (1981), *The Feeding of the Eruptive Activity of Etna Volcano: The Regional Stress Field as a Constraint to Magma Uprising and Eruption*, Bull. Volcanol. *44*, 269–282.

GRESTA, S., PERUZZA, L., SLEJKO, D., and DI STEFANO, G. (1998), *Inferences on the Main Volcano-tectonic Structures at Mt. Etna (Sicily) from a Probabilistic Seismological Approach*, J. Seismol. *2*, 105–116.

GVN (1998), *Global Volcanism Network, Bulletin of the Global Volcanism Program of the Smithsonian Institution*, Washington DC, vol. 23.

GVN (2000), *Global Volcanism Network, Bulletin of the Global Volcanism Program of the Smithsonian Institution*, Washington DC, vol. 3.

HIRN, A., NERCESSIAN, A., SAPIN, M., FERRUCCI, F., and WITTLINGER, G. (1991), *Seismic Heterogeneity of Mt. Etna: Structure and Activity*, Geophys. J. Int. *105*, 139–153.

LA DELFA, S., PATANÈ, G., CLOCCHIATTI, R., JORON, J. L., and TANGUY, J. C. (2001), *Activity Preceding the February 1999 Fissure Eruption: Inferred Mechanism from Seismological and Geochemical data*, J. Volcanol. Geoth. Res. *105*, 121–139.

LAHR, J. C. (1989), *HYPOELLIPSE/VERSION 2.0* : A Computer Program for Determining Local Earthquake Hypocentral Parameters, Magnitude, and First-motion Pattern*, U. S. Geol. Survey, Open-File Report 89/116, 81 pp.

LO GIUDICE, E. and RASÀ, R. (1986), *The Role of the NNW Structural Trend in the Recent Geodynamic Evolution of Northeastern Sicily and its Volcanic Implications in the Etnean Area*, J. Geodyn. *5*, 309–330.

PATANÈ, D., PRIVITERA, E., FERRUCCI, F., and GRESTA, S. (1994), *Seismic Activity Leading to the 1991–93 Eruption of Mt. Etna and its Tectonic Implications*, Acta Vulcanol. *4*, 47–56.

PATANÈ, D., FERRARI, F., and FERRUCCI, F. (1999), *First Application of ASDP Software: A Case Study at Mt. Etna Volcano and in the Acri Region (Southern Italy)*, Phys. Earth Planet. Inter. *113*, 75–88.

PATANÈ, D. and PRIVITERA, E. (2001), *Seismicity Related to 1989 and 1991–93 Mt. Etna (Italy) Eruptions: Kinematic Constraints by FPS Analysis*, J. Volcanol. Geotherm. Res. *109*, 77–98.

PATANÈ, D., PRIVITERA, E., GRESTA, S., ALPARONE, S., AKINCI, A., BARBERI, G., CHIARALUCE, L., COCINA, O., D'AMICO, S., DE GORI, P., DI GRAZIA, G., FALSAPERLA, S., FERRARI, F., GAMBINO, S., GIAMPICCOLO, E., LANGER, H., MAIOLINO, V., MORETTI, M., MOSTACCIO, A., MUSUMECI, C., PICCININI, D., REITANO, D., SCARFI, L., SPAMPINATO, S., URSINO, A., and ZUCCARELLO, L. (2003a), *Seismological Features and Kinematic Constrains for the July-August 2001 Lateral Eruption at Mt. Etna Volcano, Italy*, Annali di Geofisica, in press.

PATANÈ, D., DE GORI, P., CHIARABBA, C., and BONACCORSO, A. (2003b), *Magma Ascent and the Pressurization of Mount Etna's Volcanic System*, Science, *299*, 2061–2063.

PUGLISI, G., BONFORTE, A., and MAUGERI, S. R. (2001), *Ground Deformation Patterns on Mt. Etna, between 1992 and 1994, Inferred from GPS Data*, Bull. Volcanol. *62*, 371–384.

REASENBERG, P. and Oppenheimer, D. (1985), FPFIT, FPPLOT, and FPPAGE: FORTRAN Computer Programs for Calculating and Displaying Fault Plane Solutions, U.S. Geol. Surv. Open File Rep. *85/739*, 109 pp.

RESEARCH STAFF OF INGV-CT (2001), *Multidisciplinary Approach yields Insight into Mt. Etna Eruption*, EOS Trans. Am. Geophys. Uni. *82*, 653–656.

WILLIAMS, C. A. and WADGE, G. (1998), *The Effects of Topography on Magma Chamber Deformation Models: Application to Mt. Etna and Radar Interferometry*, Geophys. Res. Lett. *25*, 1549–1552.

(Received February 11, 2002, revised February 28, 2003, accepted March 14, 2003)

To access this journal online:
http://www.birkhauser.ch

Pure appl. geophys. 161 (2004) 1489–1507
0033–4553/04/071489–19
DOI 10.1007/s00024-004-2516-3

© Birkhäuser Verlag, Basel, 2004

❘ Pure and Applied Geophysics

Methods for Evaluation of Geodetic Data and Seismicity Developed with Numerical Simulations: Review and Applications

K. F. Tiampo[1], J. B. Rundle[2], J. S. Sa Martins[3],
W. Klein[4], and S. McGinnis[5]

Abstract—In this work we review the development of both established and innovative analytical techniques using numerical simulations of the southern California fault system and demonstrate the viability of these methods with examples using actual data. The ultimate goal of these methods is to better understand how the surface of the Earth is changing on both long-and short-term time scales, and to use the resulting information to learn about the internal processes in the underlying crust and to predict future changes in the deformation and stress field. Three examples of the analysis and visualization techniques are discussed in this paper and include the Karhunen-Loeve (KL) decomposition technique, local Ginsberg criteria (LGC) analysis, and phase dynamical probability change (PDPC). Examples of the potential results from these methods are provided through their application to data from the Southern California Integrated GPS Network (SCIGN), historic seismicity data, and simulated InSAR data, respectively. These analyses, coupled with advances in modeling and simulation, will provide the capability to track changes in deformation and stress through time, and to relate these to the development of space-time correlations and patterns.

Key words: Earthquake fault systems, numerical simulations, deformation patterns.

1. Introduction

In recent years, a combination of theoretical analysis and numerical simulations established the link between earthquake fault networks and the physics of critical point systems. These earthquake fault systems have been shown to be characterized by nonlinear dynamics, nonclassical nucleation, large correlation lengths, and the Gutenberg-Richter scaling law (Smalley *et al.*, 1985; Bak *et al.*, 1981;

---

[1] Cooperative Institute for Research in Environmental Sciences, University of Colorado, Boulder, CO 80309, USA and Dept. of Earth Sciences, University of Western Ontario, London, Ontario, Canada. E-mail: ktiampo@uwo.ca

[2] Center for Computational Science and Engineering, University of California, Davis, CA, U.S.A.

[3] Instituto de Fisica, Universidade Federal Fluminense, Niteroi, Brazil.

[4] Dept. of Physics, Boston University, Boston, MA, U.S.A.

[5] Cooperative Institute for Research in Environmental Sciences (CIRES), University of Colorado, Boulder, CO 80309, U.S.A.

RUNDLE, 1989; SORNETTE and SORNETTE, 1989; RUNDLE and KLEIN, 1995; RUNDLE *et al.*, 1995, 1997, 1999; MAIN, 1996; SAMMIS *et al.*, 1996; FISHER *et al.*, 1997; FERGUSON *et al.*, 1999; JAUME and SYKES, 1999). However, direct analysis of the actual earthquake fault system remains incomplete due to the inherent difficulty in sampling the continuum and its physical parameters (RICHTER, 1958; KANAMORI, 1981; TURCOTTE, 1991; GELLER *et al.*, 1997; BONNET *et al.*, 2001).

Much of the recent geophysical research associated with earthquakes themselves has centered on investigating a variety of spatial and temporal patterns that exist in local and regional deformation and seismicity data (MOGI, 1969; KEILIS-BOROK *et al.*, 1980; KANAMORI, 1981; YAMASHITA and KNOPOFF, 1989; KAGAN and JACKSON, 1992; ELLSWORTH and COLE, 1997; JONES and HAUKSSON, 1997; POLLITZ and SACKS, 1997; BOWMAN *et al.*, 1998; WYSS and WIEMER, 1999; BAWDEN *et al.*, 2001; DONG *et al.*, 2002; WATSON *et al.*, 2002). Although much of this work represents important efforts to describe the characteristic patterns in the data, none of these observations or methodologies systematically identifies all possible patterns, a necessary first step for the quantification of earthquake precursors.

Earthquake fault systems are one example of a complex nonlinear system consisting of an interacting spatial network of nonlinear cells that fire or fail as a result of a persistent driving force. Such driven systems are characterized by thresholds, residual stresses, quenched disorder, and noise, with complex spatial and temporal firing patterns that are difficult to analyze deterministically, but have been modeled using computer simulations with reasonable success (RUNDLE and KLEIN, 1995; MAIN, 1996; FISHER *et al.*, 1997; NIJHOUT *et al.*, 1997; ANGHEL and BEN-ZION, 2001; RUNDLE *et al.*, 2000). A method developed by RUNDLE *et al.* (2000), for analyzing nonlinear threshold systems based upon the principal component analysis techniques used to study another nonlinear system, the ocean-atmosphere interface (PREISENDORFER, 1988; PENLAND, 1989) was applied to numerical simulations of earthquake events. Shown in Figure 1a is the topological map view of the Virtual California fault model used to develop the methods discussed here. Virtual California is a dynamical, backslip, fault patch model of southern California in which the stress accumulation on each fault is a combination of the linear driving stress due to plate tectonics and the elastic interactions from the fault patch response, with over 300 two-dimensional patches to date (RUNDLE, 1988). All faults are vertical strike-slip faults, in a boundary element mesh of approximately 10 km horizontal and 20 km in vertical depth. Virtual California was originally used to retrieve space-time patterns in seismicity and deformation, and to compare them to the historical data for southern California (Fig. 1b).

The long-term goal of these simulations is to assist in determining the nature and value of those physical parameters that govern the fault system dynamics. Using these numerical simulations, we have discovered *systematic* space-time variations in seismicity in southern California (TIAMPO *et al.*, 2000, 2002a,b; RUNDLE *et al.*, 2000, 2002a,b,c,d). This discovery led us to investigate the feasibility of applying both established and innovative spatio-temporal analysis techniques to data collected in

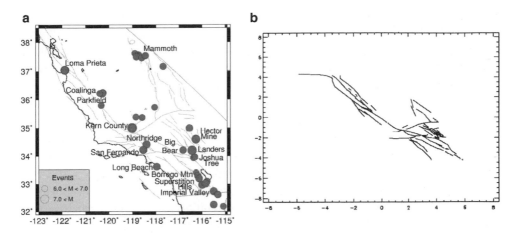

Figure 1
a) Historic events in southern California, magnitude ≥ 6.0; and b) topology (2001) of the Virtual California fault model, map view units in hundreds of kms (RUNDLE et al., 2000, 2001, 2002a).

southern California. Previously we applied two of these methods, the Karhunen-Loeve expansion (KLE) analysis and the phase dynamical probability change (PDPC) technique, to seismicity (TIAMPO et al., 2000, 2002a,b). Here we will detail the application of three analysis methods: the two mentioned above and a third that we have christened the Local Ginsberg Criteria (LGC), from their genesis in numerical simulations of fault systems to their promising application to either actual or simulated geodetic data.

2. Motivation

The Southern California Integrated GPS Network (SCIGN) is a regional network of GPS stations for the purpose of measuring the response of the southern California fault system to regional strains, to identify localized, unknown fault features and sources such as blind thrust faults, to estimate earthquake potential, and to quantify the physical parameters of the fault system itself. Important work related to this continuous array includes the calculation of station velocities (The SCIGN Project Report to NSF, 1998), the estimation of coseismic and postseismic displacements relating to both the 1992 Landers earthquake and the 1994 Northridge earthquake and their aftershocks (BOCK et al., 1997; WDOWINSKI et al., 1997), and the analysis of various noise sources (ZHANG et al., 1997; BAWDEN et al., 2001; DONG et al., 2002; WATSON et al., 2002). However, much remains to be accomplished towards the quantization of fault parameters and mechanics, and the identification of those local faults or those which are integral to the system but currently unknown, such as blind thrusts. In addition, it has been difficult to either quantitatively determine or reach agreement on the nature of the time-dependent tectonic signals and their response to local or regional strain above the noise in the data (ARGUS *et al.*, 1999; BAWDEN *et al.*, 2001; DONG *et al.*, 2002; WATSON *et al.*, 2002). The large, complex nature of the fault network precludes the simple analysis of its surficial expression, whether seismicity or deformation.

While observations of the geophysical system, including both seismicity and crustal motions, are vital to understanding the physics of the fault system and its evolution in time, a variety of data assimilation and analysis techniques and a model that accurately predicts the system dynamics are essential. For large quantities of data, such as those generated by SCIGN or InSAR, the first stage of this process is even more vital. Here we focus on that first stage using several related techniques and quantities developed with numerical simulations, that produce scaled measures of the seismicity and strain fields and show promise for specifying and quantifying the patterns of interest in the data (RUNDLE *et al.*, 2000; TIAMPO *et al.*, 2000, 2002a,b; RUNDLE and KLEIN, 2002b).

3. Methodology

3.1. Karhunen-Loeve (KL) Decomposition

Pattern evolution and prediction in nonlinear systems is complicated by nonlinear interactions and noise, however understanding such patterns, which are simply the surface expression of the underlying dynamics, is critical to understanding and perhaps characterizing the physics which control the system. In the case of a complicated fault system, while the physical parameters which control the system, such as coefficients of friction, stress drops, residual stresses, or Poisson's ratio, are often obscure or unknown, seismicity and surface deformation are both observable and measurable. RUNDLE *et al.* (2000), proposed a method of decomposing these complex spatial and temporal patterns of the physical system into their orthonormal pattern eigenstates. This KL expansion (KLE) is a technique in which a dynamical system is decomposed into a complete set of linearized, orthonormal subspaces. It has been applied to a number of other complex nonlinear systems over the last fifty years, including the ocean-atmosphere interface, turbulence, meteorology, biometrics, statistics, and solid earth geophysics (HOTELLING, 1933; PREISENDORFER, 1988; SAVAGE, 1988; PENLAND, 1989; VAUTARD and GHIL, 1989; FUKUNAGA, 1990; HOLMES *et al.*, 1996; MOGHADDAM *et al.*, 1998). The basis functions are the orthonormal decomposition of the pattern states that are the physical expression of the underlying dynamics of the system. As such, they define a unique, finite set of patterns for a given system, the eigenvalues and eigenstates of its correlation operator (FUKUNAGA, 1970). Note that the use of this decomposition implicitly assumes that one is dealing with a process that is both Markov and stationary. This implies, in the case of seismicity, that one has a complete data set that is large enough in both space and time to be effectively stationary, at least for long periods of time: a point that should be examined for each data set (MAIN, 1996; TURCOTTE, 1997).

RUNDLE et al. (2000), detail the successful application of this technique to computer simulations of the complex fault systems, identifying the spatial eigenvector patterns from a KL decomposition of fault patch simulation data, while ANGHEL (2001) and ANGHEL et al. (2003), apply a similar methodology to modeled deformation data in order to capture the coherent structures and their interactions. The method was extended to include an unequal-time correlation operator that can be used to forecast events in the simulations in the same manner as an EOF analysis is used to predict El Niño events in meteorology (PREISENDORFER, 1988; PENLAND, 1989; RUNDLE et al., 2000). A similar analysis was conducted on historic seismicity data for southern California. The data was obtained from the SCEC database, from 1932 through 1991. Relevant information consists of location in latitude and longitude and the time of occurrence. Each time step, days, is given a value of 1.0 if one or more events occur in that time period. This is done for each location time series, after which the mean for each time series is removed from the data. A KLE analysis was then performed on the entire data set. Shown in Figure 2 are results for the period 1932 through 1991. The first mode is a background seismicity mode, reflecting both the relative frequency of events and the ability to detect more events with time. The second mode delineates the occurrence of the Coalinga earthquake of 1983 as the predominant signal in the seismicity data prior to 1991, and may be thought of as a classic Mogi donut on a large spatial scale (MOGI, 1969). Note the appearance of correlations in the eighth KL mode (Fig. 2c) related to the upcoming Landers event (Fig. 1a) in 1992, despite the fact that there is no seismicity data for the Landers sequence in the analyzed data. This suggests that there exists correlated seismicity in the data prior to the upcoming event (TIAMPO et al., 2002a).

The KL procedure involves constructing a correlation operator, $C(x_i,x_j)$, for the sites that contain the spatial relationship of slip events over time. $C(x_i,x_j)$ is decomposed into the orthonormal spatial eigenmodes for the nonlinear threshold system, e_j, and their associated time series, $a_j(t)$. These spatial-temporal pattern states can be used to reconstruct the primary modes of the system, with or without noise, and quantify their relative magnitude and importance. These modes can be used to characterize the physical parameters such as stress levels and interactions that control the observable patterns. Here we demonstrate its application to one particular observable, SCIGN data.

Similar to the empirical orthogonal function (EOF) technique developed by PREISENDORFER (1988), for the atmospheric sciences, the KL expansion is obtained from the p time series that record the deformation history at particular locations in space. Each time series, $y(x_s,t_i) = y_i^s$, $s = 1, \dots p$, consists of n time steps, $i = 1, \dots n$. The goal is to construct a time series for each of a large number of locations, in this case days. These time series are incorporated into a matrix, T, consisting of time series of the same measurement for p different locations, i.e,

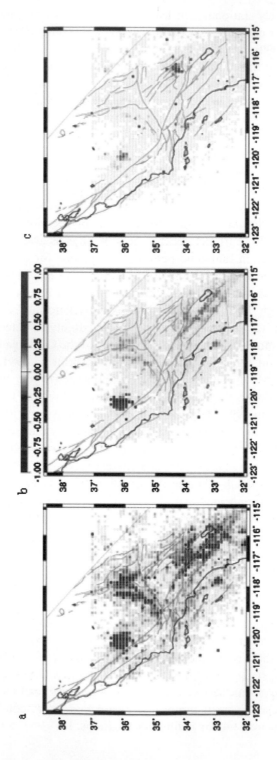

Figure 2

a) First KL mode, 1932–1991; b) Second KL eigenpattern and c) eighth KL eigenpattern. Shown are the correlations between event occurrence in each box. Here the scale represents normalized correlation, running from −1.0 to 1.0. (TIAMPO *et al.*, 2002a).

$$T = [\bar{y}_1, \bar{y}_2, ...\bar{y}_p] = \begin{bmatrix} y_1^1 & y_1^2 & \cdots & y_1^p \\ y_2^1 & y_2^2 & \cdots & y_2^p \\ \vdots & \vdots & \ddots & \vdots \\ y_n^1 & y_n^2 & \cdots & y_n^p \end{bmatrix}. \tag{1}$$

For analysis of SCIGN data, the values in the matrix T will consist of either strain or deformation measurements, horizontal or vertical. The covariance matrix, $S(x_i, x_j)$, for these events is formed by multiplying T by T^T, where S is a $p \times p$ real, symmetric matrix. $S(x_i, x_j)$, is converted to a correlation operator, $C(x_i, x_j)$, by dividing each element of $S(x_i, x_j)$, by the variance, S_p, of each time series, $y_p(x_i, t)$.

This equal-time correlation operator, $C(x_i, x_j)$, is decomposed into its eigenvalues and eigenvectors in two parts. The first employs the tri-reduction technique to reduce the matrix C to a symmetric tridiagonal matrix, using a Householder reduction. The second uses a ql algorithm to find the eigenvalues, λ_j^2, and eigenvectors, e_j, of the tri-diagonal matrix (PRESS et al., 1992). These eigenvectors, or eigenstates, are orthonormal basis vectors arranged in order of decreasing correlation that reflect the spatial relationship of events in time. If one divides the corresponding eigenvalues, λ_j^2, by the sum of the eigenvalues, the result is that percent of the correlation accounted for by that particular mode. We then reconstruct the time series associated with each location for each eigenstate, the temporal eigenvectors, $a_j(t)$, by projecting the initial data back onto these basis vectors in what is called a principal component analysis (PCA), $a_j(t_i) = \bar{e}^T \cdot T = \sum_{s=1}^{p} e_j y_i^s$, where $j, s = 1, ..., p$ and $i = 1, ..., n$ (PREISENDORFER, 1988). The spatial eigenvectors, e_j, are designated KLE modes and the associated time series, $a_j(t)$, as PC vectors, where each PC vector represents the signal associated with each particular eigenmode over time.

This preliminary success in applying the KL technique to seismicity data was the impetus for performing the same analysis on SCIGN data. For each of the existing SCIGN stations, the appropriate strain or deformation time series is identified, and a KLE analysis performed. There are several modes that one would expect to see and which are better modeled without the interaction of additional modes and noise, as this method will allow for their separation. A partial list would include plate motions, coseismic deformation, groundwater extraction, aseismic events, blind thrust faults, viscoelastic response, local variations in strain rate, and seasonal hydrologic cycles.

A simple example is given, using data obtained directly from the SCIGN website. Results are shown here for the second KL mode, using horizontal data after 1998, for only those stations located in the Los Angeles (LA) basin (Fig. 3). Note that the vertical scale on the PC time series is unitless—the surface deformation at each station is reconstituted by multiplying the time series by the eigenpattern value at that location. It should be mentioned here that the first horizontal mode, not shown, corresponds, as expected, with the plate motion velocities. This particular mode

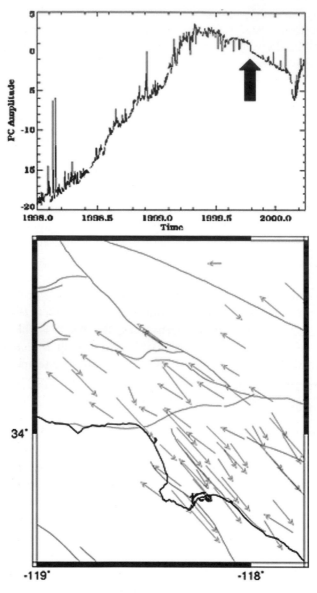

Figure 3
Second horizontal KL mode, for SCIGN 2.0 data in just the LA Basin, after 1998. At the bottom is shown the spatial eigenmode, where the arrows represent the direction of motion at each station in mms normalized to the maximm, 0.11 mm. At the top is shown the PCA signal for this eigenpattern.

appears to correspond with postseismic deformation and aftershock activity following the 1994 Northridge earthquake (see Fig. 1). Note also the jump in the PCA time series, which occurs at the time of the Hector Mine earthquake in the fall of 1999 (arrow, Fig. 3), evidence of a system-wide correlated response to that event.

The results shown in Figures 2 and 3 demonstrate the viability of applying this pattern dynamics methodology to seismicity and deformation data (TIAMPO et al., 2002a,b 2003). The data can be decomposed into its orthonormal basis functions, and the eigenstates derived from these analyses can be used to reconstruct the primary modes, either with or without associated noise. The primary modes of this pattern analysis can be used to invert for their individual sources, examples of which include but are not limited to, seasonal signals such as groundwater (BAWDEN et al., 2001; WATSON et al., 2002; TIAMPO et al., 2003), postseismic deformation (WDOWINSKI et al., 1997), or tectonic interactions such as blind faults, aseismic slip events, or time-dependent response. These modes, and their sources, will then be used to tune models such as Virtual California and its constitutive parameters, to produce strain fields and seismicity patterns on realistic spatial and temporal scales.

3.2. Phase Dynamics Probability Change (PDPC)

In phase dynamical systems, the space-time evolution of the system can be described by changes in the amplitude and phase angle of a state vector, or phase function (MORI and KURAMOTO, 1998). In such systems, space-time patterns are frequently produced as a byproduct of the dynamics (MORI and KURAMOTO, 1998; HOLMES et al., 1996; RUNDLE et al., 2000). Computer simulations strongly suggest that earthquake seismicity can be described by phase dynamics, in which the important changes in seismicity are associated only with rotations, or changes in phase angle, of the vector phase function describing the seismicity defined in a high-dimensional correlation space (RUNDLE et al., 2000; TIAMPO et al., 2002a). Realistic numerical simulations of earthquakes also suggest that space-time pattern structures are nonlocal in character; a consequence of strong correlations in the underlying dynamics (BUFE and VARNES, 1993; BOWMAN et al., 1998; BREHM and BRAILE, 1999; DODGE et al., 1996; RUNDLE et al., 1995, 1997, 2000a, 2002a; TIAMPO et al., 2002a,b). In this instance, nonlocal means that events widely separated in space and time may be correlated with each other. The PDPC procedure we discuss here is based upon the idea that seismic activity is characterized by a system state vector or phase function, $\hat{\mathbf{S}}$. Changes in the state of seismicity in space and time, corresponding to the shift of seismic activity from one location to another preceding large earthquakes, can be characterized by space-time changes in $\hat{\mathbf{S}}$.

To illustrate the method, we use results from a seismicity analysis. Define the activity rate $\psi_{obs}(x_i, t)$ as the number of earthquakes per unit time, of any size, within the box centered at x_i at time t. To construct, $\hat{\mathbf{S}}(x_i, t_0, t)$ we define the time-averaged seismicity function $\mathbf{S}(x_i, t_0, t)$ over the interval (t - t_0):

$$\mathbf{S}(x_i, t_0, t) = \frac{1}{(t - t_0)} \int_{t_0}^{t} \psi_{obs}(x_i, t) dt. \tag{2}$$

Since there are N locations for $S(x_i, t_0, t)$, and if we assume t_0 to be a fixed time, we can then consider $S(x_i, t_0, t)$ to be the i th component of a general, time-dependent vector evolving in an N-dimensional space. In previous work we showed that this N-dimensional correlation space is spanned by the eigenvectors of an $N \times N$ correlation matrix (RUNDLE *et al.*, 2000).

Denoting spatial averages over the N boxes by $<>$, the phase function $\hat{S}(x_i, t_0, t)$ is then defined to be the mean-zero, unit-norm function obtained from $S(x_i, t_0, t)$:

$$\hat{S}(x_i, t_0, t) = \frac{S(x_i, t_0, t) - < S(x_i, t_0, t) >}{\|S(x_i, t_0, t)\|}, \tag{3}$$

where $\|S(x_i, t_0, t)\|$ is the norm over all spatial boxes.

Under the assumption of phase dynamics, the important changes in seismicity will be given by $\Delta\hat{S}(x_i, t_1, t_2) = \hat{S}(x_i, t_0, t_2) - \hat{S}(x_i, t_0, t_1)$ a rotation of the N-dimensional unit vector $\hat{S}(x_i, t_0, t)$ in time. The change in probability at x_i, $\Delta P(x_i, t_1, t_2)$, from time t_1 to t_2, is then the PDPC index, $\Delta P(x_i, t_1, t_2) = \{\Delta\hat{S}(x_i, t_1, t_2)\}^2 - \mu_P$, where μ_p is the spatial mean of $\{\Delta\hat{S}(x_i, t_1, t_2)\}^2$.

Figure 4 shows a plot of $\Delta\hat{S}$ for the years 1978 to 1991. A positive (red) value represents anomalous seismic activity, and a negative (blue) value represents anomalous quiescence. The area near the upcoming Northridge event arises from anomalous quiescence, relative to the regional background rate, in the San Fernando aftershock zone, activation as one might expect. The 1983 Coalinga earthquake displays activation, as would be expected as it occurred during this time period, while the future Landers sequence is a mix of anomalous quiescence and activation. We conclude that the PDPC function does not simply identify areas associated with past events and their aftershocks, rather it quantifies the stress coherence and correlations associated with the regional seismicity.

Figure 5 illustrates that this method also represents a mapping between the measurable seismicity and the underlying stress field. On the left is shown the PDPC calculation encompassing 1996 through 1997. Here is the color scale for the PDPC is logarithmic, where the number denotes the exponent. The triangles represent the seismicity that occurs during the time period 1996–1997. The linear structure uncovered at the eastern tip of the White Wolf fault in the PDPC plot corresponds to a hidden strike-slip fault determined by BAWDEN *et al.* in 1999. While we do not propose to replace the important work necessary to quantify the mechanism and structure using relocation analysis, this does provide a potential means of identifying other similar, as yet unidentified structures for further study.

The PDPC method has been tested against catalogs that have been randomized in time, in order to determine if the measured correlations are simply a byproduct of random temporal correlations and their propagation into a fixed spatial field. The method also was tested against a second null hypothesis, the background seismicity rate. In all cases the PDPC method did significantly better when applied to the

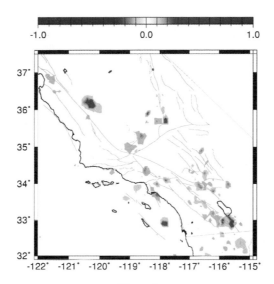

Figure 4

Map of ΔS, normalized to the maximum absolute value, for the time period 1991–1978. The color scale is linear, blue to white to red.

historic catalog than either the randomized catalogs or the background seismicity (TIAMPO *et al.*, 2002b).

Application of the PDPC methodology to surface deformation data, such as SCIGN data or InSAR images, may provide a link to earthquake deformation rates via the seismic moment; another measure that is easily incorporated into the PDPC calculation. The analysis of similar changes in the strain rate may provide insight into the variations in the underlying stress-strain field as well, hypothesized to be the physical measures of pre- and post-seismic signals.

3.3. Local Ginsberg Criteria (LGC)

We have emphasized in the prior discussions that the underlying variables of the fault system, in particular the stress field, are inherently unobservable, whereas the surface data are observable. The question we pose is whether there is some way to process surface deformation data through a mathematical mapping so that the resulting processed data would resemble the failure stress on each fault? If the earthquake cycle corresponds to an increase in the local Coulomb Failure Function (CFF) through time as a result of plate tectonic forcing, culminating in an earthquake, followed by a return to low CFF values, then we want to design a mapping that translates surface measurements to underlying CFF values. Note that in nature the fault friction strength, as well as the CFF, is inherently unobservable.

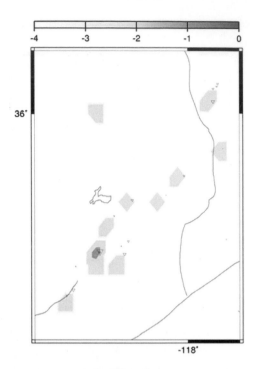

Figure 5

Anomalous seismic moment release for a hidden strike-slip structure, detected using PDPC. Shown is the PDPC calculation for the time period 1996 to 1997, at the eastern end of the White Wolf fault. Scale is logarithmic, where the triangles denote seismicity greater than magnitude 3.

Here we propose an additional decomposition technique, related to and derived from the previously described numerical simulations and pattern methods, and apply it to numerical simulation deformation data in order to determine the appropriate CFF proxy signal and wavelength for future measurement and analysis. Figure 6 is a plot of observable data from a computer simulation of southern California (RUNDLE *et al.*, 2002b,c,d). Here, simulated InSAR observations are shown. Figure 6 shows that, in theory, it is possible to observe pre- and post-differences in this type of surface deformation measurements by differencing the images (RUNDLE *et al.*, 2002b,c,d). The appropriate mapping for this surface change into the fault system stress is called the "Local Ginzberg Criterion" (LGC) (RUNDLE and KLEIN, 2002b) after a similar method of analysis in the theory of critical phenomena (MA, 1976). The method consists of defining a normalized,

▶

Figure 6

a) InSar C-band image corresponding to a simulated earthquake event; b) deformation for the five years prior to the simulated event shown in a; c) InSar image of the deformation for the five years after the same simulated event; d) the differenced image of b and c, the pre- and post- InSar images (RUNDLE *et al.*, 2002b,c).

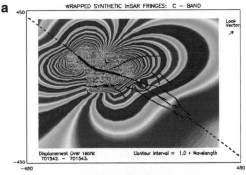

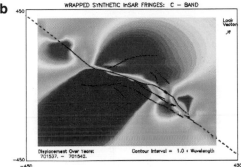

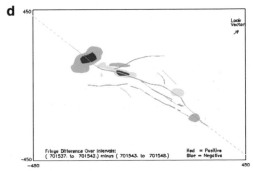

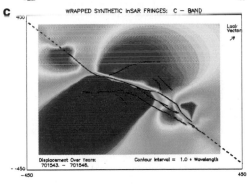

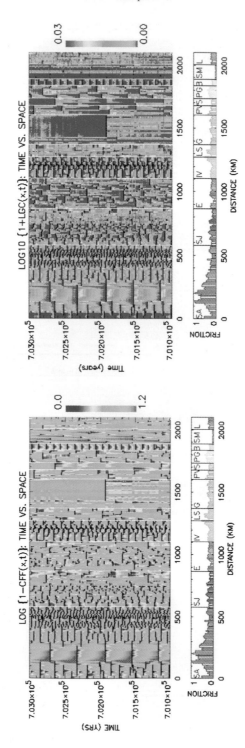

Figure 7

On the left is shown a plot of unobservable CFF, while on the right is the observable LGC, both from numerical simulation data generated by the Virtual California fault model (RUNDLE *et al.*, 2002b,c).

squared, time-dependent strain rate for each fault so that all locations can be treated as having the same strain rate statistics. In the theory of critical phenomena, sudden changes in the dynamical state of the system are preceded by large fluctuations in the order parameter. For earthquake fault systems, the order parameter is the slip deficit on a fault, fluctuations of which correspond to fluctuations in strain rate along the fault. The "Ginzberg Criteria" (GC) is defined to be the variance in the order parameter divided by the square of the mean. When this parameter becomes large, on the order of 10% or more, sudden changes in the order parameter of the system are imminent.

We therefore define a related quantity, a fluctuation metric, for each earthquake fault segment *as a function of time*. We define the LGC as the surface shear strain rate at time t and position x divided by the time average of surface shear strain rate at x over the interval $[0,t]$, squared,

$$LGC = \left(\frac{\dot{e}(t)}{\bar{\dot{e}}(t)} \right)^2 . \tag{4}$$

The LGC can vary over several orders of magnitude, so that values of the quantity $\psi = \log_{10}[1 + LGC]$ compare favorably with values of CFF in numerical simulations. Figure 7 shows the observable LGC, seen to be a close representation of the unobservable CFF. By applying this technique to actual SCIGN data, along with InSAR data, it may be possible to not only identify areas of high stress, but to define other such mappings as well (RUNDLE *et al.*, 2002b,c,d).

4. Conclusions

We have demonstrated here the importance of employing large-scale simulations not only to study the limits and capabilities of future data networks and to better delineate associated research priorities, but also to develop the data assimilation and analysis techniques necessary to understand the earthquake fault system and its underlying dynamics. In addition, we have detailed three different techniques developed using these numerical and demonstrated their applicability to the southern California fault system that shows promise for the analysis and understanding of earthquakes and their precursors. While the tests shown here have been primarily retrospective, and, as such, are subject to the possibility of unforeseen and unintentional biases, whether they lie in the models or data (MULARGIA, 2001), future validation studies have begun on at least one of the techniques, the PDPC method (RUNDLE *et al.*, 2002a; TIAMPO *et al.*, 2002c). Additional work includes their application to a variety of available data sets, analysis of their sensitivity to the data error and completeness, and the study of the validity and limitation of each method.

Acknowledgments

Research by JBR was funded by USDOE/OBES grant DE-FG03-95ER14499 (theory), and by NASA grant NAG5-5168 (simulations). Research by KFT was conducted under NASA grant NGT5-30025. This research was also supported by the Southern California Earthquake Center. SCEC is funded by NSF Cooperative Agreement EAR-0106924 and USGS Cooperative Agreement 02HQAG0008. The SCEC contribution number for this paper is 704. Research by WK was supported by USDOE/OBES grant DE-FG02-95ER14498 and W-7405-ENG-6 at LANL. WK would also like to acknowledge the hospitality and support of CNLS at LANL. JdsM was supported as a Visiting Fellow by CIRES/NOAA, University of Colorado at Boulder, and SM was supported by a NASA student fellowship. We are grateful to Drs. Ian Main and Franceso Mulargia, and an anonymous reviewer, for their critical review of an earlier version of this work. KFT would also like to thank the Instituto de Astronomía y Geodesia, Universidad Complutense de Madrid for its hospitality during the completion of this work.

REFERENCES

ANGHEL, M., and BEN-ZION, Y. (2001), *Nonlinear System Identification and Forecasting of Earthquake Fault Dynamics Using Artificial Neural Networks*, EOS Trans., AGU *82*, F571.

ANGHEL, M., BEN-ZION, Y., and MARTINEZ, R. R. (2003), *Dynamical System Analysis and Forecasting of Deformation Produced by an Earthquake Fault*, Pure Appl. Geophys., in press.

ARGUS, D. F., HEFLIN, M. B., DONNELLAN, A., WEBB, F. H., DONG, D., HURST, K. J., JEFFERSON, D. C., LYZENGA, G. A., WATKINS, M. M., and ZUMBERGE, J. F. (1999), *Shortening and Thickening of Metropolitan Los Angeles Measured and Inferred by Using Geodesy*, Geology *27*, 703–706.

BAK, P., TANG, C., and WEISENFIELD, K. (1987), *Self-organized Criticality: An Explanation of the 1/f Noise*, Phys. Rev. Lett. *59*, 381–384.

BAWDEN, G. W., MICHAEL, A. J., and KELLOGG, L. H. (1999), *Birth of a Fault: Connecting the Kern County and Walker Pass, California*, Geology *27*, 601–604.

BAWDEN, G. W., THATCHER, W., STEIN, R. S., HUDNUT, K. W., and PELTZER, G. (2001), *Tectonic Contraction Across Los Angeles after Removal of Groundwater Pumping Affects*, Nature *412*, 812–815.

BOCK, Y., WDOWINSKI, S., FANG, P., ZHANG, J., WILLIAMS, S., JOHNSON, H., BEHR, J., GENRICH, J., DEAN, J., VAN DOMSELAAR, M., AGNEW, D., WYATT, F., STARK, K., ORAL, B., HUDNUT, K., KING, R., HERRING, T., DINARDO, S., YOUNG, W., JACKSON, D., and GURTNER, W. (1997), *Southern California Permanent GPS Geodetic Array: Continuous Measurements of Crustal Deformation between the 1992 Landers and 1994 Northridge Earthquakes*, J. Geophys. Res. *102*, 18,013–18,033.

BONNET, E., BOUR, O., ODLING, N. E., DAVY, P., MAIN, I., COWIE, P., and BERKOWITZ, B. (2001), *Scaling of Fracture Systems in Geological Media*, Rev. Geophys. *39*, 347–384.

BOWMAN, D. D., OUILLON, G., SAMMIS, C. G., SORNETTE, A., and SORNETTE, D. (1998), *An Observational Test of the Critical Earthquake Concept*, J. Geophys. Res. *103*, 24,359–24,372.

BREHM, D. J. and BRAILE, L. W. (1999), *Intermediate-term earthquake prediction using the modified time-to-failure method in Southern California*, Bull. Seismol. Soc. Am., *89*, 275–293.

BUFE, C. G., and VARNES, D. J. (1993), *Predictive modeling of the sesimic cycle of the greater San Francisco bay region*, J. Geophys. Res., *98, 9871–9883*.

DODGE, D. A., BEROZA, G. C., and ELLSWORTH, W. L. (1996), *Detailed observations of California Foreshok Sequences: Implications for the earthquake initiation process*, J. Geophys. Res., *101*, 22,371–22,393.

DONG, D., FANG, P., BOCK, Y., CHENG, M. K., and MIYAZAKI, S. (2002), *Anatomy of Apparent Seasonal Variations from GPS-derived Site Position Time Series*, J. Geophys. Res. *107*, doi:10.1029/2001JB000573.

ELLSWORTH, W. I., COLE, A. T., and DIETZ, L. (1998), *Repeating Earthquakes and the Long-Term Evolution of Seismicity on the San Andreas Fault near Bear Valley, California*, Seis. Res. Lett. *69*.

FERGUSON, C. D., KLEIN, W., and RUNDLE, J. B. (1999), *Spinodals, Scaling, and Ergodicity in a Threshold Model with Long-range Stress Transfer*, Phys. Rev. E *60*, 1359–1373.

FISHER, D. S., DAHMEN, K., RAMANATHAN, S., and BEN-ZION, Y. (1997), *Statistics of Earthquakes in Simple Models of Heterogeneous Faults*, Phys. Rev. Lett. *78*, 4885–4888.

FUKUNAGA, K., *Introduction to Statistical Pattern Recognition* (Academic Press 1970).

GELLER, R. J., JACKSON, D. D., KAGAN, Y. Y., and MULARGIA, F. (1997), *Enhanced: Earthquakes Cannot be Predicted*, Science *275*, 1616–1617.

HOLMES, P., LUMLEY, J. L., and BERKOOZ, G., *Turbulence, Coherent Structures, Dynamical Systems, and Symmetry* (Cambridge University Press, Cambridge, UK 1996).

HOTELLING, H. (1933), *Analysis of a complex of statistical variables into principal components*, J. Educ. Psych., *24*, 417–520.

JAUME, S. C., and SYKES, L. R. (1999), *Evolving Towards a Critical Point: A Review of Accelerating Seismic Moment/Energy Release prior to Large and Great Earthquakes*, Pure Appl. Geophys. *155*, 279–306.

JONES, L. M., and HAUKSSON, E. (1997), *The Seismic Cycle in Southern California: Precursor or response?*, Geophys. Res. Lett. *24*, 4, 469–472.

KAGAN, Y. Y., and JACKSON, D. D. (1992), *Seismic Gap Hypothesis, Ten Years After*, J. Geophys. Res. *96*, 21,419–21,431.

KANAMORI, H., *The nature of seismicity patterns before large earthquakes. In Earthquake Prediction. An International Review*, AGU Monograph (AGU, Washington, D. C. 1981). pp. 1–19.

KEILIS-BOROK, V. I., KNOPOFF, L., and ALLEN, C. R. (1980), *Long-term Premonitory Seismicity Patterns in Tibet and the Himalayas*, J. Geophys. Res. *85*, 813–830.

MA, S. K. (1976), *Renormalization Group by Monte-Carlo Methods*. Phys. Rev. Lett. *37*, 461–464.

MAIN, I. (1996), *Statistical Physics, Seismogenesis, and Seismic Hazard*, Rev. Geophys. *34*, 433–462.

MOGI, K. (1969), *Some Features of Recent Seismic Activity in and near Japan, (2) Activity before and after Great Earthquakes*, Bull. Earthquake Res. Inst., Tokyo Univ. *47*, 395–417.

MORI, H., and KURAMOTO, Y., *Dissipative Structures and Chaos* (Springer-Verlag, Berlin 1998).

MULARGIA, F. (2001), *Retrospective Selection Bias (or the Benefit of Hindsight)*, Geophys. J. Int. *146*, 489–496.

NIJHOUT, H. F., *Pattern formation and biological systems*. In *Pattern Formation in the Physical and Biological Sciences*, Lecture Notes V, SFI (Addison Wesley, 1997) pp. 269–298.

PENLAND, C. (1989), *Random Forcing and Forecasting Using Principal Oscillation Pattern Analysis*, Mon. Weath. Rev. *117*, 2165–2185.

POLLITZ, F. F., and SACKS, I. S. (1997), *The 1995 Kobe, Japan, Earthquake: A Long-delayed Aftershock of the Offshore 1944 Tonankai and 1946 Nankaido Earthquakes*, BSSA *87*, 1–10.

PREISENDORFER, R. W., *Principle Component Analysis in Meteorology and Oceanography* (ed. C. D. Mobley) Develop. Atm. Sci. 17 (Elsevier, 1988).

PRESS, W., TEUKOLOSKY, S., VETTERING, W., and FLANNERY, B., *Numerical Recipes in C: The Art of Scientific Computing* (Cambridge University 1992).

RICHTER, C. F., *Elementary Seismology* (Freeman, San Francisco 1958).

RUNDLE, J. B. (1988), *A Physical Model for Earthquakes: 2. Applications to Southern California*, J. Geophys. Res. *93*, 6255–6274.

RUNDLE, J. B. (1989), *Derivation of the Complete Gutenberg-Richter Magnitude-frequency Relation Using the Principle of Scale Invariance*, J. Geophys. Res. *94*, 12,337–12,342.

RUNDLE, J. B., and KLEIN, W. (1995), *New Ideas about the Physics of Earthquakes*, Rev. Geophys. Space Phys. Suppl. (July) *283*, 283–286.

RUNDLE, J. B., GROSS, S., KLEIN, W., FERGUSON, C., and TURCOTTE, D. L. (1997), *The Statistical Mechanics of Earthquakes*, Tectonophysics *277*, 147–164.

RUNDLE, J. B., KLEIN, W., and GROSS, S. (1999), *Physical Basis for Statistical Patterns in Complex Earthquake Populations: Models, Predictions and Tests*, Pure Appl. Geophys. *155*, 575–607.

RUNDLE, J. B., KLEIN, W., TIAMPO, K. F., and GROSS, S. (2000), *Linear Pattern Dynamics of Nonlinear Threshold Systems*, Phys. Rev. E *61*, 2418–2432.

RUNDLE, J. B., KLEIN, W., TIAMPO, K. F., and SÁ MARTINS, J. S. (2002a), *Self-organization in Leaky Threshold Systems: The Influence of Near-mean Field Dynamics and its Implications for Earthquakes, Neurobiology, and Forecasting*, Proc. Nat. Acad. Sci. U.S.A., Suppl. 1, *99*, 2514–2521.

RUNDLE, J. B., and KLEIN, W. (2002b), *Towards a Forecast Capability for Earthquake Fault Systems: Integrating NASA Space Geodetic Observations with Numerical Simulations of a Changing Earth*, Third ACES Conference Proceedings.

RUNDLE, J. B., RUNDLE, P. B., KLEIN, W., SÁ MARTINS, J. S., TIAMPO, K. F., DONNELLAN, A., and KELLOGG, L. H. (2002c), *GEM Plate Boundary Simulations for the Plate Boundary Observatory: A Program for Understanding the Physics of Earthquakes on Complex Fault Networks via Observations, Theory and Numerical Simulation*, Pure Appl. Geophys. *159*, 2357–2381.

RUNDLE, J. B., KELLOGG, L. H., and DONNELLAN, A. (2002d), *Observables and Mapping of Fault Networks Developed Using Large-scale Earthquake Fault Simulations*, GESS Conference Proceedings.

SAMMIS, C. G., SORNETTE, D., and SALEUR, H., *Complexity and earthquake forecasting. In Reduction and Predictability of Natural Disasters*, SFI Series in the Science of Complexity, XXV (Addison-Wesley, Reading, MA 1996).

SAVAGE, J. C. (1988), *Principal Component Analysis of Geodetically Measured Deformation in Long Valley Caldera, Eastern California, 1983–1987*, J. Geophys. Res. *93*, 13,297–13,305.

SMALLEY, R. F., TURCOTTE, D. L., and SOLLA, S. A. (1985), *A Renormalization Group Approach to the Stick-slip Behavior of Faults*, J. Geophys. Res. *90*, 1894–1900.

SORNETTE, A., and SORNETTE, D. (1989), *Self-organized Criticality and Earthquakes*, EPL *9*, 197–202.

TIAMPO, K. F., RUNDLE, J. B., KLEIN, W., McGINNIS, S., and GROSS, S. J., *Observation of systematic variations in non-local seismicity patterns from southern California. In Geocomplexity and the Physics of Earthquakes*, AGU Monograph (AGU, Washington, 2000).

TIAMPO, K. F., RUNDLE, J. B., GROSS, S. J., and KLEIN, W. (2002a), Eigenpatterns in southern Califorina Seismicity, J. Geophys. Res. *107*, doi:10.1029/2001JB000562.

TIAMPO, K. F., RUNDLE, J. B., McGINNIS, S., GROSS, S., and KLEIN, W. (2002b), *Mean-field Threshold Systems and Phase Dynamics: An Application to Earthquake Fault Systems*, EPL *60*, 481–487.

TIAMPO, K. F., RUNDLE, J. B., KLEIN, W., BEN-ZION, Y., and McGINNIS, S. (2003), *Using Eigenpattern Analysis to Constrain Seasonal Signals in Southern California*, Pure Appl. Geophys., in press.

TURCOTTE, D. L. (1991), *Earthquake Prediction*, Ann. Rev. Earth Planet. Sci. *19*, 263–281.

TURCOTTE, D. L., *Fractals and Chaos in Geology and Geophysics*, 2nd ed. (Cambridge University Press 1997).

VAUTARD, R., and GHIL, M. (1989), *Singular Spectrum Analysis in Nonlinear Dynamics, with Applications to Paleodynamic Time Series*, Physica D *35*, 395–424.

WATSON, K. M., BOCK, Y., and SANDWELL, D. T. (2002), *Satellite Interferometric Observations of Displacements Associated with Seasonal Groundwater*, J. Geophys. Res. *107*, doi:10.1029/2001JB000470.

WDOWINSKI, S., BOCK, Y., ZHANG, J., and FANG, P. (1997), *Southern California Permanent GPS Geodetic Array: Spatial Filtering of Daily Positions for Estimating Coseismic and Postseismic Displacements Induced by the 1992 Landers Earthquake*, J. Geophys. Res. *102*, 18,057–18,070.

WYSS, M., and WIEMER, S. (1999), *How Can One Test the Seismic Gap Hypothesis? The Case of Repeated Ruptures in the Aleutians*, Pure Appl. Geophys. *155*, 259–278.

YAMASHITA, T., and KNOPOFF, L. (1989), *A Model of Foreshock Occurrence*, Geophys. J. *96*, 389–399.

ZHANG, J., BOCK, Y., JOHNSON, H., FANG, P., GENRICH, J., WILLIAMS, S., WDOWINSKI S., and BEHR, J. (1997), *Southern California Permanent GPS Geodetic Array: Error Analysis of Daily Position Estimates and Site Velocities*, J. Geophys. Res. *102*, 18,035–18,055.

(Received February 22, 2002, revised February 7, 2003, accepted May 13, 2003)

 To access this journal online:
http://www.birkhauser.ch

Pure appl. geophys. 161 (2004) 1509–1517
0033–4553/04/071509–9
DOI 10.1007/s00024-004-2517-2

© Birkhäuser Verlag, Basel, 2004

▌Pure and Applied Geophysics

A Free Boundary Problem Related to the Location of Volcanic Gas Sources

JESÚS ILDEFONSO DÍAZ[1] and GIORGIO TALENTI[2]

Abstract — A mathematical model of general gas emitting systems is derived, and a sample of relevant mathematical results is offered. The present paper indicates that shallow subsurface gas sources in typical volcanic areas can be located if appropriate physico-chemical measurements are made on the Earth's surface and put to use.

Key words: Subsurface source, surface measurement, partial differential equations of parabolic type, free boundary, obstacle problem, *a priori* bound.

1. Introduction

Locating gas sources ranks among high priority goals of volcanic surveillance. In this work we explore the potential of making the grade via physico-chemical surface methods. Roughly speaking, we address ourselves to the following question. Assume that gas moves out of some extended underground source and travels towards the soil with vertical velocity through some homogeneous porous-permeable medium — much as it happens in typical volcanic areas. Assume that both physical and chemical observations are made on the soil: namely, bulk gas flow is sampled at the Earth's surface over time and source gas concentration is measured at and beneath the Earth's surface. Can the gas source be located?

This paper demonstrates that — under suitable hypotheses, which are detailed below—the answer is *yes*, in principle.

Locating subsurface gas sources by surface methods results in a typical *inverse problem*. Inverse problems are currently intensively worked on both in mathematics and in geophysics, and occur in such areas of geophysics as, e.g., locating masses and heat sources based on gravity and temperature fields (see, e.g., GLASKO, 1984; GOSH ROY, 1991; MENKE, 1984; TARANTOLA, 1987). However, the present case has received no attention thus far, to the best of the authors' knowledge.

---

[1] Departamento de Matemática Aplicada, Universidad Complutense de Madrid, Madrid, Spain. E-mail: ji_diaz@mat.ucm.es

[2] Dipartimento di Matematica U. Dini, Università di Firenze, Italy. E-mail: talenti@math.unifi.it

Our work was spurred by the alert condition declared during springtime 1988 over the Island of Vulcano and the ensuing Summer 1988 Crash Programme by the Italian Gruppo Nazionale di Vulcanologia (see, e.g., SUNDRY AUTHORS, 1991).

2. Derivation of a Mathematical Model

We model the affairs as follows. The natural system of concern is idealized as a layer of some homogeneous porous-permeable medium sandwiched from above and from below by two bodies of different gases: the Earth's atmosphere, consisting of air, and an extended gas reservoir beneath, containing mainly carbon dioxide. The upper boundary of the layer is a reference plane that faces air and represents the Earth's surface; the lower boundary is a geometric two-dimensional surface that may uprise and subside over time, and represents the roof of the gas reservoir—i.e., the subsurface gas source. The former is bound to host data, the latter must be determined. The layer itself includes no sources or sinks of gas, and is filled up by a mixture of air and subsurface gas with varying composition—air percolates downward; the subsurface gas flows upward; both slowly move through the medium.

We assume the system is isothermal and the equation of perfect gases is in force. We choose to give prominence to diffusion and advection, and to ignore any extra process that might occur in fixing the configuration of our system.

We call *time t*, call space coordinates x, y and z, and let x stand for *depth* throughout. As usual, $\nabla = $ *gradient* with respect to x, y and z; $\cdot = $ scalar product of vectors; div $= \nabla \cdot$, the divergence operator; $\Delta = $ div ∇, Laplace operator.

Let P and u denote the *total gas pressure* and the *concentration of the subsurface gas*, respectively. For convenience, we assume P is the ratio between the actual total pressure and the atmospheric pressure—dimensionless.

Let ρ, ρ_1 and ρ_2 denote the total gas density, the density of air and the density of the subsurface gas, respectively. We have

$$\rho_1 = \rho(1-u) \quad \text{and} \quad \rho_2 = \rho u \tag{1}$$

since (concentration of air) + (*concentration* of the subsurface gas) = 1.

A form of *Fick's law* ensures that

$$\text{diffusive flow rate of constituent no.} i = -(D/P)\nabla \rho_i \tag{2}$$

for some positive constant D (which depends on the medium). On the other hand,

$$\text{advective flow rate of constituent no.} i = \rho_i \mathbf{W} \tag{3}$$

provided $\mathbf{W} = $ *bulk gas velocity*. Therefore the flow rate of constituent no i, caused by diffusion and advection, amounts to $-(D/P)\nabla \rho_i + \rho_i \mathbf{W}$. The conservation of mass implies

$$\frac{\partial \rho_i}{\partial t} = \operatorname{div}\left(\frac{D}{P}\nabla\rho_i - \rho_i\mathbf{W}\right). \tag{4}$$

We deduce successively

$$\frac{\partial \rho}{\partial t} = \operatorname{div}\left(\frac{D}{P}\nabla\rho - \rho\mathbf{W}\right) \tag{5}$$

and

$$\rho\frac{\partial u}{\partial t} = D\operatorname{div}\left(\frac{\rho}{P}\nabla u\right) + \left(\frac{D}{P}\nabla\rho - \rho\mathbf{W}\right)\cdot\nabla u. \tag{6}$$

As the temperature is constant, ρ is a constant multiple of P. Hence

$$\frac{\partial P}{\partial t} = \operatorname{div}\left(\frac{D}{P}\nabla P - P\mathbf{W}\right) \tag{7}$$

and

$$P\frac{\partial u}{\partial t} = D\Delta u + \left(\frac{D}{P}\nabla P - P\mathbf{W}\right)\cdot\nabla u. \tag{8}$$

Darcy's law tells us that

$$-\mathbf{W} = \Pi\nabla P \tag{9}$$

for some positive constant Π (which depends on the medium and includes permeability, porosity, tortuosity and viscosity).

The following equations

$$\frac{\partial P}{\partial t} = \Delta\left(D\ln P - \frac{\Pi}{2}P^2\right), \tag{10}$$

$$P\frac{\partial u}{\partial t} = D\Delta u + \nabla\left(D\ln P + \frac{\Pi}{2}P^2\right)\cdot\nabla u \tag{11}$$

are established. They encode the essentials of *gas transport* within the considered system — the former describes changes in bulk gas distribution; the latter accounts for chemical composition.

Boundary conditions can be appended. First, our model applies in the absence of gas surges, i.e., when total gas pressure near the ground is close to atmospheric pressure and bulk gas flow is weak enough to enable atmospheric circulation disposal. Therefore

$$P = 1 \quad \text{at} \quad x = 0, \tag{12}$$

and

$$u = 0 \quad \text{at} \quad x = 0. \tag{13}$$

Secondly, we ask that gas composition be analyzed at diverse depths in the subsoil. Therefore we let

$$[0, T] = \text{life span of the relevant observation} \tag{14}$$

and

$$\frac{\partial u}{\partial x} = \text{a given datum for } x = 0 \text{ and } 0 \leq t \leq T. \tag{15}$$

Thirdly, we think of the subsurface gas reservoir as the place where no air is present and pure subsurface gas occurs. In other words, the subsurface gas source plays the role of a *free boundary* in the present framework, and

$$u = 1 \text{ at the subsurface gas source.} \tag{16}$$

In case suitable extra conditions are met, the classical WKBJ method (see LUNENBURG, 1964; or VAINBERG, 1989, for instance) enables us to simplify the model in hand considerably. Surmise the medium is *fairly permeable* and the subsurface gas source is *shallow*. Then Π is considerably larger than D, total gas pressure can be regarded as nearly constant, and the following arguments apply.

Let

$$\lambda = D/\Pi. \tag{17}$$

Let D remain constant,

$$\lambda \to 0 \tag{18}$$

and the following asymptotic expansions

$$\begin{aligned} P &\simeq 1 + \lambda P_1 + \lambda^2 P_2 + \dots \\ u &\simeq u_0 + \lambda u_1 + \lambda^2 u_2 + \dots \end{aligned} \tag{19}$$

hold—the relevant coefficients are bound not to depend on λ. These expansions, (10) and (11) return a system of equations that determines $P_1, P_2, \dots$ and $u_0, u_1, \dots$ recursively; the beginning of such a system reads

$$\text{div}(D\nabla P_1) = 0 \tag{20}$$

and

$$\frac{\partial u_0}{\partial t} = D\Delta u_0 + (D\nabla P_1) \cdot \nabla u_0. \tag{21}$$

The same Ansatz and Darcy's law tells us that bulk gas flow $\mathbf{W}$ obeys

$$-\mathbf{W} \simeq (D\nabla P_1) + \lambda(D\nabla P_2) + \dots; \tag{22}$$

in particular, $-\mathbf{W}$ approaches $(D\nabla P_1)$ asymptotically.

$$\text{div } \mathbf{W} = 0; \tag{23}$$

concentration of subsurface gas obeys

$$\frac{\partial u}{\partial t} = D\Delta u - \mathbf{W} \cdot \nabla u. \tag{24}$$

Equations (23) and (24) can be conveniently treated under the additional hypothesis that *one-dimensional geometry prevails*, i.e., bulk gas velocity is purely vertical and both bulk gas velocity and the concentration of subsurface gas are invariant under horizontal space translations. In fact, such a hypothesis and (23) imply that the horizontal components of **W** vanish and

$$-(\text{vertical component of } \mathbf{W}) = F, \tag{25}$$

a function of time only. As a consequence, bulk gas velocity becomes an *observable* in the present setting—it takes at any depth the same value that it takes *at the Earth's surface.* Moreover, (24) can be recast thus

$$\frac{\partial u}{\partial t} = D\frac{\partial^2 u}{\partial x^2} - F(t)\frac{\partial u}{\partial x}. \tag{26}$$

In conclusion, *equation (26) governs gas composition within the considered system* under suitable hypotheses. Under the same hypotheses F represents bulk gas flow at the Earth's surface, hence it can be viewed as a datum. Assembling equation (26) and boundary conditions (13), (15) and (16) results in a mathematical problem that will be examined in the next section.

3. Analysis of the Mathematical Model

Motivated by foregoing arguments, in the present section we are concerned with the mathematical problem of determining a function

$$L : [0, T] \to]0, +\infty[\tag{27}$$

(depending upon time), with the physical meaning

$$L = \text{depth of the subsurface gas source}, \tag{28}$$

and a sufficiently smooth map

$$u : \{(t, x) \mid 0 \le t \le T, 0 \le x \le L(t)\} \to [0, 1] \tag{29}$$

(depending upon both time and depth) that obey the following conditions

$$(P)\begin{cases} u_t = Du_{xx} + F(t)u_x & \text{for } 0 < t < \text{T and } 0 < x < \text{L(t)}, \\ 0 < u(t, x) < 1 & \text{for } 0 < t < \text{T and } 0 < x < \text{L(t)}, \\ u(t, 0) = 0 & \text{for } 0 < t < \text{T}, \\ u_x(t, 0) = g(t) & \text{for } 0 < t < \text{T}, \\ u(t, L(t)) = 1 & \text{for } 0 < t < \text{T}. \end{cases} \tag{30}$$

where D, T, F, g are given (D and T are positive constants, F and g are sufficiently smooth functions of time).

Recall that D is a diffusion coefficient, T is the time span of observations, F stands for bulk gas flow, and u stands for concentration of subsurface gas.

Problem (P) can be advantageously approached by the following recipe. Let G be the *maximal monotone graph* defined by

$$G(r) = \begin{cases}]-\infty, 0] & \text{if } r=0, \\ \{0\} & \text{if } 0 < r < 1, \\ [0, +\infty[& \text{if } r=1, \\ \text{the empty set} & \text{if either } r<0 \text{ or } r>1. \end{cases} \tag{31}$$

Consider the problem of determining a sufficiently smooth function

$$u : \{(t,x)|0 \le t \le T, 0 \le x < \infty\} \to]-\infty, \infty[\tag{32}$$

such that

$$(OP) \begin{cases} u_t = Du_{xx} + F(t)u_x + G(t) \ni 0 & \text{for } 0 < t < T \text{ and } 0 < x < \infty, \\ u(t,0) = 0 & \text{for } 0 < t < T, \\ u_x(t,0) = g(t) & \text{for } 0 < t < T. \end{cases} \tag{33}$$

Let u be a solution to (OP) that develops *a coincidence set* , i.e., obeys

$$\{(t,x)|0 \le t \le T, 0 \le x < \infty, u(t,x) = 1\} \text{ is not empty,} \tag{34}$$

and define L by

$$L(t) = \inf\{x \in [0, +\infty[| u(t,x) = 1\} \tag{35}$$

for every t from $[0, T]$. Then the pair comprised of L and a self-evident restriction of u is a solution to (P).

The first line in (OP) measures up to a *differential inclusion*. (OP), resembls the so-called *obstacle problems* that are studied in wide-spread mathematical literature (see, for instance, RODRIGUES, 1987). However, (OP) departs from standards, since it involves a pair of Cauchy conditions on the boundary segment where $x = 0$ and involves no initial condition.

The following result gives conditions regarding physical and chemical data, ensuring that a subsurface gas source is present. Its proof will appear in a forthcoming paper by the authors.

Theorem 1. *Assume*

$$F(t) \le M = \text{ constant} \tag{36}$$

and

$$g(t) \ge N = \text{constant} \tag{37}$$

for every t from $[0, T]$; *assume M and N satisfy*

$$N > \max(M, 0) + (\pi DT)^{-1/2}\exp\left(-\frac{M^2}{4D^2}T\right).$$ (38)

Then any solution u to problem (OP) develops a coincidence set.

A feature which problem (P) shares with other problems for partial differential equations of parabolic type is *ill-posedness*. First, a small perturbation of data need not result in a small perturbation of solutions; secondly, stability can be restored if solutions themselves are suitably constrained *a priori*. A relevant result can be found in BACCHELLI (1997); the following result is a consequence of FRANCINI (2000).

Theorem 2. *Let* g_1 *and* g_2 *be two copies of* **g**, *let* (u_1, L_1) *and* (u_2, L_2) *be the corresponding solution pairs to problem* (P) *above. Assume* M, α, β, γ *are positive constants; assume*

$$|F(t)| \leq M$$ (39)

for every t from $[0, T]$, *and*

$$\alpha \leq L_i(t) \leq \beta, \quad |L_i'(t)| \leq \gamma$$ (40)

for every t from $[0, T]$ *and* $i = 1, 2$. *Let* $0 < a < b \leq T$. *Then a positive constant C exists such that*

$$\sup\{|L_1(t) - L_2(t)| : a \leq t \leq b\} \leq C\left(\log\frac{1}{\varepsilon}\right)^{-0.1}$$ (41)

provided

$$\sup\{|g_1(t) - g_2(t)| : 0 \leq t \leq T\} \leq \varepsilon$$ (42)

and ε *is sufficiently small.*

A detailed analysis of problem (P) was made in SGHERI *et al.* (1993), and TALENTI and TONANI (1995), in the case where depth L of the subsurface gas source happens to be constant in time and can be summarized thus. Let a function

$$U : \{(l, t, x)|0 < l < \infty, 0 \leq t \leq T, 0 \leq x \leq l\} \rightarrow]-\infty, \infty[$$ (43)

(depending upon time, depth and an extra parameter l) obey

$$\begin{cases} U_t = DU_{xx} + F(t)U_x & \text{for } 0 < t < T \text{ and } 0 < x < l, \\ U(l, t, 0) = 0 & \text{for } 0 < t < T, \\ U(l, t, l) = 1 & \text{for } 0 < t < T, \\ u(l, 0, x) = x/l & \text{for } 0 < x < l, \end{cases}$$ (44)

a standard boundary value problem for a standard partial differential equation. Such a function U can be computed and plotted via available algorithm and *FORTRAN* code. *Ad hoc* arguments show that $U_x(l, t, 0)$ decreases strictly from $+\infty$ to 0 as l increases from 0 to $+\infty$ and t is fixed. Consequently, the following equation

$$U_x(l, T, 0) = g(T) \tag{45}$$

has exactly one positive root l which can be routinely computed. The root in hand is an estimate of the source depth which becomes increasingly accurate as the time span of the observation enlarges. The following result appears in the papers quoted above.

Theorem 3. *Let*

$$|F(t)| \leq M = \text{constant} \tag{46}$$

for every t from $[0, T]$ *let problem* (**P**) *have a solution pair such that*

$$L = \text{ constant}, \tag{47}$$

and let l be the unique root of (45). Assume $L_{\max}$ *is an upper bound for both L and l. Then*

$$\frac{|l - L|}{L_{\max}} \leq 147 \exp\left(-\frac{\pi^2 \cdot D}{L_{\max}} T + 2.3 \frac{L_{\max}}{D} M \right), \tag{48}$$

provided T is large enough.

4. Conclusions

The mathematical model proposed here applies in circumstances (such as those met in specimen areas of Vulcano Island) where *one-dimensional geometry* prevails, the considered layer is *fairly permeable*, the subsurface gas source is *shallow*, and bulk gas velocity is *low*. Our results can be summarized accordingly;

(i) Simple conditions on the data are available, allowing one to predict whether a subsurface gas source does exist.

(ii) Detecting a subsurface gas source via surface measurements is an *ill-conditioned* problem, i.e., the solution is extremely sensitive toward errors on data. However, stability can be restored, i.e., gross data can be potentially handled, provided reasonable hypotheses are made and pathological configurations of the gas source are ruled out.

(iii) An easy algorithm is available in the case where the subsurface gas source is presumed to be constant in time.

Acknowledgments

The research of J.I. Díaz was partially supported by project REN2000-0766 of the DGES (Spain) and the RTN HPRN-CT-2002-00274 of the EC.

REFERENCES

BACCHELLI, V. (1997), *Stability of Level Lines of Solutions of a Linear Parabolic Equation* Rend. Ist. Lombardo, *131*.

FRANCINI, E. (2000), *Stability Results for Solutions of a Linear Parabolic Noncharacteristic Cauchy Problem*, J. Inv. Ill-Posed Problems *8*(3), 255–272.

GLASKO, V. B., *Inverse problems of Mathematical Physics* (Amer. Inst. of Phys. 1984).

GOSH ROY, D. N., *METHODS OF INVERSE PROBLEMS IN PHYSICS* (CRC PRESS 1991).

LUNENBURG, R. K., *Mathematical Theory of Optics* (Univ. of California Press 1964).

MENKE, W., *Geophysical Data Analysis: Discrete Inverse Theory* (Academic Press 1984)

RODRIGUES, J. F., *Obstacle Problems in Mathematical Physics* (Noth–Holland, Amsterdam 1987).

SGHERI, L., TALENTI, G., and TONANI, F., *Complementary Techniques and Results on Soil Gas Distribution* (Consiglio Nazionale delle Ricerche, Istituto di Analisi Globale e Applicazioni, Firenze, 1993).

SUNDRY AUTHORS (1991), *Collected Papers on Unrest at Vulcano*, Acta Vulcanologica *1*, 93–254.

TALENTI, G., and TONANI, F. (1995), *Sounding for Underground Gas Sources by Surface Physico-chemical Methods: A Mathematical Basis*, Inverse Problems *11*, 6, 1265–1297.

TARANTOLA, A., *Inverse Problem Theory* (Elsevier, 1987).

VAINBERG, B. R., *Asymptotic methods in equations of mathematical physics* (Gordon and Breach, 1989).

(Received March 13, 2002, Revised May 19, 2003, Accepted May 20, 2003)

 To access this journal online:
http://www.birkhauser.ch

Pure appl. geophys. 161 (2004) 1519–1532
0033–4553/04/071519–14
DOI 10.1007/s00024-004-2518-1

© Birkhäuser Verlag, Basel, 2004

❘ Pure and Applied Geophysics

High CO_2 Levels in Boreholes at El Teide Volcano Complex (Tenerife, Canary Islands): Implications for Volcanic Activity Monitoring

V. SOLER[1], J. A. CASTRO-ALMAZÁN[1], R. T. VIÑAS[1], A. EFF-DARWICH[2],
S. SÁNCHEZ-MORAL[3], C. HILLAIRE-MARCEL[4], I. FARRUJIA[5], J. COELLO[2],
J. DE LA NUEZ[2], M. C. MARTÍN[2], M. L. QUESADA[2], and E. SANTANA[6]

Abstract — Emissions of CO_2 have been known for more than a hundred years as fumarolic activity at the terminal crater of El Teide volcano and as diffuse emissions at numerous water prospection drillings in the volcanic island of Tenerife. Large concentrations of CO_2 (> 10% in volume) have been found inside galleries, long horizontal tunnels excavated for water mining. However, CO_2 concentrations of only 2900 ppm have been observed at the surface of the central region of the island (Las Cañadas del Teide caldera). In this work we analysed CO_2 concentrations in the subsurface of Las Cañadas caldera, in an attempt to study the vertical distribution of carbon dioxide and, in particular, the low emissions at the surface. This has been done through a series of 17 vertical profiles in two deep boreholes excavated in the Caldera.

We found high levels of CO_2, varying in time from 13 vol% up to 40 vol% in different profiles directly above the water table, while no significant concentrations were detected above the thermal inversion that takes places in both boreholes at approximately 100 m from the water table. Water analyses also showed high dissolved CO_2 levels in equilibrium with the air, and an average $\delta^{13}C$ value in DIC of $+4.7\%$ (PDB), apparently induced by fast CO_2 degassing in the bicarbonated water.

Key words: Cañadas caldera, CO_2, ^{222}Rn, Ground temperature, Isotope fractionation.

1. Introduction

In past years numerous works have dealt with the quantification and characterisation of gaseous emissions in active volcanic areas, in the context of monitoring volcanic activity and pre-eruptive processes (e.g., TEDESCO, 1995, for a review of the topic). A commonly used species to study degassing processes is CO_2, the most

[1] Estación Volcanológica de Canarias, IPNA-CSIC, Aptdo. Correos 195, 38206 La Laguna, Spain. E-mail: vsoler@ipna.csic.es
[2] Departmento de Edafología y Geología, Universidad de La Laguna, La Laguna, Spain.
[3] Departamento de Geología, Museo Nacional de Ciencias Naturales, CSIC. 28006 Madrid, Spain.
[4] Université du Québèc à Montréal, GEOTOP-UQAM, H3C 3P8 Montréal, Canada.
[5] Consejo Insular de Aguas (CIA), Cabildo de Tenerife, Santa Cruz de Tenerife, Spain.
[6] Servicio Electrónico, Universidad de La Laguna, La Laguna, Spain.

abundant gas, after water vapour, in the volatile phase of magmas. It has low solubility in silicate melts, especially in basalts (STOLPER and HOLLOWAY, 1988) and hence is usually one of the first gases to be released from ascending magma. Even in quiescent conditions, it is usually the major constituent of the dry gas emitted by active volcanoes. It is therefore essential to quantify the amount of CO_2 released by magma in order to better understand the dynamics of volcanoes. Moreover, carbon dioxide in large concentrations is a toxic gas and therefore it could be dangerous to humans if variations of volcanic activity release large amounts of this gas, as occurred in the Lake Nyos event (CHEVRIER, 1990) and others recorded in central Italy (CHIODINI and FRONDINI, 2001) .

An important factor that must be taken into account in the case of volcanic gas studies is the presence of cold-water aquifers (T < 30 °C), common to many volcanic regions (HERNÁNDEZ *et al.*, 2000; CHIODINI and FRONDINI, 2001), since some carbon dioxide generated by magma may dissolve in these reservoirs and hence, the actual degassing rate could be underestimated (D'ALESSANDRO *et al.*, 1997).

Carbon dioxide measurements may be complemented by data from a non-reactive species only affected by physical transport processes. Among them, ^{222}Rn, a noble radioactive ultratrace gas, has been associated with magmatic activity (FLEROV *et al.*, 1986; CONNOR *et al.*, 1996). When radon is released from rocks, it escapes through pores and cracks towards the surface. However, the distance it can travel by molecular diffusion does not exceed a few metres as a result of having a short half-life of 3.8 days. Hence, large amounts of radon not explained by the emanating power of rocks could only be transported by gas and/or water flows.

Tenerife is an active volcanic island where CO_2 emissions have been known for more than a hundred years as fumarolic activity at the terminal crater of El Teide volcano (TD in Fig. 1) and as diffuse emissions at numerous water prospection drillings. In recent works, large concentrations of CO_2 (up to 10 vol%) have been found inside water galleries, horizontal tunnels several kilometres long, that have been excavated for water mining (VALENTÍN *et al.*, 1989; HERNÁNDEZ *et al.*, 1998). These concentrations have been responsible for several fatal accidents inside galleries in recent decades. However, HERNÁNDEZ *et al.* (2000) found average CO_2 concentrations of only 2900 ppm in most (nearly 93%) of the 350 analysed samples collected at the surface of the central region of the island (Las Cañadas caldera). Larger values were found at the summit fumaroles of El Teide volcano with carbon dioxide concentrations reaching 29 vol%.

In this work we analysed CO_2 concentrations in the subsurface of Las Cañadas caldera, Tenerife, directly above the local aquifer, in an attempt to study the vertical distribution of carbon dioxide and, in particular, the low emissions at the surface. For this purpose 17 vertical profiles of CO_2 were obtained at two boreholes (505-m and 400-m deep, respectively) over a period of 20 months, together with continuous measurements (over 36 hours) of CO_2 and ^{222}Rn emissions. Isotopic determinations of $\delta^{13}C$ in CO_2 were also carried out in water and gas samples collected at the

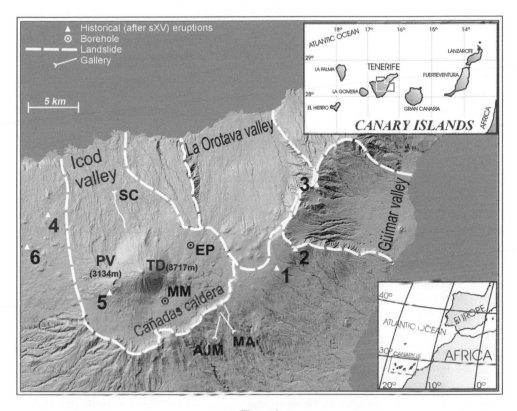

Figure 1

Location map: PV: Pico Viejo; TD: Teide; EP: El Portillo; MM: Montaña Majúa; MA: Madre del Agua; AJM: Ancón de Juan Marrero; SC: Saltadero de Las Cañadas. Historical eruptions: 1, 2, 3: Siete Fuentes Eruptions (1704–05); 4: Montaña Negra (1706); 5: Chahorra (1798); 6: Chinyero (1909).

bottom of the boreholes. Finally, water samples were also taken out to determine the chemical composition.

2. Description of Monitoring Sites and Geological Context

Tenerife is the largest island (2078 km^2) of the Canarian Archipelago and the world's second largest oceanic island, after Hawaii. It is located between longitudes 28–29N and latitudes 16–17W, 280 km from the African coast (see Fig. 1). The central part of the island is occupied by Las Cañadas del Teide caldera, a depression covering an area of 16 × 9 km that constitutes the base of the El Teide-Pico Viejo stratovolcano, a 3717-m high volcanic complex whose last eruption in the terminal crater took place in 1400 AD (QUIDELLEUR *et al.*, 2001). The Caldera is closed by a wall to the south and it is open to the sea on the northern face through a landslide valley (BRAVO, 1962; NAVARRO and COELLO, 1989), hidden by the accumulation of eruptive material. Although the eruptive events associated with the construction of the stratovolcano Teide-Pico Viejo were gradually spaced in time (GILLOT *et al.*, 2001), a high level of historical and sub-historical activity persists in the flanks of the main complex (CARRACEDO *et al.*, 2003). This activity is also evident in the diffuse and fumarolic gaseous emissions, the high level of microseismicity (MEZCUA *et al.*, 1992) and the presence of thermal and chemical anomalies (CARRACEDO and SOLER, 1983).

Experimental data were collected at two boreholes named El Portillo (hereafter EP) and Montaña Majúa (MM), percussion-drilled by the Water Authority of Tenerife (Consejo Insular de Aguas, hereafter CIA) for piezometric studies and water quality control. MM (where 10 profiles have been carried out) is located in the central area of the Caldera, 2264 m above sea level, reaching a depth of 505 m, well below the water level (at a depth of 446 m), while EP (7 profiles) is located in the northern area of the Caldera, 7.1 kilometres from the first borehole, 2133 m above sea level and it reaches a depth of 400 m (water level is at a depth of 370 m) (Figs. 1 and 2).

EP has an initial drilling diameter of 700 mm and there is an internal pipe with a diameter of 500 mm that is slotted in the last 100 m. The initial drilling diameter at MM is also 700 mm and its internal pipe is 450 mm in diameter and is also slotted in the last 100 m. The internal pipe in MM contains 13 additional pipes installed for different water monitoring programs. We used the largest pipe, 115 mm in diameter, to carry out our measurements. All of the pipes are opened at both ends and they allow air circulation under stable conditions.

There is a good correlation between the stratification found in both boreholes. The lithology corresponds to an upper layer of phonolitic and pumiceous rocks followed by basaltic lavas and some intermediate trachybasaltic flows. This stratification belongs to three subunits of the Post-Caldera formations, namely the initial basaltic sequence, Teide-Pico Viejo trachybasalts and Teide-Pico Viejo phonolitic flank vents, respectively, with materials gradually more differentiated (NAVARRO, 2000; FARRUJIA *et al.*, 2001).

3. Observational Procedure

The experimental methodology is based on sensors that were pulled down into the boreholes to obtain vertical profiles of carbon dioxide and temperature. The design of these instruments is constrained by the spatial limitations of the boreholes, taking into account that the pipe at MM was 115 mm in diameter.

Sensors were placed on a stainless steel plate that was enclosed within an open PVC tube, 75 mm in diameter, sufficiently narrow to fit in the boreholes. We developed a specific data acquisition system only 6-cm wide and based on a PIC (16C774) microcontroller with 8 different 12 bit, 0-5 V channels. The power supply system consisted of 8 alkaline "D" 1.5 V batteries and a small control unit.

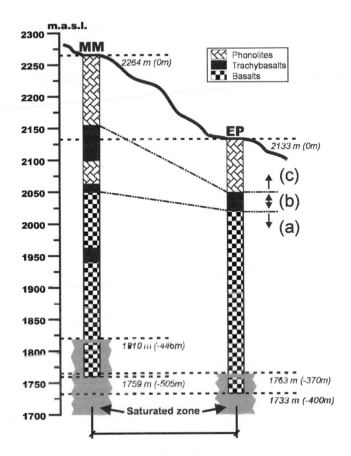

Figure 2

Schematic lithology, absolute height, relative position and geologic units of Montaña Majúa (MM) and El Portillo (EP) boreholes. Relative depths are shown into brackets. (a): Initial basaltic sequence, (b): Teide-Pico Viejo trachybasalts and (c): Teide-Pico Viejo phonolitic flank vents (NAVARRO, 2000, FARRUJIA et al., 2001). m.a.s.l. = meters above sea level.

Carbon dioxide concentrations were obtained with a double beam non-dispersive infrared spectrometer (NDIR) sensitive to values from 0 to 50% (model Gasmitter ATME 13). The pump used by the sensor has an independent power supply and was isolated with a foam capsule to avoid perturbations to the remainder of the detectors. Temperature was measured by a PT-100 type sensor (0–30 °C in 0–5 V with an accuracy of 0.01 °C) equipped with a high accuracy amplifier INA-125. The PT-100 sensor was suspended at a distance of 10 m from the main unit to minimise perturbations.

Data collection continued while slowly lowering the sensors (programmed to record data every 5 seconds) through the borehole and stopping approximately 30 seconds every 25 metres to obtain measurements under stable conditions. The measuring process never exceeded 30 minutes.

Data obtained with profiles were complemented with continuous measurements of CO_2 and ^{222}Rn, that were carried out at the bottom of MM (−440 m) over 36 hours (starting on 23 January 2002), using the NDIR sensor for CO_2 and a silicon diode α-particle detector (Alphanuclear 611) for ^{222}Rn. Sampling intervals were 30 minutes for carbon dioxide and 1 hour for radon data.

We also performed for the first time isotopic analyses of CO_2 in the air and water of the boreholes, complemented with air samples collected in three galleries. The latter were done to compare with measurements reported earlier by other authors. The isotopic air samples were collected by emptying a tube containing distilled water with a delayed electromagnet system. The water samples were collected by sinking a tube until the water pressure opened a small pressure valve (a depth of approximately 2 m sufficed), filling the tube. The analyses were performed using an Isocarb^{TM} preparation device in line with a triple collector VG-Prism^{TM} mass spectrometer. Results are expressed in δ units calculated against the V-PDB standard value (COPLEN-TYLER, 1995).

Water samples for chemical analyses were taken at both boreholes in 200 ml bottles which had been pre-washed in double-distilled water. Some parameters such as the water temperature, electrical conductivity, pH and CO_2-HCO_3^--CO_3^{2-} contents must be measured *in situ* with portable instruments, including a carbon dioxide gas-sensing electrode. Complete chemical analyses have been performed in the laboratory by atomic absorption spectrometry and ionic capillary electrophoresis technique. Chemical speciation and geochemical calculations were performed with speciation models and chemical mass transport codes (e.g., PHRQPITZ, PLUMMER *et al.*, 1988).

4. Results and Discussion

Typical vertical profiles obtained at MM and EP for CO_2 and temperature are shown in Figure 3. The shape of the temperature profiles can be split in two main trends. The first decreases with depth and this is due to the difference between the atmospheric temperature and the temperature of the rock in the center of the boreholes. Here the temperature is about 5.5 °C at MM and 6.5 °C at EP and it may be induced by the percolation of meteoric snow melt. The second trend is determined by the water table temperature (18.4 °C and 14.7 °C at MM and EP, respectively) higher than the rock temperature above it. Water temperature has been independently measured for more than 4 years in MM by CIA, giving temporal variations lower than 0.2 °C per year. The profiles repeated in different periods of the year produced similar results.

Carbon dioxide profiles are similar in the two boreholes, ranging from approximately 13 to 40 vol% in the different profiles, contrasting with the low concentrations reported at the surface (HERNÁNDEZ *et al.*, 2000). The vertical distribution shows that CO_2 is mainly concentrated in the region between the water

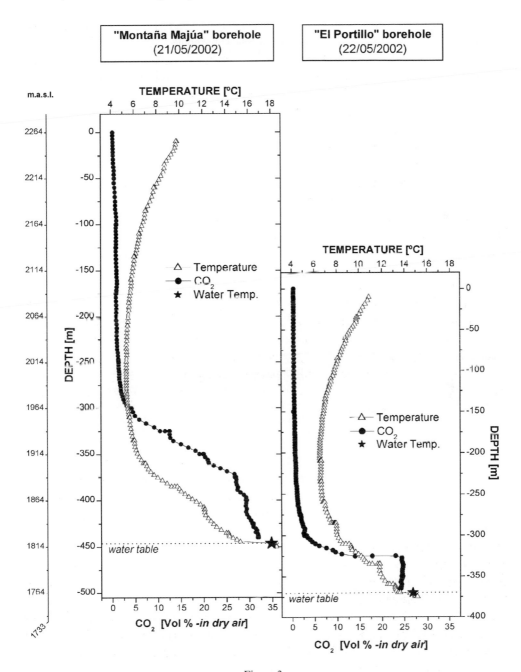

Figure 3

Typical vertical distribution of CO₂ and temperature at MM and EP boreholes. Water temperature was also independently measured and it is represented by the star point in each profile. Absolute height is shown on the left axis. m.a.s.l. = meters above sea level.

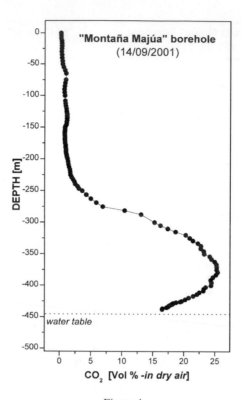

Figure 4
Vertical distribution of CO_2 at MM borehole when the maximum on concentration is not in direct contact with the water table. m.a.s.l. = meters above sea level.

table and the level where the change in the temperature gradient takes place. This distribution may be induced by the thermal inversion occurring above the water table. Warmer CO_2 flows up from the top of the aquifer while the effect of gravity and the cooling effect of the rock refrain this gas from flowing upward.

Variations in the position of the maximum on the concentration of CO_2 have been observed in several profiles in MM. Indeed, the maximum of CO_2 concentration is shifted from the water table (Fig. 4) when there is a decrease in barometric pressure, as illustrated in the left panel of Figure 5. However, when barometric pressure gradient is small, the maximum of CO_2 is in contact with the water table (see right panel of Fig. 5 and Fig. 3). Nonetheless it is essential to carry out extended continuous monitoring to find a conclusive explanation.

Although temperature profiles are very stable in time, we looked for short-term temporal variations of carbon dioxide. For this purpose, continuous measurements of CO_2 and ^{222}Rn were carried out at –440 m in MM (see Fig. 6). They exhibited similar temporal variations that are well correlated to temporal changes in barometric pressure. EFF-DARWICH *et al.* (2002) showed that the observed behaviour

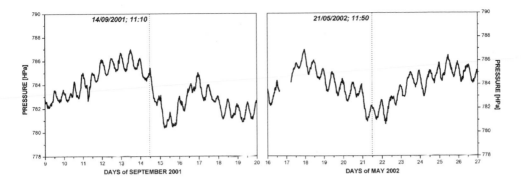

Figure 5

Temporal distribution of barometric pressure for a ten-day interval centered at the moment when MM profiles shown in Figure 3 (21/05/2002) and Figure 4 (14/09/2001) were carried out (shown in dotted lines).

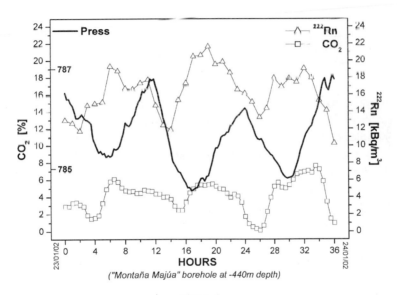

("Montaña Majúa" borehole at -440m depth)

Figure 6

Temporal variation of CO$_2$, ^{222}Rn and pressure (in HPa) in MM at a depth of 440 m below the topographic surface over 36 hours, starting on January 23, 2002.

of the radon signal is explained by the pumping effect of temporal changes in barometric pressure. Considering that CO$_2$ and Radon present similar behaviour (see Fig. 6), temporal changes in CO$_2$ should be induced by barometric pressure variations.

The origin of the carbon dioxide was studied on the basis of isotopic determinations of δ^{13} C. The results obtained in water and air at the two boreholes range from +3.5‰ to +4.9‰, as shown in Table 1. Typical values for magmatic gas samples range from −4‰ to −9‰ (HOEFS, 1997) and specifically from −7‰ to −8‰ in

Table 1

δ^{13} C isotopic relations of water and air samples. See Figure 1 for location.

Location	Type	δ^{13} C (air)	δ^{13} C (DIC in water)
MA	g	−10.2 (~1.32%)	–
		−10.5 (~1.32%)	–
AJM	g	−10.7 (~0.61%)	–
		−10.5 (~0.61%)	–
SC	g	−10.6 (~2.24%)	–
		−10.6 (~2.24%)	–
		−9.4 (~2.85%)	–
		−9.3 (~2.85%)	–
EP	b	+3.5 (~20%)	4.7
		+4.2 (~20%)	4.8
MM	b	–	4.5
		–	4.9

g: gallery   In brackets: CO_2 concentration at the moment of sampling
b: borehole δ^{13} C -CO_2 : [‰ PDB]

oceanic islands (MOORE *et al.*, 1977). In the Island of Tenerife, previous works (VALENTÍN *et al.*, 1989; ALBERT *et al.*, 1989) reported values between −10.8‰ and −5‰ in samples collected in galleries; the most negative results being interpreted as biogenic "contamination," and those greater than −7‰ as the result of the precipitation of carbonates in the gallery. Our results in galleries (Table 1) agree with these previous works, while the samples in the boreholes show high δ^{13} C-values observed for both DIC (Dissolved Inorganic Carbon) and gaseous CO_2. A plausible explanation for that high δ^{13} C-values would involve fast CO_2-degassing and carbonate precipitation under pressure release, both compounds being 13-depleted vs. isotopic equilibrium conditions, due to kinetic effects. Kinetic fractionation results from irreversible, i.e., one-way physical or chemical processes (VOGET *et al.*, 1970). The residual DIC would then get progressively enriched in heavy carbon, following a Rayleigh enrichment process (fractionation factor < 1; MOOK *et al.*, 1974; MOOK, 1986). Such mechanisms have often been observed for travertine deposition, for example, and evoked for a hydrothermal system (e.g., RIHS *et al.*, 2000).

    Water chemical analyses (Table 2) yielded a sodium bicarbonated composition with high amounts of dissolved CO_2. The dissolved constituents of groundwater could be explained by the reaction of CO_2-charged water with plagioclase feldspar and minor amounts of magnesium silicates and K-feldspars, as major constituents of volcanic rock formation. The theoretical concentration of carbon dioxide in air in equilibrium with water was also calculated (Table 3) ranging from 9.5 vol% (EP) to 29.8 vol% (MM). In spite of the strong variations detected in the contents of dissolved CO_2, the concentration in calcium stays constant and the state of

Table 2

Chemical composition of water samples from MM and EP boreholes. Results in ppm. CO_2 dissolved in water in mg/l. MM-1 correspond to a sample taken at the surface of the aquifer and MM-2 to a sample taken at the bottom of the borehole

Sample	CO_2	T (°C)	pH	HCO_3^-	NO_3^-	SO_4^{2-}	Cl^-	Ca^{2+}	Mg^{2+}	Na^+	K^+	Si	F
MM-1	402	18.4	6.63	1402.4	27.1	30.0	8.0	46.2	64.8	302.8	84.9	28.80	2.43
MM-2	518	18.7	6.60	1414.0	26.4	30.5	10.1	47.7	68.6	314.4	88.8	29.90	2.11
EP	163	17.0	6.95	1128.0	4.3	13.5	4.1	45.7	44.9	247.8	75.6	25.00	5.72

Table 3

State of mineral saturation of the water samples, regarding the main carbonate minerals. CO_2 Partial pressure and saturation index calculated with PHRQPITZ (PLUMMER et al., 1988)

Sample	Aragonite	Calcite	Dolomite	P_{CO_2} water	Equilibrium Air CO_2 (%)
MM-1	−0.42	−0.22	+0.06	$10^{-0.65}$ bar	22.5
MM-2	−0.50	−0.31	−0.08	$10^{-0.54}$ bar	29.6
EP	−0.21	−0.02	+0.30	$10^{-1.04}$ bar	9.5

saturation of the water regarding the main carbonated minerals remains very close to the equilibrium state (Table 3). Therefore, a long residence time of water in contact with the rock is expected, being the precipitation and dissolution processes of calcium carbonated minerals an essential factor in the control of the contents and emissions of CO_2 as well as of the isotopic composition both of air and water. The long residence time and especially the CO_2 confinement at the bottom of the boreholes may explain the low emission rates of carbon dioxide (HERNÁNDEZ et al., 2000) in soil gases at Las Cañadas del Teide caldera.

Temporal variations on the total concentration of CO_2 in the air-water system may depend on changes of atmospheric pressure on the aquifer, as well as changes in water temperature and/or emissions of CO_2 arising from the volcanic activity. Theoretical models (PLUMMER et al., 1988) show that an increase of 1 °C in the water temperature is followed by an increase of the CO_2 in equilibrium in air of 0.5 vol% due to the increase of the water CO_2 partial pressure (see Fig. 7). Continuous monitoring of carbon dioxide and atmospheric pressure, together with the working monitoring system of physical-chemical characteristics of water, could supply immediate information of changes occurring in the aquifer, as well as to differ if these changes are related with processes of volcanic reactivation. Moreover, provided that carbon dioxide is a toxic gas, it is advisable to monitor this gas as a result of the high levels found in this work.

Since the aquifer covers the entire area of the caldera and it is in an intermediate position between the magma and the surface, boreholes such as EP and MM provide

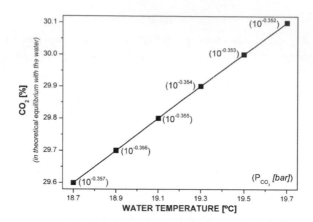

Figure 7

Increase of equilibrium air CO_2 as a result of a theoretical 1°C water heating in the aquifer. In brackets is the CO_2 partial pressure for each point. CO_2 partial pressure and saturation index calculated with PHRQPITZ (PLUMMER *et al.*, 1988).

exceptional access to monitor volcanic activity of extensive active volcanic regions such as Las Cañadas caldera.

Acknowledgements

This study was supported by the Spanish Project PB98-0643 and the European Union Project EVR1-1999-00017. We are grateful to the Water Authority of Tenerife (Consejo Insular de Aguas, Cabildo de Tenerife) for providing access to the boreholes. We also would like to thank Mr. Abelardo Díaz Torres and Mr. Santiago Acosta Rodríguez for their helpful contributions in hardware development. Finally, we thank Dr. David Pyle and another anonymous referee for their careful reading and valuable suggestions.

REFERENCES

ALBERT, J. F., DIEZ-GIL, J. L., VALENTIN, A., GARCÍA DE LA NOCEDA, C., and ARAÑA, V., *El sistema fumaroliano del Teide*, In *Los Volcanes y la Caldera del Parque Nacional del Teide (Tenerife, Islas Canarias)* (eds. Araña, V., and Coello, J.) (Ed. ICONA, serie técnica n° 7, Madrid, 1989) pp. 347–358.
BRAVO, T. (1962), *El circo de Las Cañadas y sus dependencias*, Bol. R. Soc. Esp. Hist. Nat. *40*, 93–108.
CARRACEDO, J. C., and SOLER, V. (1983), *Anomalías térmicas asociadas al volcanismo en las islas Canarias*, V Asamblea Nacional de Geodesia y Geofísica. Madrid. *Vol. Com.* pp. 2351–2363.
CARRACEDO, J. C., GUILLOU, H., PATERNE, M., PÉREZ-TORRADO, F. J., PARIS, R., and BADIOLA, E. R. (2003), *Carbon-14 Ages of the Past 20 Ka of Eruptive Activity of Teide volcano, Canary Islands*, Geophys. Res. Abstracts, EGS-AGU-EUG Joint Assembly, vol. 5.
CHEVRIER, R. M. (1990), *Lake Nyos : Phenomenology of the Explosive Event of December 30, 1986*, J. Volcanol. Geotherm. Res. *42*–4, 387–390.

CHIODINI, G., and FRONDINI, F. (2001), *Carbon Dioxide Degassing from the Albani Hills Volcanic Region, Central Italy*, Chem. Geology *177*, 67–83.

CONNOR, C., HILL, B., LAFEMINA, P., NAVARRO, M., and CONWAY, M. (1996), *Soil ^{222}Rn Pulse during the Initial Phase of the June- August 1995 Eruption of Cerro Negro, Nicaragua*, J. Volcanol. Geotherm. Res. *73*, 119–127.

COPLEN-TYLER, B. (1995), *Discontinuance of SMOW and PDB*, Nature *375*, 285–288.

D'ALESSANDRO, W., GIAMMACO, S., PARELLO, F., and VALENZA, M. (1997), *CO$_2$ Output and δ^{13} C(CO$_2$) from Mount Etna as Indicators of Degassing of Shallow Asthenosphere*, Bull. Volcanol. *58*, 455–458.

EFF-DARWICH, A., Martín-Luis, C., QUESADA, M., DE LA NUEZ, J., and COELLO, J. (2002), *Variations on the Concentration of ^{222}Rn in the Subsurface of the Volcanic Island of Tenerife, Canary Island*, Geophs. Res. Lett. 29, 22.

FARRUJIA, I., BRAOJOS, J., and FERNÁNDEZ, J. (2001), *Ejecución de dos sondeos profundos en Las Cañadas del Teide*, VII Simposio de Hidrogeología de la Asociación Española de Hidrogeólogos. Murcia. *XXIII*, 661–672.

FLEROV, G. N., CHIRKOV, A. M., TRETYAKOVA, S. P., DZHOLOS, L. V., and MERKINA, K. I. (1986), *The Use of Radon as an Indicator of Volcanic Processes*, Earth Phys. *22*, 213–216.

GILLOT, P-Y., SOLER, V., and QUIDELLEUR, X., *Piling rate and magmatic evolution through time of the Teide volcano (Tenerife, Canary Islands)*. In *EUG XI Meeting* (T. Nova, 2001).

HERNÁNDEZ, P., PÉREZ, N., SALAZAR, J., NOTSU, K., and WAKITA, H. (1998), *Diffuse Emission of Carbon Dioxide, Methane and Helium-3 from Teide Volcano, Tenerife, Canary Islands*, Geophys. Res. Lett. 23, 3311–3314.

HERNÁNDEZ, P., PÉREZ, N., SALAZAR, J., SATO, M., NOTSU, K., and WAKITA, H. (2000), *Soil CO$_2$, CH$_4$ and H$_2$ Distribution in and around Las Cañadas Caldera, Tenerife, Canary Islands, Spain*, J. Volcanol. Geotherm. Res. *103*, 425–438.

HINKLE, M. E. (1994), *Environmental Conditions Affecting Concentrations of He, CO$_2$, O$_2$ and N$_2$ in Soil Gases*, Appl. Geochem. *9*, 53–63.

HOEFS, J., *Stable Isotope Geochemistry* (Springer, Berlin 1997).

MEZCUA, J., BUFORN, E., UDÍAS, A., und RUEDA, J. (1992), *Seismotectonics of the Canary Islands* Tectonophysics, *208*, 447–452.

MOOK, W. G., BOMMERSON, J. C., and STAVERMAN, W.H. (1974), *Carbon Isotope Fractionation between Dissolved Bicarbonate and Gaseous Carbon Dioxide*, Earth Planet. Sci. Lett. *22*, 169–176.

MOOK, W. G. (1986), *^{13}C in Atmospheric CO$_2$*. Neth. J. Sea Res. 20 (2/3), 211–223.

MOORE, J. G., BATCHELDER, J. N., and CUNNINGHAM, C. G. (1977), *CO$_2$ Filled Vesicles in Mid-oceanic Basalt*, J. Volcanol. Geotherm. Res. 2, 309–327.

NAVARRO, J. M., and COELLO, J. (1989), *Depressions Originated by Landslide Processes in Tenerife*, ESF Meeting on Canarian Volcanism. Lanzarote, 150–152.

NAVARRO, J. M., *Geología del Parque Nacional del Teide*, In *El Parque Nacional del Teide* (Estagnos, 2000) pp. 19–72.

PLUMMER, L. N., PARKHURST, D. L., FLEMING, G. W., and DUNKLE, S. A. (1988), *PHRQPITZ, A Computer Program Incorporating Pitzer's Equations for Calculation of Geochemical Reactions in Brines*, U.S. Geol. Surv. Water Res. Inv. Report 88–4153, 310 pp.

QUIDELLEUR, X., GILLOT, P.-Y., SOLER, V., and LEFÈVRE, J.-C. (2001), *K/Ar Dating Extended into the Last Millennium: Application to the Youngest Effusive Episode of the Teide Volcano (Spain)*, Geophys. Res. Lett. 28–16, 3067–3070.

RIHS, S., CONDOMINES, M., and POIDEVIN, J.-L. (2000), *Long-term Behaviour of Continental Hydrothermal Systems: U-series Dating of Hydrothermal Carbonates from the French Massif Central (Allier Valley)*, Geochimica et Cosmochimica Acta *64*, 3189–3199.

STOLPER, E., and HOLLOWAY, J. R. (1988), *Experimental Determination of the Solubility of Carbon Dioxide in Molten Basalt at Low Pressure*, Earth Planet. Sci. Lett. *87*, 397–408.

TEDESCO, D., *Monitoring fluids and gases at active volcanoes*. In *Monitoring Active Volcanoes* (eds. McGuire, B., Kilburn, C., and Murray, J.) (UCL, London 1995) pp. 315–345.

VALENTÍN, A., ALBERT, J., DÍEZ-GIL, J. L., and GARCÍA DE LA NOCEDA, C., *Emanaciones magmáticas residuales en Tenerife*. In *Los Volcanes y la Caldera del Parque Nacional del Teide (Tenerife, Islas Canarias)* (eds. Araña, V. and Coello, J.) (ed. ICONA, serie técnica n° 7, Madrid, 1989) pp. 299–310.

VOGEL, J. C., GROOTES, P. M., and MOOK, W. G. (1970). *Isotope Fractionation between Gaseous and Dissolved Carbon Dioxide*, Z. Phys. *230*, 225–238.

(Received February 28, 2002, revised April 29, 2003, accepted May 15, 2003)

 To access this journal online:
http://www.birkhauser.ch

Pure appl. geophys. 161 (2004) 1533–1547
0033–4553/04/071533–15
DOI 10.1007/s00024-004-2519-0

❚ Pure and Applied Geophysics

Simulation of the Seismic Response of Sedimentary Basins with Vertical Constant-Gradient Velocity for Incident *SH* Waves

F. Luzón[1,2], L. Ramírez[3], F. J. Sánchez-Sesma[3],
and A. Posadas[1,2]

Abstract — The simulation of the seismic response of heterogeneous sedimentary basins under incident plane waves is computed using the Indirect Boundary Element Method (IBEM). To deal with these kinds of basins we used approximate analytical expressions for the two-dimensional Green's functions of a medium with constant-gradient wave propagation velocity. On the other hand, for the homogeneous half space underlying the sedimentary basin, the full space Green's functions were used. The response of semi-circular heterogeneous basins under incident *SH* waves is explored by means of the displacements in the frequency-space diagrams and synthetic seismograms. Moreover, we compared these results with those obtained for other homogeneous semi-circular models. The principal differences among them are pointed out. This simulation provided interesting results that displayed a complex amplification pattern in a rich spectrum of frequencies and locations. The maximum amplitudes levels were found around the edges of the heterogeneous sedimentary basins. In time domain some features characterize the seismic response of the basin which include enhanced generation and trapping of surface waves inside the sediments, and the reduced emission of seismic energy to the hard rock. In the heterogeneous models the lateral reflections of surface waves greatly influence the total displacements at the free surface in comparison with the homogeneous models where the displacements have a shared influence among both vertical and lateral reflections.

Key words: Boundary element method, Green's function, vertical heterogeneity, sedimentary basins, strong ground motion, seismology.

1. Introduction

Local effects in sedimentary basins can lead to significant spatial and temporal variations of seismic ground motions. This may induce large structural damage when strong and moderate earthquakes occur. Local site response has received significant attention in the past two decades and its characterization has been dealt with using both experimental and numerical analysis (see e.g., AKI, 1988 and SÁNCHEZ-SESMA,

[1] Departamento de Física Aplicada. Universidad de Almería. Cañada de San Urbano s/n. 04120, Almería, Spain. E-mail: fluzon@ual.es; aposadas@ual.es
[2] Instituto Andaluz de Geofísica y Prevención de Desastres Sísmicos.
[3] Instituto de Ingeniería, Universidad Nacional Autónoma de México, Cd. Universitaria, Coyoacán 04510, México D.F., México. E-mail: sesma@serridor.unam.mx; lrag@pumas.iingen.unam.mx

1996, respectively). A profusion of theoretical works on site effects has been focused on study of the seismic response of different configurations subjected to specified incoming wavefields. Usually plane waves are assumed and homogeneous sedimentary basins have been considered (e.g., LUCO *et al.*, 1990; PAPAGEORGIOU and KIM, 1991; SÁNCHEZ-SESMA *et al.*, 1993; LUZÓN *et al.*, 1995). The numerical methods that have been widely used to calculate the propagation of elastic waves inside of sedimentary inclusions can be classified in domain, boundary, and asymptotic methods. In the first group, where the discretization of the media is required, techniques are included as the finite difference (see e.g., ALTERMAN and KARAL, 1968) or the finite-element method (e.g., LYSMER and DRAKE, 1972). On the other hand, in the boundary methods, where only the discretization of the boundaries is required, there are two main approaches: one is based on the use of complete systems of solutions (HERRERA, 1980), and the other, on the use of boundary integral equations (BREBBIA, 1978). When the problem of propagation of waves has interest in the high frequency band and the diffraction can be ignored, the asymptotic methods can be useful. Ray theories are based on the asymptotic behavior of the wave equation for the wavelength $\lambda \approx 0$, and have been used to study the ground motion in sedimentary inclusions (RIAL, 1984) or dipping layers (ZIEGLER and PAO, 1984). On the other hand, in their pioneering work, AKI and LARNER (1970) introduced a method based on a discrete superposition of plane waves, which was used to deal with 2-D (e.g., BARD and BOUCHON, 1980) and 3-D sedimentary basins (HORIKE *et al.*, 1990).

The most realistic simulations of elastic wave propagation to date have been done by using the finite-difference method. The recent work of OLSEN (2000), where the 3-D response of the Los Angeles basin was calculated, illustrates this fact well. However, these kinds of methods require considerable computational time and large memory supercomputers. Boundary element methods have gained increasing popularity due to their numerical advantages. They reduce by one the dimensionality of the problem, avoid the introduction of fictitious boundaries, and provide highly accurate results. In recent years numerical codes using these techniques have been developed to deal with elastic wave propagation in 3-D alluvial basins (SÁNCHEZ-SESMA and LUZÓN, 1995) and 3-D topographies (LUZÓN *et al.*, 1997, 1999). These methods are formulated in terms of the Green's function of the problem, which represents the solution for a unit force applied in a point within a certain domain of interest. The availability of these functions is a severe limitation to the boundary element methods. In fact, the Green's functions can be easily obtained only for homogeneous unbounded medium. This has constrained the study of the seismic response of sedimentary basins to homogeneous models. However, there is a wide class of problems for which it is reasonable to assume an increase of wave propagation velocities with depth (see e.g., BARD and GARIEL, 1986; VRETTOS, 1990).

In this paper we analyze the ground motion in semi-circular sedimentary inclusions with velocity varying linearly with depth by using an approximate analytic

Green's function calculated recently (SÁNCHEZ-SESMA *et al.*, 2001). We illustrate the effects of these sedimentary basins by means of frequency-space plots, synthetic seismograms on the free surface, and snapshots at different times of the motion in depth under incident *SH* plane waves.

2. The Method and the Green's Function

Consider a 2-D alluvial valley on the surface of a half space which is named the exterior region E. The alluvial basin occupies region R as depicted in Fig. 1, and assume the material of the valley has shear-wave velocity increasing linearly with depth. To compute the seismic response of such alluvial valleys, the Indirect Boundary Element Method is used. Assuming harmonic excitation, the refracted displacement inside the basin can be expressed as

$$u_i^r(\mathbf{x}) = \int_{S_R} \phi_j^R(\xi) G_{ij}^R(\mathbf{x}, \xi)\, dS_\xi, \tag{1}$$

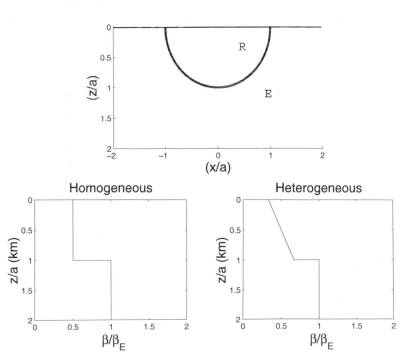

Figure 1

Top: Semi-circular sedimentary basin of radius a (region R) underlying a half space (region E). Bottom: Velocity profiles in depth for $x = 0$ for both, the homogeneous model (*S*-wave velocity $\beta_R = \beta_E/2$), and the heterogeneous model ($\beta_R(0) = \beta_E/3$, $\beta_R(a) = 2\,\beta_E/3$). The mass density for both models is $\rho_R = 2\rho_E\,/\,3$ ($n = 0$ of equation (4)).

where $G_{ij}^R(\mathbf{x}, \xi)$ = Green's function of the basin region R (the Green's function is the displacement produced in the direction i at $\mathbf{x}$ due to the application of a harmonic unit line force at the point ξ in the direction j), $\phi_j^R(\xi)$ = force density in the direction j at ξ and S_R = boundary of region R. Equation (1) can be obtained from Somigliana's identity (see SÁNCHEZ-SESMA and CAMPILLO, 1991). On the other hand, the total wavefield in the half space is the sum of the so-called *free-field* and *diffracted* wave,

$$u_i(\mathbf{x}) = u_i^{(0)}(\mathbf{x}) + u_i^{(d)}(\mathbf{x}) = u_i^{(0)}(\mathbf{x}) + \int_{S_E} \phi_j^E(\xi) G_{ij}^E(\mathbf{x}, \xi)\, dS_\xi, \qquad (2)$$

where $u_i^{(0)}(\mathbf{x})$ = displacement associated with the free-field, i.e., the motion in the absence of any irregularity, which includes incident and reflected waves by the half space free surface, $u_i^{(d)}(\mathbf{x})$ = displacement of diffracted waves, and $G_{ij}^E(\mathbf{x}, \xi)$ = Green's function of the region E. Tractions can be obtained by direct application of Hooke's law and by equilibrium considerations around an internal neighborhood of the boundary. When the boundary conditions are imposed, that is zero tractions on the free surface of the basin and of the half space, and continuity of displacements and tractions at the common interface of both media, the discretized versions of tractions and displacements give rise to a system of linear equations where $\phi_j^{E,R}(\xi)$ are the unknowns that can be used to compute the diffracted and the refracted displacements. For further details on the method and in the discretization process of the surfaces involved we refer to SÁNCHEZ-SESMA et al. (1993).

Consider an elastic isotropic sedimentary basin with velocity varying linearly with depth. In such medium the S-wave velocity and the mass density inside the basin can be expressed as:

$$\beta(z) = \beta_0 \frac{1 + \gamma z}{1 + \gamma z_0} = \beta(0) \cdot (1 + \gamma z), \qquad (3)$$

$$\rho(z) = \rho_0 \left(\frac{1 + \gamma z}{1 + \gamma z_0}\right)^n = \rho(0) \cdot (1 + \gamma z)^n, \qquad (4)$$

where β = S wave velocity, β_0 = S wave velocity at source depth, $\gamma = 1/h$, where $\beta(-h) = 0$, ρ = mass density, ρ_0 = mass density at source depth and $n \geq 0$. The approximate analytic 2-D Green's function for SH waves in this case can be expressed (SÁNCHEZ-SESMA et al., 2001) as:

$$G_{yy}^R(\omega, x) = \Lambda \frac{i}{4\mu_0} H_0^{(1)}(\omega \tau), \qquad (5)$$

where

$$\Lambda = \left(\frac{1 + \gamma z_0}{1 + \gamma z}\right)^{\frac{n+2}{2}} \sqrt{\frac{\beta_0 \tau}{R_w}},$$

$i = \sqrt{-1}$, μ_0 = shear modulus at the source level, $H_0^{(1)}$ = Hankel's function of the first kind and zero order, ω = circular frequency, τ = travel time, and $R_w = (1 + \gamma z_0) \sinh(\gamma \beta(0) \tau)$ is the radius of the wave front. In equation (5) and hereafter the explicit time dependence $e^{-i\omega t}$ is omitted. Note that in the homogeneous case, that is, when the velocity is constant as occurs on the half space, the Green's function is

$$G_{yy}^E(\omega, x) = \frac{i}{4\mu_0} H_0^{(1)}(\omega\tau), \tag{6}$$

therefore equation (6) can be observed as the limit case in which the homogeneous medium is considered, in which case the factor Λ tends to 1.

3. Results for Inhomogeneous Basins

3.1. Testing the Numerical Approach

In order to test this numerical approach, using the Green's functions for an inclusion with velocity varying linearly with depth under incident SH waves, we deal with two cases. First is the limit case where $\gamma = 0$ in equation (5), which corresponds with that of a homogeneous basin, and second is the case in which an inhomogeneous inclusion is considered. Our results are compared with those presented by previous authors using incident SH waves in a semi-circular basin. The response has been computed for the normalized frequency $\eta = 1$, being $\eta = a\omega/\pi \beta_E$ with a = radius of the sedimentary inclusion, and β_E = velocity of S waves in the half space. The physical properties of the basin in the homogeneous case studied here are (see Fig. 1): S-wave velocity $\beta_R = \beta_E/2$, and a mass density of $\rho_R = 2\rho_E/3$. The discretization has been extended horizontally up to a radius of $2a$. Figure 2 displays the surface displacements for incident SH plane waves with an angle of $30°$ with respect to the vertical. These amplitudes are plotted along the x-axis from $x = -2a$ to $x = +2a$. In the same plot the results of SÁNCHEZ-SESMA et al. (1993) are reproduced by symbols using the indirect boundary element method and the exact solution provided by TRIFUNAC (1971). The agreement among three results is excellent.

In the heterogeneous case the material properties considered are (see Fig. 1): $\beta_R(0) = \beta_E/3$, $\beta_R(a) = 2\beta_E/3$, and a mass density of $\rho_R = 2\rho_E/3$ ($n = 0$ of equation (4)). In Figure 3 are presented the surface amplitudes of the transfer function, for $\eta = 1$, caused by vertical incident SH plane waves calculated in this work, and by those symbols computed by BENITES and AKI (1994) using a boundary integral-Gaussian beam method, and those computed by the same authors with the finite-element method. The results of the three techniques used independently have some small discrepancies, although the overall agreement is quite good for all the cases.

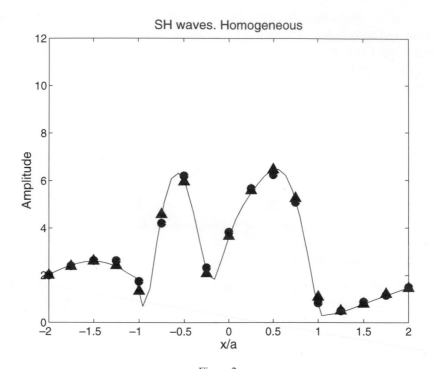

Figure 2

Amplitudes of horizontal antiplane displacement for incident *SH* plane waves with 30°, with respect to the vertical upon a homogeneous semi-circular sedimentary inclusion. Normalized frequency $\eta = 1$. Solid line corresponds to results obtained in the present study, while solid triangles and circles correspond to the solution of SÁNCHEZ-SESMA *et al.* (1993) and TRIFUNAC (1971), respectively.

3.2. *Examples*

In order to analyze the principal characteristics of the displacements produced inside sedimentary inhomogeneous basins and to underscore the differences with the homogeneous problem, we deal with four models of different physical properties with semi-circular geometry and with a radius equal to 1 km. Two of these inclusions are homogeneous and the other two are heterogeneous, in such a way that different velocity contrasts are considered. In all of the models the *S* velocity for the half space has been considered $\beta_E = 3$ km/s, and the density inside the sediments is $\rho_R = 2\rho_E/3$ in the four cases. Figure 4 presents the velocity profiles for the four cases. The velocities inside the basins are valid for the range $0 \leq z \leq 1$ km. Therefore these profiles correspond to the vertical models in $x = 0$. The models homogeneous A and homogeneous B (HO-A and HO-B) can be observed as sedimentary basins with high and low contrast-velocity, respectively. The heterogeneous A (HE-A) is a model in which the basin has the same *S*-wave velocity in the free surface as the model HO-A, and the same velocity of HO-B in $z = 1$ km. In this way the contrast velocity between the basin and the half space is 2/3 at this depth. On the other hand the

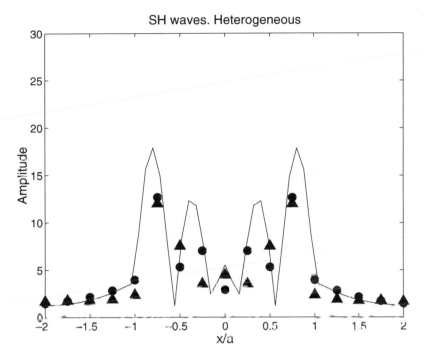

Figure 3

Amplitudes of horizontal antiplane displacement for vertical incident *SH* plane waves upon a heterogeneous semi-circular sedimentary inclusion. Normalized frequency $\eta = 1$. Solid line corresponds to results obtained in the present study, while solid triangles and circles correspond to the solution provided by BENITES and AKI (1994) using a boundary integral-Gaussian beam method and the finite-element method, respectively.

sedimentary inclusion of model heterogeneous B (HE-B) does not have contrast in $z = 1$ km with respect to the bedrock, whereas its velocity at the free surface is equal to that in the HE-A at this position. Note that in the case of HE-B the difference of velocity between $z = 0$ and $z = 1$ km is greater than the difference in HE-A between the same depths.

We computed the displacement amplitudes under vertical incident *SH* waves, in the frequency domain, for 128 frequencies from 0 to 2 Hz on 51 receivers located at the free surface from $x = -2$ km up to $x = 2$ km (the separation between each station is 80 m). The zero frequency corresponds with the case in which no inclusion exists, that is, the solution in the free surface of the half space. In Figure 5 these results are presented by means of *f-x* plots for the four models. In this way one can analyze the transfer function (relative to the amplitude of incident waves) along the *x* axis and also as a function of the frequency. This provides patterns of amplification which can be related with the position in the basin. The global resonances observed in these plots are characteristics of the behavior of these 2-D structures and cannot be predicted using 1-D models of the site. On the other hand, it is noticed that the

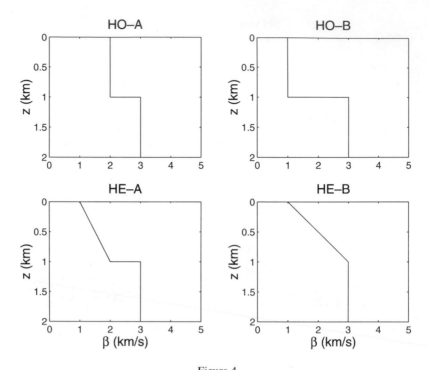

Figure 4
Velocity profiles in depth at $x = 0$ for four semi-circular sedimentary basins of radius $a = 1$ km. The models presented at the top correspond to homogeneous basins (HO-A and HO-B), whereas the models shown at the bottom correspond to heterogeneous models (HE-A and HE-B). Note that all models have the same geometry and the same S velocity for the half space $\beta_E = 3$ km/s.

response is completely different in each example, in the frequencies where resonances are produced and in the amplitude levels, even in the case, as presented here, where all models have the same geometry and the same properties of the half space. For example, if the results of models HO-A and HO-B are compared, the fundamental resonances for most positions at the surface of the basin are around 0.7 Hz and 0.35 Hz, respectively. The differences in the amplitude levels are important too, whereas the maximum in HO-A is around 5 times next to the center of the basin, the HO-B model has amplitudes that reach 16 times the incident one. In the higher resonant modes of both basins the divergence of the results is notable as well. These effects are well known and defined. As has been pointed out by previous authors (see for example, BARD and BOUCHON, 1980; SÁNCHEZ-SESMA *et al.*, 1993), when high-impedance contrast exists, as in HO-B, surface waves are very efficiently generated and reflected at the edges of the basin, creating a considerable part of the differences observed among HO-A and HO-B. But let us observe the *f-x* plot in Figure 5 of the heterogeneous HE-A model. Again, these results are different in the frequencies where the vibration modes and their amplitude levels appear, when these are compared with the results of the homogeneous models. However if we fix our

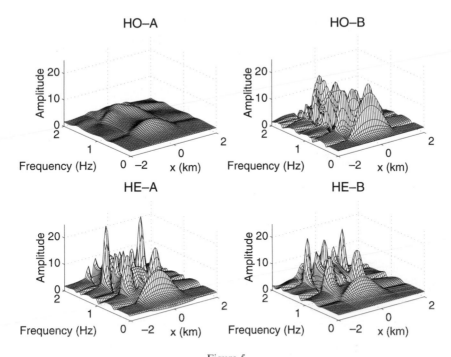

Figure 5
Amplitude of the displacements in the frequency-space domain (*f-x*) for the vertical incidence of *SH* waves of each of the four sedimentary inclusions (see Fig. 4) considered in the present study.

attention on the fundamental mode of the resonance of HE-A, we can observe that it is located around 0.5 Hz, which is approximately the average value of the fundamental modes produced in the models HO-A and HO-B. Moreover, the displacement's amplitude levels, about 10 in the centered positions of the basin, are also approximately the average value of the amplitudes of the homogeneous models. This behavior is consistent with the fact that HE-A is like an *average* model of HO-A and HO-B, in the sense that it has the same *S*-wave velocity in the free surface as the model HO-A, and the same velocity of HO-B at its maximum depth $z = 1$ km. Regardless, these results are only observed for the fundamental resonance. For the higher modes this comportment is not present at all, and the heterogeneous basin has its own characteristics. In fact, whereas the homogeneous models have approximately the same amplitude level at all the positions over the surface of the basin in each of their resonant modes, the HE-A model has substantial differences, depending on the surface location. This is well observed, for example, at the frequency 1.5 Hz, where the amplitude reaches the value of 24 on the corners of the basin at the positions $x \approx \pm 1$ km, whereas in other positions more centered on the basin the displacements arrive in some cases to 12. This phenomena was also observed by BARD and GARIEL (1986), who studied the case of sedimentary basins in which the shear-wave velocity of the sediments increased linearly with depth. This effect, which characterizes the behavior of the HE-A, is present too in the HE-B basin. However, in the HE-B model this difference between the amplitudes on the edges of the basin and the locations near its center is not as important as in the case of HE-A for 1.5 Hz. This is due to the fact that its velocity gradient is greater and possibly this difference on the amplitudes in different locations may be greater at higher frequencies.

In order to analyze the effects of these inhomogeneous basins we simulated the displacements in the time domain inside and outside the sedimentary inclusions by means of synthetic seismograms. Using the transfer functions in the frequency domain and the Fast Fourier Transform algorithm we computed the simulated motion produced in each station with a Ricker wavelet. These calculations were performed considering a characteristic period of $t_p = 1.\hat{6}$ s, and a time-lag of $t_s = 5$ s. For the sake of comparison of the different results in each of the four sedimentary basins presented in this work, the displacements in time are displayed in Figure 6. The principal variations which appear among the results of the different models belong to the motion produced at the stations located inside the basins. The propagation of *SH* energy along the half space refracted from inside the basins is well observed, and has similar amplitude levels in models HO-B, HE-A and HE-B. In the

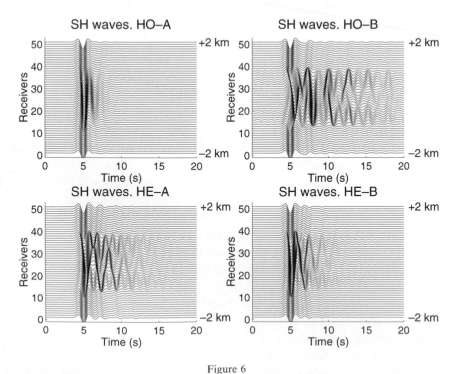

Figure 6

Synthetic seismograms for the horizontal antiplane displacement produced at the free surface of the four models (see Fig. 4) considered in the present study, under vertical incidence of *SH* waves. The incident time signal is a Ricker pulse with a characteristic period of $t_p = 1.\hat{6}$ s.

inclusion of HO-A these kinds of waves are practically negligible. This is due to the low velocity contrast of the model which not produces in efficient Love waves inside the basin (BARD and BOUCHON, 1980). On the contrary, at the models in which higher impedance contrast exists, these surface waves are more efficiently generated and their reflections at the edges of the sedimentary inclusion contribute to produce the longer duration of the signals registered on the receivers of the basin. The pattern of displacements on the surface of the sediments is very complex and is composed principally of the interferences between the different travelling waves that rebound up and down, and the Love waves that propagate laterally. In the homogeneous basin HO-A, due to the low contrast at the interface, a substantial part of the energy is reflected on the free surface of the basin and is lost outside to the interior of the half space. Oppositely, in the model HO-B the energy that goes to the half space is less and a sizable part remains inside producing multiple reflections in all the basin half space interfaces as also can be observed in Figure 7. In this plot the snapshots of the

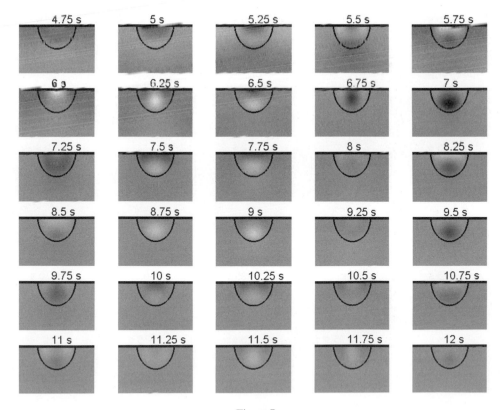

Figure 7
Snapshots at different times showing the horizontal antiplane displacement produced by the vertical incidence of a Ricker pulse (*SH* wave) with a characteristic period of $t_p = 1.\hat{6}$ s in the homogeneous model HO-B.

displacement for 30 different times are shown, with an increment of time for each correlative snapshot of 0.25 s. However, in the heterogeneous models the impedance contrast at their interfaces, and therefore the refraction coefficients are not constant and depends on its depth. In this way, for zones next to the free surface the contrast is bigger that in those locations with more depth of the inclusion, providing this situation has more lateral reflections than vertical. This is well observed in Figure 8, where snapshots of the displacement for the same times as in Figure 7 are depicted for the model HE-A. In this model the lateral reflections of surface waves greatly influence the total displacements at the free surface in comparison with the homogeneous model in which the displacements have a shared influence among both vertical and lateral reflections. This fact influences the time duration of the signals inside the basin HE-A, which is less than the duration produced in HO-B, where the refraction coefficients are the same along the entire interface. Conversely, a notable difference over the duration of the displacements inside heterogeneous basins with

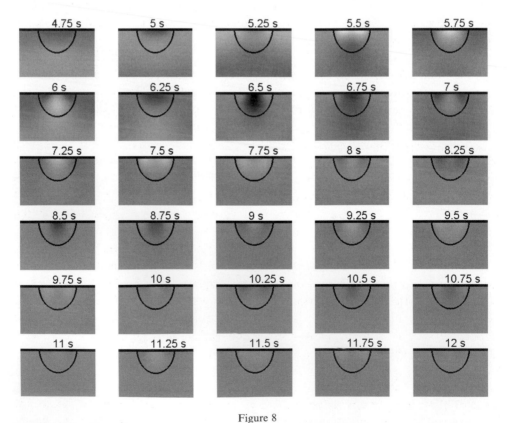

Figure 8
Snapshots at different times showing the horizontal antiplane displacement produced by the vertical incidence of a Ricker pulse (*SH* wave) with a characteristic period of $t_p = 1.\hat{6}$ s in the heterogeneous model HE-A.

the same velocity along the free surface is also observed when the inclusions HE-A and HE-B are compared independently in figure 6. This is because for a given depth $z \neq 0$, the velocity contrast between basin half space in HE-B is smaller than in HE-A, in such a way that when we consider $z = 1$ km there is no contrast in the former model. This causes that vertical reflections to be more effective in the model HE-A than in the HE-B.

4. Conclusions

We have computed the seismic response of different homogeneous and heterogeneous models of sedimentary basins using the indirect boundary element method (IBEM) for vertical incidence of SH plane waves only, however it is possible to deal with other irregular geometries and other incident angles as well. Moreover, the case of incident P and SV waves can be easily treated using the appropriate Green's functions. We have analyzed the results in both frequency and time domains. These show very interesting aspects from the view point of both the physical phenomena that occur on these situations, and their implications on earthquake engineering and microzonation studies.

The fundamental mode of resonance of the heterogeneous model HE-A is approximately the average value of the fundamental modes produced by the homogeneous inclusions HO-A and HO-B. This is consistent with the fact that basin HE-A has the same wave velocity in the free surface as the model HE-A, and the same velocity in the maximum depth $z = 1$ km as HE-A. For the higher modes of vibration this average comportment is not present, and the heterogeneous basin has its own characteristics of vibration. Nonetheless, these effects must be known by those microzonation studies in which only the fundamental frequency of the site is required, and suggest that not only necessary is the knowledge of the value of S-wave velocity at the free surface, but also the depth distribution of velocities extending to the basement. On the other hand, it has been observed that in the heterogeneous models lateral reflections of surface waves have enormous influence on the total displacements at the free surface in comparison with the homogeneous models in which the displacements have a shared influence among both, vertical and lateral reflections. This behavior produces large differential motions, depending on the location in the free surface. This has been shown by means of the amplitude of displacements in frequency-space diagrams and by the analysis in time domain of the soil motion. This effect must be considered in earthquake engineering studies which deal with heterogeneous sedimentary basins as analyzed here, because on the contrary a significant underestimation of surface amplification, particularly at the edges of the basin, can be obtained.

Acknowledgments

We thank M. Bouchon and R. Madariaga for their helpful suggestions and reviews of the manuscript. This work was partially supported by CICYT, Spain, under Grants REN2002-04198-C02-02/RIES and REN2001-2418-C04-02/RIES, by the European Community with FEDER, by the research team of *Geofísica Aplicada* (RNM194) of Junta de Andalucía, Spain, and by DGAPA-UNAM, Mexico, under Project IN104998.

REFERENCES

AKI, K. *Local site effects on strong ground motion*. In *Earthquake Engineering and Soil Dynamics II-Recent Advances in Ground Motion Evaluation* (ed. J. Lawrence von Thun) (American Society of Civil Engineering, 1988) pp. 103–155.

AKI, K., and LARNER, K. L. (1970), *Surface Motion of a Layered Medium Having an Irregular Interface due to Incident Plane SH Waves*, J. Geophys. Res. *75*, 933–954.

ALTERMAN, Z., and KARAL, F. C. (1968), *Propagation of Elastic Waves in Layered Media by Finite Difference Methods*, Bull. Seismol. Soc. Am. *58*, 367–398.

BARD, P.-Y., and BOUCHON, M. (1980), *The Seismic Response of Sediment-filled Valleys. Part 1. The Case of Incident SH Waves*, Bull. Seismol. Soc. Am. *70*, 4, 1263–1286.

BARD, P.-Y., and GARIEL, J.-C. (1986), *The Seismic Response of Two-dimensional Sedimentary Deposits with Large Vertical Velocity Gradients*, Bull. Seismol. Soc. Am. *76*, 343–366.

BENITES, R., and AKI, K. (1994), *Ground Motion at Mountains and Sedimentary Basins with Vertical Seismic Velocity Gradient*, Geophys. J. Int. *116*, 95–118.

BREBBIA, C. A., *The boundary Element Method for Engineers* (Pentech Press, London 1978).

HERRERA, I. (1980), *Variational Principles for Problems with Linear Constraints, Prescribed Jumps and Continuation Type Restrictions*, J. Inst. Maths. and Applics. *25*, 67–96.

HORIKE, M., UEBAYASHI, H., and TAKEUCHI, Y. (1990), *Sesimic Response in Three-dimensional Sedimentary Basin due to Plane S-wave Incidence*, J. Phys. Earth *38*, 261–284.

LUCO, J. E., WONG, H. L., and DE BARROS, F. C. P. (1990), *Three-dimensional Response of a Cylindrical Canyon in a Layered Half Space*, Earthq. Eng. Struct. Dyn. *419*, 799–817.

LUZÓN, F., AOI, S., FÄH, D., and SÁNCHEZ-SESMA, F. J. (1995), *Simulation of the Seismic Response of a 2-D Sedimentary Basin: A Comparison between the Indirect Boundary Element Method and a Hybrid Technique*, Bull. Seismol. Soc. Am. *85*, 1501–1506.

LUZÓN, F., SÁNCHEZ-SESMA, F. J., RODRÍGUEZ-ZÚÑIGA, J. L., POSADAS, A. M., GARCÍA, J. M., MARTÍN, J., ROMACHO, M. D., and NAVARRO, M. (1997), *Diffraction of P, S and Rayleigh Waves by Three-dimensional topographies*, Geophys. J. Int. *129*, 571–578.

LUZÓN, F., SÁNCHEZ-SESMA, F. J., GIL, A., POSADAS, A., and NAVARRO, M. (1999). *Seismic Response of 3-D Topographical Irregularities under Incoming Elastic Waves from Point Sources*, Phys. Chem. Earth (A), *24*(3) 231–234.

LYSMER, J., and DRAKE, L. A., *A finite element method for seismology*, In *Methods in Computacional Physics, Vol. 11 Seismology* (ed. Bolt B. A.) (Academic Press, New York 1972).

OLSEN, K. B. (2000). *Site Amplification in the Los Angeles Basin from Three-dimensional Modeling of Ground Motion*, Bull. Seismol. Soc. Am. *90*(6B), S77–S94.

PAPAGEORGIOU, A. S., and KIM, J. (1991), *Study of the Propagation and Amplification of Seismic Waves in Caracas Valley with Reference to the 29 July 1967 Earthquake: SH Waves*, Bull. Seismol. Soc. Am. *481*, 2214–2233.

RIAL, J. A. (1984), *Caustics and Focusing Produced by Sedimentary Basins. Application of Catastrophe Theory to Earthquake Seismology*, Geophys. J. R. Astr. Soc. *79*, 923–938.

SÁNCHEZ-SESMA, F. J., *Strong ground motion and site effects*. In *Computer Analysis of Earthquake Resistant Structures* (eds. D. E. Beskos and S. A. Anagnostopoulos) (Comp. Mech. Publications, Southampton 1996) pp. 200–229.

SÁNCHEZ-SESMA, F. J., and CAMPILLO, M. (1991), *Diffraction of P, SV and Rayleigh Waves by Topographic Features: A Boundary Integral Formulation*, Bull. Seismol. Soc. Am. *481*, 2234–2253.

SÁNCHEZ-SESMA, F. J., and LUZÓN, F. (1995), *Seismic Response of Three-dimensional Alluvial Valleys for Incident P, S and Rayleigh Waves*, Bull. Seismol. Soc. Am. *85*, 269–284.

SÁNCHEZ-SESMA, F. J., MADARIAGA, R., and IRIKURA, K. (2001), *An Approximate Elastic 2-D Green's Function for a Constant-Gradient Medium*, Geophys. J. Int. *146*, 237–248.

SÁNCHEZ-SESMA, F. J., RAMOS-MARTÍNEZ, J., and CAMPILLO, M. (1993), *An Indirect Boundary Element Method Applied to Simulate the Seismic Response of Alluvial Valleys for Incident P, S and Rayleigh Waves*, Earthq. Eng. Struct. Dyn. *422*, 279–295.

TRIFUNAC, M. D. (1971), *Surface Motion of a Semi-cylindrical Alluvial Valley for Incident Plane SH Waves*, Bull. Seismol. Soc. Am. *61*, 1755–1770.

VRETTOS, Ch. (1990), *In-plane Vibrations of Soil Deposits with Variable Shear Modulus: II. Line Load*, Intl. J. Numer. Anal. Meth. Geomech. *14*, 649–662.

ZIEGLER, F., and PAO, Y.-H. (1984), *Transient Elastic Waves in a Wedge-shaped Layer*, Acta Mechanica *52*, 133–163.

(Received February 15, 2002, revised/accepted September 24, 2002)

 To access this journal online:
http://www.birkhauser.ch

Pure appl. geophys. 161 (2004) 1549–1559
0033–4553/04/071549–11
DOI 10.1007/s00024-004-2520-7

© Birkhäuser Verlag, Basel, 2004

❙ Pure and Applied Geophysics

The Use of Ambient Seismic Noise Measurements for the Estimation of Surface Soil Effects: The Motril City Case (Southern Spain)

Z. Al Yuncha[1,2], F. Luzón[1,2], A. Posadas[1,2], J. Martín[1,2],
G. Alguacil[2], J. Almendros[2], and S. Sánchez[3]

Abstract—This paper presents a study of the ambient seismic noise features in Motril City. The predominant resonant period at each site was determined using the Horizontal to Vertical Noise Ratio (*HVNR*) of the microtremors records; it was applied to 91 points which cover the city and we present the distribution of the periods over a map which shows important characteristics requiring consideration in a study focussing on the mitigation of the seismic risk. Using these results, two zones can be distinguished: north of the GG' profile with predominant periods smaller than 0.25 sec which corresponds with a hard soil, and south of GG' with periods ranging from 0.25 to 0.6 indicating a medium-soft soil. Microtremors are a preliminary useful tool that compensates for the lack of seismic recorded data, and that can complement other valuable information (borehole logs, geology, etc.) to evaluate the seismic risk on places as Motril City where some of this information is unavailable. Through the predominant period the *HVNR* indicates the kind of soil we are dealing with.

Key words: Microtremors, spectral ratio, seismic microzonation, ambient seismic noise.

1. Introduction

Local site response can lead to large variations of seismic ground motion which would cause severe structural damage even when moderate earthquakes occur. During recent decades numerical and theoretical analyses have been done in order to understand the physics and nature of the problem. It has been shown in particular, that resonant phenomena often appear in unconsolidated deposits (SÁNCHEZ-SESMA, 1996; LUZÓN *et al.*, 2002). For these, the amplitude and duration over certain period bands may be several times larger than at sites located on rock. It is of interest that the near-surface impedance contrast, between the sediment deposits and hard rock, can significantly affect the amplitude content and duration of the earthquake ground motion. One example is the January 17th, 1994, Northridge California earthquake (TENG and AKI, 1996), registering a magnitude $M_w = 6.7$, that shook heavily the

---

[1] Departamento de Física Aplicada, Universidad de Almería, La Cañada de S. Urbano s/n, 04120 Almería, Spain.
[2] Instituto Andaluz de Geofísica y Prevención de Desastres Sísmicos. Granada, Spain.
[3] Departamento de Edafología y Química Agrícola, Universidad de Almería, Spain.

communities throughout the San Fernando Valley and Simi Valley. The epicenter was located 20 miles from Los Angeles and the focal depth at 19 km. This earthquake was the costlier disaster in United States history, with estimated losses of 900 billion dollars (damage and replacement value), 11 major freeways were damaged as far as 32 km from the epicenter; closure of 11 major roads to downtown Los Angeles; 9 bridges collapsed; 2500 dwellings uninhabitable, 7000 buildings unsafe to occupy, the number of deaths 57 and total injured 9000. One year later, on the same day of January, the Hyogo-ken Nambu earthquake with a magnitude $M_w = 6.8$ struck the City of Kobe, Japan. The epicenter was located north of Awaji Island and the focal depth was about 16 km. The earthquake caused over 15100 deaths and were injured 30,000, the estimated total losses were 10 times higher than the losses from the Northridge earthquake (SOMERVILLE, 1995), 103,520 buildings were destroyed and 300,000 people left homeless.

Southern Spain, located in the Eurasian and African tectonic interaction zone, is the most seismically dangerous region in Spain (Fig. 1, top). It is considered as a zone of high-moderate seismic activity, and rated with an *MSK VIII* intensity for a return period of 100 years by the Spanish Seismic Code (NCSE-94, 1995).

The City of Motril is located in the southern part of Granada province and has a population of about 60,000 people. From a geological point of view Motril is placed within the frame of the Betic Mountain range. This range constitutes the western end of the Alpine chains (FALLOT, 1948). There are two domains to be recognized; the internal and the external one. The section under study is included in the internal zone which has been affected by alpine metamorphism (Fig. 1, bottom) (SANZ DE GALDEANO, 1992, 1993). They are formed by paleozoic and triasic materials, and constitute three complexes which are tectonically structured as superimposed thrust layers. These complexes are, from floor to ceiling, the Nevado-Filábride, the Alpujárride and the Maláguide; the first two present paleozoic and triasic sequences strongly affected by the metamorphic alpine stages. The city is located between the profiles AA' and KK'; one part of the city is located on the northern sector of the GG' profile, with paleozoic materials within the Nevado-Filábride complex; they are formed by micaschists, quartzites and marbles. The southern sector is formed by quaternary materials constituted by detritic formations of conglomerates and alluvial sands (Fig. 2) (IGME, 1974).

The microzonation studies are very important to accomplish seismic disaster mitigation for a city and, as is well known, an estimation of the site response is crucial in such studies. With the goal of performing this kind of study a great variety of empirical techniques have been used. KANAI *et al.* (1954) considered the possibility of estimating the dominant period and the amplification level of soft sediments by obtaining the Fourier spectrum of the microtremors horizontal components. NAKAMURA (1989) proposed a technique involving the record of the three components of motion in a site and their Horizontal to Vertical Noise Ratio (*HVNR*). The method has been used to obtain the predominant period of a site and in many cases it has been observed as a good

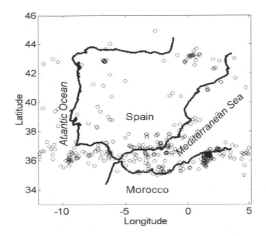

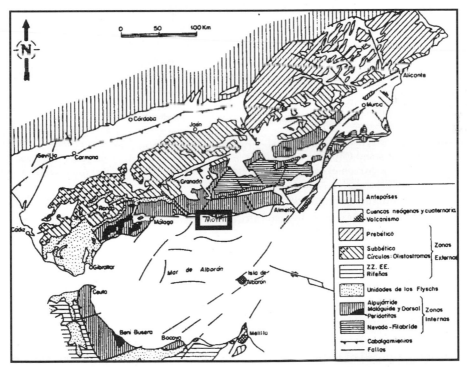

Figure 1

Top: Map showing the epicenters (circles) of the region for earthquakes with magnitudes ≥ 4 during the period 1970–2001. Bottom: Geological map showing the Betic Cordillera, Rif and Motril location (adapted from SANZ DE GALDEANO, 1992).

correlation with its geology. During recent years several microzonation studies have been carried out on the main cities of this area, as for example Granada City (VIDAL et al., 1998), Almería (NAVARRO et al., 2001). Here we obtain a map of the distribution

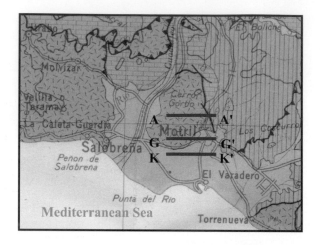

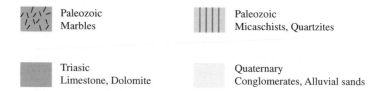

Figure 2

Geological map of Motril City surroundings. We can see the first profile AA' of our grid, the last one KK', and the one like a division of paleozoic and quaternary materials, GG' (adapted from IGME, 1974).

of the predominant resonant periods of microtremors in the City of Motril and we check our results with the geological information available.

2. *The HVNR Technique and its Application in Motril City*

NAKAMURA (1989) considered that the spectral amplification at a site could be obtained by evaluating the *HVNR* recorded at the site. The technique (LERMO and CHÁVEZ-GARCÍA, 1994; DRAVINSKI *et al.*, 1996) implies that microtremors are primarily composed of Rayleigh waves propagating in a single layer over a half-space; a further assumption is that microtremors motion is due to very local sources. Nakamura claims that the motion at the interface between the surface layer and the half-space is not affected by local sources. With the additional hypothesis that the horizontal and vertical motions at this interface are equal, he stated that site effects $S(f)$ can be estimated by the spectral ratio of the horizontal and vertical motions; that is,

$$S(f) = \frac{H(f)}{V(f)}. \tag{1}$$

KONNO and OHMACHI (1998) developed Nakamura's technique, extending the problem to a multilayered system. These authors reinforced the technique, which up to that moment did not have a very clear theoretical explanation. Several studies have shown that the *HVNR* can reveal with some confidence the fundamental resonant frequency at a site (e.g., LERMO and CHÁVEZ-GARCÍA, 1994; FIELD and JACOB, 1995; LACHET *et al.*, 1996; SEEKINS *et al.*, 1996; COUTEL and MORA, 1998). Some works have dealt with the applicability of the technique by simulation of seismic waves in specific structures such as sedimentary basins (AL YUNCHA and LUZÓN, 2000; DRAVINSKI *et al.*, 1996). LUZÓN *et al.* (2001), by a numerical study in a 2-D flat sedimentary inclusion, concluded that *HVNR* technique can also predict, reasonably well, the fundamental local frequency when there is a high-impedance contrast between the sedimentary basin and the bedrock. This technique has become extensively utilized worldwide, basically due to two aspects: a) it is not necessary the occurrence of an earthquake, and b) the analysis of the data is relatively easy. Many works have determined the predominant periods of different regions with Nakamura's *HVNR*, as IBS-VON SEHT and WOHLENBERG (1999) in the Lower Rhine Embayment (Germany), DELGADO *et al.* (2000) in the Segura River valley (Spain), or NAVARRO *et al.* (2001) in Almería City (Spain).

The ambient seismic noise measurements in the Motril urban area were performed during July 1998. The total number of points considered for the records was 91. The distribution of the points was selected taking into account a rectangular grid that covered the entire city area. The grid size is approximately 200 m × 200 m. We used three instrumental sets; each one composed of a three-component high-sensitive seismometer (with natural period of one second), a digital recorder of 24 bits technology to ensure a good quality of data in a laptop computer, and a *GPS* module to localize the spatial position. To see the signal-noise ratio, three minutes of instrumental noise was recorded: a) blocking the seismometers and b) removing them. Figure 3 shows an example of instrumental and soil motion noise spectral amplitudes recorded at a site. As can be observed, the amplitude of the soil motion is large enough to consider the instrumental noise as negligible.

It is known that microtremor spectra can be affected by some disturbances such as the weather and very nearby sources (traffic and other urban facilities conditions). A stationary test was carried out to check the amplitude and period stability in the area. Three-component records (for 3 minutes) were obtained every hour, during an interval of 24 hours in four selected sites of the city. Figure 4 shows an example at one of the sites; it can be seen that the periods and amplitudes of the microtremors (0.1 s < T < 1.0 s) are not very stable. Therefore we tried to avoid the possible spurious local disturbances by recording during the night, when nearby sources from human activity are minima.

The horizontal and vertical components of ambient noise were recorded at each site during three minutes (with a sample frequency of 100 Hz), see Figure 5 (top), and several 20 sec windows of the record were selected for the analysis. The signal

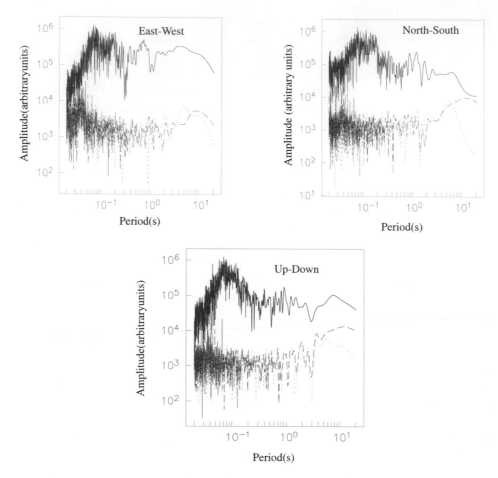

Figure 3

Fourier spectra of ambient noise (solid line) and instrumental noise when the sensors are blocked (dotted line) and when they are removed (discontinuous line).

was Fourier transformed and smoothed using a 0.7 Hz Hanning windows. The horizontal to vertical spectral ratio was obtained using this equation:

$$S(f) = \frac{\sqrt{H_{NS}^2(f) + H_{EW}^2(f)}}{V(f)}, \tag{2}$$

where $H_{NS}(f)$ and $H_{EW}(f)$ are the North-South and East-West horizontal components of the displacement, respectively. The method used here consists of the application of the last equation to successive data windows along the traces. This procedure yields several *HVNR* functions that can be represented as a two-dimensional contour plot versus frequency and time, as proposed by ALMENDROS *et al.* (2002). This plot, which we name ratiogram, represents the evolution of the

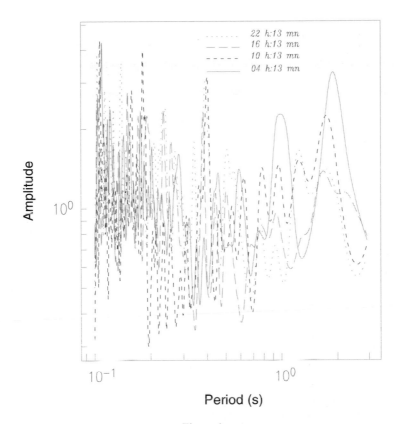

Figure 4

Horizontal to vertical noise ratio, at a selected point, from records obtained at different hours of the same day.

HVNR with time and frequency. Figure 5, bottom left side presents an example of the application of the *HVNR* using the above methodology.

3. Results and Conclusions

Various microzonation studies deal with the estimation of the amplification and the predominant period for a specific site in future earthquakes. Nakamura's technique, as used here, is a good tool for identifying the predominant period. One way to characterize the area under study is by means of a map depicting the characteristic magnitudes which evolved in the problem. Figure 6 presents the distribution of the predominant resonant periods of microtremors for Motril City.

The surface geology in Motril City north of the GG' profile (see Fig. 2) can be considered as hard soil and the predominant periods found are smaller than 0.25 sec. No evidence of soft soil has been observed. This could be due to the geological

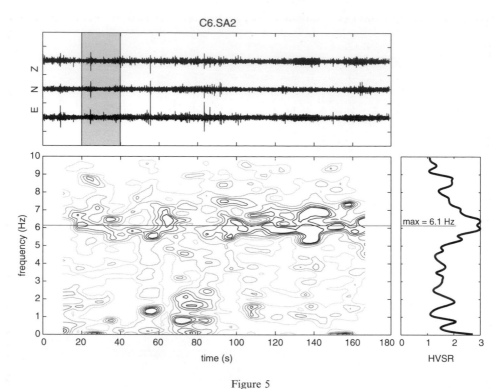

Figure 5

Top: Example of a three-component record (180 sec) of microtremors. *Bottom left*: Ratiogram representing the *HVNR* as a function of frequency and time (for the 180 sec record). *Bottom right*: *HVNR* amplitude versus frequency (for the window from 20 to 40 sec, of the 180 sec record).

features of the area in which the paleozoic materials are dominant. However, towards the west of the city (see Fig. 6) some highly predominant periods appear; there is a dry river bed which could have a high potential of sediments. The southern part of the city towards the sea, south of the GG' profile, is characterized by the presence of soft sediments (quaternary conglomerates and alluvial sands, see Fig. 2). The predominant periods range from 0.25 to 0.60 sec advocating for a medium soft soil. Some sites in this part of the city have small periods; this could be due to very local special conditions, as for example the existence of an outcrop.

In general we observe a good correlation between the surface geological conditions and the predominant periods obtained. This good agreement is similar as mentioned in previous papers (see. e.g., NAVARRO *et al.*, 2001). As a last conclusion we can asses that the results we present in this paper, integrated with the geological information, borehole logs, site effects observed for strong earthquakes, realistic numerical modeling of signals, etc., can represent a useful tool for accomplishing a C level microzonation of a city.

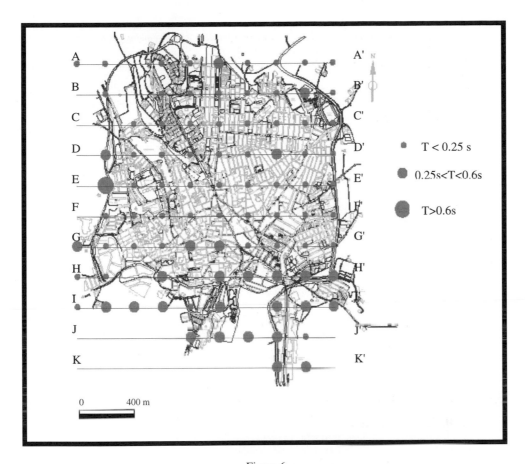

Figure 6
Distribution of predominant periods of the microtremors. The figure shows the profiles AA', GG' and KK', which also appear in Figure 2.

Acknowledgements

We appreciate the suggestions offered by two anonymous reviewers, which aided in improving this paper. This work was partially supported by Spanish Grants, REN2001-2418-C04-02/RIES, REN2002-04198-C02-02/RIES, by the European Community with FEDER, by Plan Propio de la Universidad de Almería and by the research team RNM-194 of Junta de Andalucía, Spain.

REFERENCES

AL YUNCHA, Z. and LUZÓN, F. (2000), *On the Horizontal-to-vertical Spectral Ratio in Sedimentary Basins*, Bull. Seismol. Soc. Am. *90*, 1101–1106.

ALMENDROS, J., LUZÓN, F., and POSADAS, A. (2002), *Microtremors Analysis at Teide Volcano (Canary Island, Spain): Assessment of Natural Frequencies of Vibration Using Time-dependent Horizontal-to-vertical Spectral Ratios*, Pure Appl. Geophys., accepted.

COUTEL, F. and MORA, P. (1998), *Simulation-based Comparison of Four Site-response Estimation Techniques*, Bull. Seismol. Soc. Am. *88*, 30–42.

DELGADO, J., LÓPEZ-CASADO, C., ESTÉVEZ, A. C., GINER, J., CUENCA, A., and MOLINA, S. (2000), *Mapping Soft Soils in the Segura River Valley (SE Spain): A Case Study of Microtremors as an Exploration Tools*, J. Appl. Geophy. *45*, 19–32.

DRAVINSKI, M., DING, G., and WEN, K.-L. (1996), *Analysis of Spectral Ratios for Estimating Ground Motion in Deep Basin*, Bull. Seismol. Soc. Am. *86*, 646–654.

FALLOT, P. (1948), *Les Cordilléres Bétiques*, Est. Geol. *8*, 83–172.

FIELD, E. H. and JACOB, K. H. (1995), *A Comparison and Test of Various Site-response Estimation Techniques, Including Three that are not Reference-site Dependent*, Bull. Seismol. Soc. Am. *85*, 1127–1143.

IBS-VON SEHT, M. and WOHLENBERG, J. (1999), *Microtremor Measurements Used to Map Thickness of Soft Sediments*, Bull. Seismol. Soc. Am. *89*, 250–259.

IGME (1974), *Mapa Geológico de Motril*, 1055 - (19–44). Esc. 1:50000. Serv. Publi. Ministerio de Industria, p.15.

KANAI, K., TANAKA, T., and OSADA, K. (1954), *Measurement of Microtremor, I*, Bull. Earthq. Res. Institute *32*, 199–209.

KONNO, K. and OHMACHI, T. (1998), *Ground-motion Characteristics Estimated from Spectral Ratio between Horizontal and Vertical Components of Microtremor*, Bull. Seismol. Soc. Am. *88*, 228–241.

LACHET, C., HAZFELD, D., BARD, P.-Y., THEODULIDIS, N., PAPAIOANNOU, C., and SAVVAIDIS, A. (1996), *Site Effects and Microzonation in the City of Thessaloniki (Greece) Comparison of Different Approaches*, Bull. Seismol. Soc. Am. *86*, 1692–1703.

LERMO, J. and CHÁVEZ-GARCÍA, F. J. (1994), *Are Microtremors Useful in Site Response Evaluation?* Bull. Seismol. Soc. Am. *84*, 1350–1364.

LUZÓN, F., AL YUNCHA, Z., SÁNCHEZ-SESMA, F. J., and ORTIZ-ALEMÁN, C. (2001), *A Numerical Experiment on the Horizontal to Vertical Spectral Ratio in Flat Sedimentary Basins*, Pure appl. Geophys. *158*, 2451–2461.

LUZÓN, F., PALENCIA, V. J., MORALES, J., and SÁNCHEZ-SESMA, F. J. (2002), *Evaluation of Site Effects in Sedimentary Basins*, Física de la Tierra *14*, 183–214.

NAKAMURA, Y. (1989), *A Method for Dynamic Characteristics Estimation of Subsurface Using Microtremor on the Ground Surface*, Report Railway Tech. Research Institute *30*, 1, 25–33.

NAVARRO, M., ENOMOTO, T., SÁNCHEZ, F. J., MATSUDA. I., IWATATE, T., POSADAS A., LUZÓN, F., and SEO, K. (2001), *Surface Soil Effects Study Using Short-period Microtremor Observations in Almería City, Southern Spain*, Pure Appl. Geophys. *158*, 2481–2497.

NCSE-94. (1995), *Normativa de Construcción Sismorresistente Española NCSE-94*. Real Decreto 2543/94, B.O.E. 33, 8 de Febrero 1995.

SÁNCHEZ-SESMA, F. J. (1996), *Strong ground motion and site effects. In Computer Analysis and Earthquake-resistant Structures* (D. E. Beskos and S. A. Anagnostopoulos, eds.), Computational Mechanics Publications, Southampton, pp. 200–229.

SANZ DE GALDEANO, C. (1992), *Some Geological Problems of the Betic Cordillera and Rif*, Física de la Tierra. 4, Editorial Complutense, Madrid. pp. 11–40.

SANZ DE GALDEANO, C. (1993), *Principal Geological Characteristics of the Betic Cordillera*, Some Spanish Karstic Aquifers. Uni. de Granada. pp. 1–17.

SEEKINS, L. C., WENNERBERG, L., MARGHERITI, L., and LIU, H. P. (1996), *Site Amplification at Five Locations in San Francisco, a Comparison of S Waves, Codas, and Microtremors*, Bull. Seismol. Soc. Am. *86*, 627–635.

SOMERVILLE, P. (1995), *Geotechnical Reconnaissance of the Effects of the January 17, 1995, Hyogo-ken Nanbu Earthquake, Japan*. Technical Report UCB/EERC-95/01, Earthquake Engineering Research Center, University of California, Berkely, California.

TENG, T.-L. and AKI, K. (1996), *Preface to the 1994 Northridge Earthquake Special Issue*, Bull. Seismol. Soc. Am. *86*, S1–S2.

VIDAL, F., ALGUACIL, G., MORALES, J., FERICHE, M., MOURABIT, T., CHEDDADI, A., SEO, K., YAMANAKA, H., KURITA, K., SUZUKI, K., KOBAYASHI, H., and SAMANO, T. (1998), *Study on Seismic Risk Assessement in Granada City, Spain*. As an activity in the Spanish-Japanese joint research on seismic microzonation in basin areas. The effects of Surface Geology on Seismic Motion. Recent Progress and New Horizon on ESG Study. Yokohama, 1–3 December, 1998.

(Received April 2, 2002, revised February 28, 2003, accepted April 3, 2003)

 To access this journal online:
http://www.birkhauser.ch

Pure appl. geophys. 161 (2004) 1561–1578
0033–4553/04/071561–18
DOI 10.1007/s00024-004-2521-6

© Birkhäuser Verlag, Basel, 2004

| Pure and Applied Geophysics

Results of Analysis of the Data of Microseismic Survey at Lanzarote Island, Canary, Spain

A. V. Gorbatikov[1], A. V. Kalinina[1], V. A. Volkov[1],
J. Arnoso[2], R. Vieira[2], and E. Velez[2]

Abstract—From the data of a microseismic survey of the Lanzarote Island territory (Canary Archipelago) we obtained a microseisms amplitude distribution in the frequency range 0.3–12.5 Hz. We found a distinguished anomaly such as an amplitude depression, whose size and magnitude depend on the frequency. After studying the statistical and polarization properties of microseism signals we proposed a model explaining this depression, based on the presence of a rigid intrusive body in the center of the island. Results of our survey coincided well with independent detailed gravity survey results whose interpretation also implies the presence of intrusion. We estimated shear-wave velocity for the rocks of intrusion and for surrounding rocks using microseismic data.

Key words: Microseisms, survey, amplitude anomaly, intrusion, Lanzarote, Canary.

1. Introduction

Using microseisms – the Earth's surface weak background oscillations – to gain information regarding peculiarities in the structure and mechanical parameters of subsurface zones is very attractive. First, microseism signals are always present at any point on the Earth's surface. Second, the measurements themselves both in the methodical and expense aspect are much easier, as a rule, compared to other seismological methods. That is why many authors developed the methods utilizing microseisms as a sounding signal. More or less established terminology has appeared in this area. The distribution between long-period (T > 1 s) and short-period (T < 1 s) corresponds to the traditional distinction between "microseisms" with natural origin and "microtremors" with artificial origin (BARD, 1999). Microseisms are widely used for site response effect studies. It is recognized as an important factor to be considered in seismic microzonation.

[1] Joint Institute of Physics of the Earth, Russian Academy of Sciences, B. Gruzinskaya, 10, Moscow 123995, Russia. E-mails: avgor70@mail.ru; kalinina@uipe-ras.scgis.ru; volkov@uipe-ras.scgis.ru

[2] Instituto de Astronomía y Geodesia (CSIC-UCM), Universidad Complutense de Madrid, Facultad de Matemticas, 28040 Madrid, Spain. E-mails: vieira@iagmat1.mat.ucm.es; arnoso@iagmat1.mat.ucm.es; emilio_velez@mat.ucm.es

Microseisms research began as far back as the early works of OMORI (1908) and was developed both by the study of the microseism sources nature and their applications for local site response problems and engineering purposes. Concerning the microseisms origin, we know already since GUTENBERG (1927) that they appear as a result of atmospheric perturbations transmission for the waters of oceans and propagate over continental surface as Love and Rayleigh type waves. A high efficiency of their propagation is explained by their surface nature.

Many authors studied the microseisms nature. For example, AKI (1957); LACOSS *et al.* (1969), OMOTE and NAKAJIMA (1973), ASTEN, (1978), IRIKURA and KAWANAKA, (1980), SATO *et al.* (1991), HOUGH *et al.*, (1992), GORBATIKOV and BARABANOV (1994) studied the structure of short-period microseisms, their connection with geological structure and its condition, using seismic arrays and separate instruments. SAKAJIRI (1982); HORIKE (1985) on the other hand investigated the nature of long-period microseisms and their connection with the subsurface structure.

A solid majority of papers devoted to microseisms utilization, could be classified into three groups by their methodical approaches. This classification was suggested by LERMO and CHAVEZ-GARCIA (1994). The first group implies direct interpretation of Fourier spectra (e.g., KANAI and TANAKA, 1954; KATZ and BELLON, 1978); the second one – calculation and study of spectral relations between reference and studied site (e.g., OHTA *et al.*, 1978; KAGAMI et al., 1986; FIELD *et al.*, 1990); the third group includes determination of spectra relations between horizontal and vertical spatial components (NAKAMURA, 1989).

We use microseisms as a sounding signal in our study as well. In contrast to the majority of researchers studying the properties and response of sediment layer lying on some rigid basement, the specificity of our subject is that we are interested in a further degree to know the properties of the basement in comparison to the sediments. This paper could be referred to as the second group of the adduced classification in a methodical aspect. We study the spatial distribution of spectral amplitudes of microseisms ranging 0.3–12.5 Hz.

Customarily researchers look for additional information for the verification of the results of microseismic study when microseisms are used for microzonation. In one case (UDWADIA and TRIFUNAC, 1973; CHAVEZ-GARCIA *et al.*, 1995) a zonation map was either confirmed with real accelerograms of a strong earthquake, or with the map of intensities built on results of the investigation. In a different case (MATSUSHIMA and OKADA, 1990) the microseismic zonation map found its confirmation in drilling data along some profile with subsequent verifying calculated evaluations.

In our paper we use independently obtained detailed gravity survey data as comparative information (CAMACHO *et al.*, 2000), and endeavor to make a complex interpretation of these two fields.

2. Geophysical Observations at Lanzarote Island

The Lanzarote Island belongs to the Canary Archipelago which is located at the edge of the West African Continental Margin. The island has an elongated shape and is situated at the northeast side of the archipelago. Its orientation has a SW-NE direction. It is approximately 55 km in length and 20 km in width. Lanzarote is defined by a shallow basement, probably about 4 5-km thick as deduced from seismic profiles (BANDA *et al.*, 1981). The basement is formed by a group of sedimentary rocks (quartzite and shale), plutonic rocks (basic and ultra-basic), and sub-volcanic rocks (basaltic and rachitic dikes) with an abundance of xenoliths of quartzite and sandstone emitted by its volcanoes. A solid majority of earlier geophysical research deals with extended surrounding oceanic areas and islands. We concentrate our study on the Lanzarote Island only and aim at revealing the peculiarities concerning its upper crust.

We should note here that the Institute of Astronomy and Geodesy of Madrid University has long conducted stationary observations at the underground geodynamic observatory, placed in the lava tunnel Cueva de los Verdes at the northeast edge of the island. The observatory is equipped with a number of instruments for observing the gravity field tidal variations, mareographs, meteorological instruments and others. On the west side of the island in Timanfaya National Park a measurement of microseismicity is being conducted as well as a measurement of heat flow and gas analysis (VIEIRA, 1991; ARNOSO, 2001). Since 1995 a tiltmeter station based on Ostrovsky pendulums and seismic station has been installed in frames of cooperation between Madrid University and the Institute of Physics of the Earth, Russian Academy of Sciences.

3. Microseisms Observations

Developing seismic observations on the island is of great interest, especially concerning registration of the local weak seismicity for its comparison with geodynamic processes. It was the challenge of searching for the appropriate places for future long-term regional seismic stations which iniated the plan for a microseism survey.

We conducted observations during two cycles – in summer 2000 and in summer 2001. The second cycle had a specifying character and was organized based on results of the first one. For the first cycle we chose 30 observational sites which were regularly distributed along the island territory (if the opportunity was offered). One of the sites was organized in the tunnel 18 deep from the day surface. During the second cycle we conducted measurements not along the surface but along a definite profile crossing an anomalous zone revealed during the first cycle. The profile

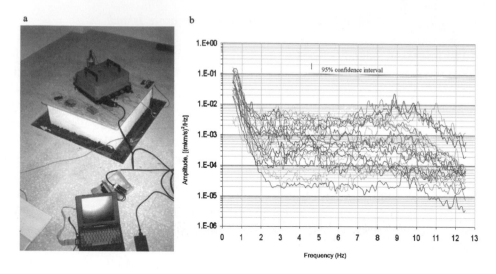

Figure 1

(a) Seismometer and acquisitioning system used in microseismic survey. (b) All measured (64 times stacked) spectra obtained at Lanzarote during microseismic survey, summer 2000 campaign.

contained 10 measurement sites. Figure 4 shows the locations of sites in the first cycle and Figure 5a shows the sites in the second cycle profile.

For observations we applied a movable station with control from a book-type computer, with power supply from an accumulator battery and with one 3-component seismometer. Our registering system provided 3 channels of velocimeter type with a frequency band 0.3–12.5 Hz, maximum sensitivity 80000 Volt*sec/meter with the possibility of gradual attenuation. The ADC had a resolution of 16 bits and the sampling rate was 100 samples per second by channel (see Fig. 1a).

Examples of fragments of microseisms are presented in Fig. 2a. The figure demonstrates the possibility of visually evaluating amplitudes and the character of microseisms on the island. We can see that a low frequency component with character periods of about 1 second prevails in the signal. The amplitude of those oscillations in displacements achieves 1 micron, which is an enormous value. For high frequencies beginning from 2 Hz, amplitudes by visual estimations achieve 0.1–0.01 microns. Such amplitudes are not unusual but typical for many places on the Earth.

During the survey we measured microseismic signal power spectra and stacked them 64 times at each site. For individual power spectra we used signal fragments of 20 seconds, with their preliminary multiplying by a Hunning window. We will discuss

▶

Figure 2

(a) Fragment of microseisms signal at Lanzarote both in displacement and velocity. (b) Examples of particle motion projections obtained at site 14. The projections are given in relative units.

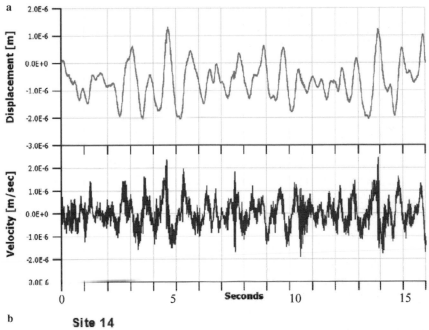

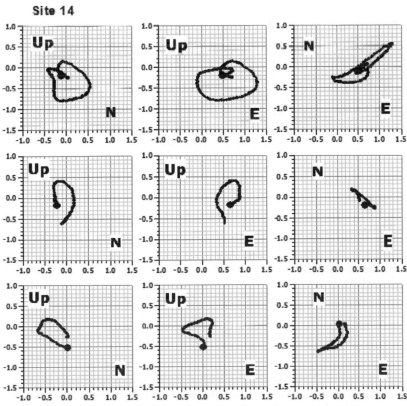

later a question concerning the number of stacking (N = 64) chosen as optimum. We registered microseisms with three reciprocal components; however for further processing we took only vertical component records to date. Thus, all the following speculations in this paper will concern the vertical component data only.

Figure 1b shows all measured stacked power spectra. A significant difference between individual spectra attracts the attention. The difference between spectral amplitudes achieves a magnitude of two orders in the frequency band 2–12 Hz. Regarding the low frequency part, considerable variations in amplitudes could be seen here as well. It is interesting to see that spectra exist with high low-frequency and low high-frequency components, however, one can find spectra with the inverse behavior. To provide correct interpretation of the observations it is important to answer two essential questions: 1. To what degree are the differences between the measured spectra significant? Or in other words, are the microseisms signal stationary and in which interval. After answering the question we would be able to compare the spatially separated sites where measurements were done asynchronously. 2. What type of waves compose the microseismic signal on the island in main? This aspect will be important when we evaluate the medium based on the received data.

4. Signal Properties Study

We conducted two special experiments in order to evaluate the stationary properties. In the first experiment we installed the microseismic station in one of the network sites for a period slightly exceeding one day (26 hours). Similar to all the other sites of the network, the measurement consisted of 64 times power spectra stacking, but was periodically repeated in the 26-hour period of the experiment. The resulting stacked spectra are exhibited as a surface on Figure 3b. Here the vertical axis is marked in logarithms of spectral amplitudes, the X-axis means frequency and the Y-axis means time marked from right to the left. We can see from the figure that the experiment started on 31.05.2001 at 20:09 local time and concluded late evening the next day. A night period can be well distinguished on the surface. Amplitudes smoothly decrease nearly ten times less from about 3–4 a.m. and then rather sharply increase from 8–9 a.m. However these variations do not concern the entire spectra but only frequencies ranging higher than 1.5 Hz. The lower frequencies are not exposed to this daily influence. We can see this from the behavior of isoline parallel to X-axis at frequencies lower than 1.5 Hz. In fact the applied approach is reminiscent of those

▶

Figure 3
(a) Illustration for the stationarity properties study. The examples of standard deviation behaviour in increasing time window are presented for four chosen sites. (b) The 64 times stacked spectrum behaviour in time during a one-day period.

a

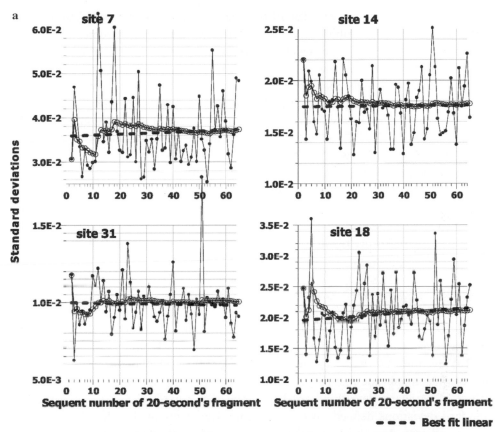

site 7

site 14

Standard deviations

site 31

site 18

Sequent number of 20-second's fragment   Sequent number of 20-second's fragment

- - - - Best fit linear

b

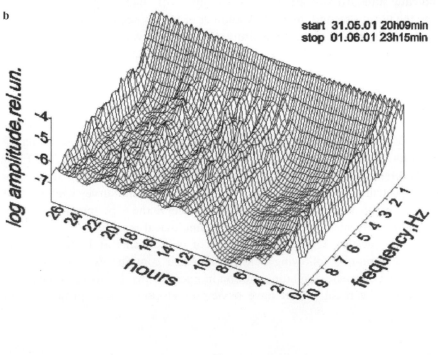

start 31.05.01 20h09min
stop 01.06.01 23h15min

log amplitude,rel.un.

hours

frequency,Hz

proposed by SEO *et al.* (1996) for distinguishing microseisms from microtremors where significant daily amplitude variations clearly indicate microtremors.

One more analogous experiment was conducted at another site. We measured a stacked power spectrum at one and the same place but in days with different weather conditions. We purposely chose a windy day and a day with very calm weather. The result was similar to the previous one. Wind influence occurred only on frequencies higher than 1.5 Hz. Thus we can preliminarily conclude that for frequencies lower than 1.5 Hz we can make asynchronous measurements and compare separated in space sites with each other.

The question about stationarity concerned the observational period optimization. We had to determine the optimal number of stacking while measuring the power spectrum at each site. Assuming from previous data that the microseism signal is a random Gaussian process with its mean value equal to zero, we could consider applicable the following relation between dispersion and spectral density:

$$\sigma_x^2 = 2 \int_0^\infty S_x(f)df \ ,\tag{1}$$

where $x(t)$ is a random microseismic process, σ_x^2 and $S_x(f)$ is its dispersion and spectral density, correspondingly (BENDAT and PIERSOL, 1966).

Taking into account (1), let us consider the behavior of dispersions of microseisms amplitude measured at four different sites (Fig. 3a). On all figures the values of standard deviations are laid along the Y-axis. The large dots indicate standard deviations defined on twenty-second fragments of the record. The empty circles indicate standard deviations in a window with increasing size, which is divisible by the twenty-second interval. A number of twenty-second intervals are laid along the X-axis, and they in turn constitute one long interval. We can well see that the standard deviations remain more or less equal from the beginning to the end of the experiment within each twenty-second interval. However a standard deviation in increasing window comes to a certain stable value after several oscillations at the beginning. The best-fit lines built on the clouds of short interval standard deviations are shown by dashes. We see that for all four exhibited cases the standard deviation in the increased window coincides with the best fit line after the long window length increases up to about 40 short intervals. The deflection does not exceed 3%. From this figure and relation (1) we can conclude that after 40 times stacking already the microseisms spectrum becomes stationary. For the microseismic survey itself we took 64 as a number of stacking. It exceeds with a 30% reserve the experimentally defined boundary and is convenient for a number of technological reasons.

The solution question regarding a content of microseisms at Lanzarote leads towards two illustrations following below. Figure 2b contains examples of three projections for 3-D particle oscillating motion in microseisms. In the presented cases projections on the vertical planes have a view of ellipsoids, and projections on horizontal plane have a linear form. Here the beginning of the oscillation phase is indicated with a large dot. We can trace the motion direction in the oscillation. We see that the oscillation character corresponds to the motion in the Rayleigh-type surface waves. From these examples we can also estimate a direction to the source. The projections on horizontal plane make it possible to perceive that the source for the given oscillating phases is not the only one. We analyzed many more examples than presented here. But due to size limitation of this paper we do not illustrate all the cases. From our analysis we saw that there is no singled out source of the microseisms on the island. The source looks like a distributed one. This is not surprising considering that the island has limited sizes and its shores are permanently subjected to the ocean surf impact. We also concluded that microseisms' energy is defined mainly by the contribution of Rayleigh-type waves.

To verify this we conducted an independent experiment which consisted in comparatively of the microseisms amplitude at frequencies 0.5–1.5 Hz measured both at a site on the day surface and on the depth 18 meters beneath the chosen point. We used a fortunate circumstance for our measurements – the presence of a natural lava tunnel, and found an attenuation of signal with the depth. The attenuation was frequency-dependent and increased in magnitude the higher the frequency was. Besides the amplitude, dependence on the frequency coincided with that calculated for the Rayleigh wave, under the assumption of shear-wave, velocity at that point had a value $V_S \cong 100–200$ m/s. Thus we obtained indirect proof that microseisms ranging from 0.5–1.5 Hz are presented with surface waves of Rayleigh type. Resultingly Figure 1b shows that the main part of energy originates from the frequency range 0.5–1.5 Hz and we can conclude that microseisms are mainly presented with Rayleigh waves.

5. Results of Observations

The results of our observations are shown in two figures – Figure 4 and Figure 5b. Figure 4 illustrates the first cycle observation results. Figure 5b contains results of the second cycle. Figure 4 has six sheets, with each corresponding to some frequency indicated at the top-left corner of the sheet. Each sheet represents a surface of experimental amplitude's distributions. Amplitudes in turn were selected from the row of the stacked spectra in Figure 1b. All sheets contain the island contour and observational sites are indicated with empty triangles. It is necessary to mention that average amplitudes for the different frequencies in spectra differ (Fig. 1b). Thus, to achieve maximum contrast we normalized the surface images independently – each to its own amplitude. We observe that an area with depressed amplitude values can be distinguished. This area located in the central part of the island and is slightly shifted to its east coast. The nearest settlement to the zone is named San-Bartolome.

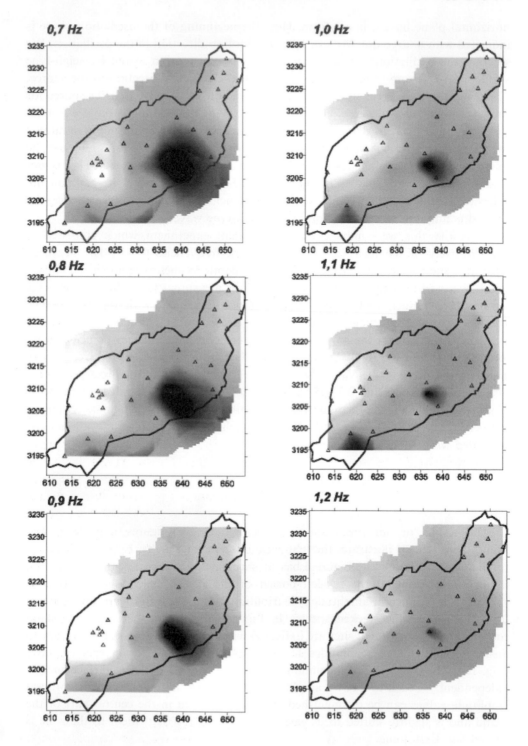

The size of this area and the depth of the depression depend on frequency. Figure 4 contains surfaces for the frequencies from 0.7 Hz onwards. We do not adduce here the lower frequencies, as they do not differ considerably from the 0.7 Hz sheet. The depressed amplitude area decreases in size and concentrates around a certain point when the frequency increases from 0.7 to 1.2 Hz. From 1.3 Hz on this area mostly cannot be seen on the background.

We assume that this dependence can be explained only under the assumption that microseisms are present with the surface–type waves. Furthermore, it is necessary to assume the presence of geological inclusion more rigid by its mechanical properties than surrounding rocks. It is necessary to presume also that this inclusion does not appear at the surface, but rather is hidden under a certain sediment layer where the amplitude depression is disappearing beginning from the frequency 1.3 Hz. Under those assumptions the depression forming looks like the following. A surface Rayleigh wave propagates across the island and if it has enough length and thus enough deep penetration to the depth, it runs against the rigid inclusion. When propagating across the inclusion the wave amplitude decreases. If the wavelength is not long enough and penetration is not deep then the wave does not touch the inclusion and pass above it.

The fact of stable amplitude loss at a definite area is interesting because microseisms amplitude growth can be observed in practice very often. Besides, one can find numerous reasons for amplitudes growth than for the loss. It could be the influence of cultural sources for example.

We decided to verify this result again, and as we stated before, during next field season we conducted an additional cycle of microseismic survey. This survey consisted of the observations along the profile. Figure 5a shows the distribution of the observational sites relative to the island contour and the revealed zone of anomalous amplitudes depression. Observational results are presented in Figure 5b as dependence of microseisms' amplitudes (indicated by inking) both on the frequency (along the vertical axis) and on the coordinate in the profile (along the horizontal axis). The observational sites were placed imprecisely along a straight line and thus Figure 5b relates to a conditional line indicated on the figure by dashes.

We observe that the first cycle results are being proved. As in the first case we find the depression above the revealed zone at frequencies 0.5–0.8 Hz. As in the first case we found the depression size and depth decrease with the frequency growth. Starting from the approximate frequency 1.25 Hz, this depression is no longer noticeable on the background. Additionally we see on Figure 5b that with frequency growth from 1.3 Hz, the amplitude distribution character in the profile changes to the opposite

◀

Figure 4

Amplitude distribution of indicated frequency components along the territory of the island. The inking of each pattern is normalized by the average amplitude of the indicated frequency in the experimental spectra. Empty triangles show the places of the survey network for the summer 2000 campaign.

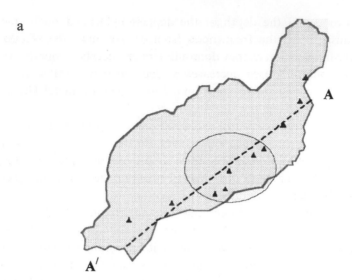

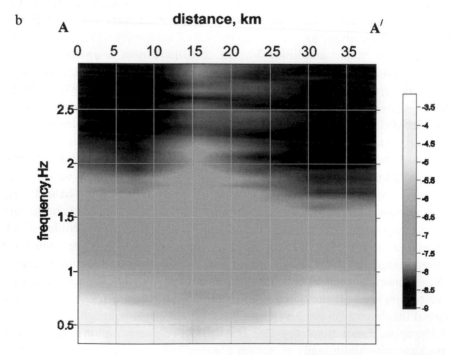

Figure 5
(a) The observational sites, positions in the profile during the summer 2001 campaign. The ellipse with increased density indicates the anomalous zone revealed in summer 2000 campaign. (b) Distribution of stacked spectra amplitudes versus frequency and site position in the profile. Inking indicates the logarithmic amplitudes of microseisms.

one. In the zone where we observed a depression we see amplitude growth. Here we just note this interesting fact. It is probably also in connection with peculiarities of the subsurface geological structure at the given place of the island. However we will not consider and analyze the phenomenon models in this paper.

6. Discussion

In spite of the possibility to explain the observed loss in amplitudes aided by the presence of subsurface rigid inclusion, we made our best effort to attract independent data on other geophysical fields for the same territory.

Thus recently CAMACHO et al., (2001) presented results of modelling the crust of anomalous masses of Lanzarote Island, which were based on application of a new version for inverse gravimetry problem solution and data which were obtained in a detailed gravimetry survey of the island in 1988. Figure 6a was adopted from the paper and contains a map of the Bouguer gravimetric anomalies distribution. Dots indicate a network of gravity stations (totally 296 stations). A bright positive anomaly in the center of the island rises above the background near San-Bartolome. The transversal size of this zone is about 13 km. Its mass anomaly constitutes $582*10^{12}$ kg. CAMACHO et al. (2001) interpret this anomalous zone as an intrusive body, reaching with its roots to a depth of 15 km, where it almost achieves the Moho boundary alleged here. Up to him the depth of the intrusion is still under discussion. According to MACFARLANE and RIDLEY (1969) this body corresponds to a major igneous center within the crust, it was important at least during the early sub-aerial growth of the island, and may be formed initially at the intersections of fundamental NNE- and SW- trending fault systems. However, while large earth movements and an erosion process may have disrupted this structure until now, only the ridges remain as horst blocks, the central cone having subsided between them and having been covered by younger volcanic eruptions. Moreover, the detailed survey of the island revealed a number of smaller and shallower positive bodies, interpreted as less-developed magmatic intrusions. Unfortunately density of our microseismic network made if impossible for us to distinguish these minor structures, though we can expect to distinguish them in the microseismic field as well if the experiment is properly arranged.

Comparing Figure 4 with Figure 6a we note a remarkable coincidence of anomalous zones revealed in the microseismic and gravimetrical fields. The interpretations of both anomalies independently imply the analogous reason of their formation and namely the presence of an intrusive body in the area of San-Bartolme. We consider this fact a ponderable mutual proof of these interpretations.

Based on the inverse gravimetry problem solution CAMACHO et al. (2001) estimated density of the intrusive body. We also endeavored to estimate elastic parameters of this body based on the microseisms amplitudes distribution

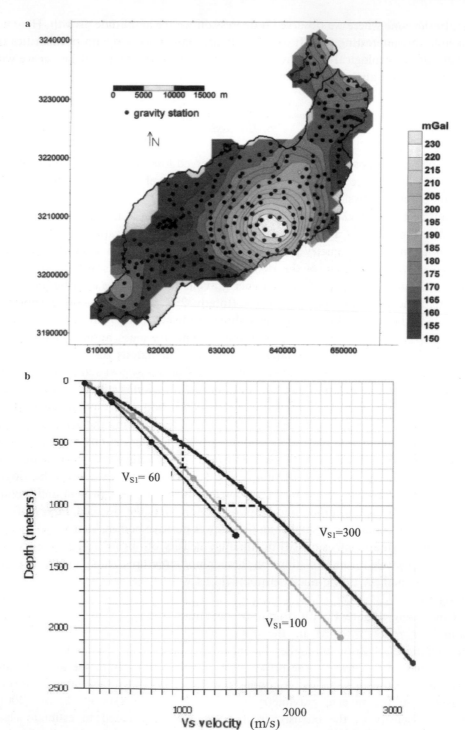

information. For this we employed a simple elastic model of two half spaces with different values of densities and elastic parameters. According to LEVSHIN et al. (1992), a dependence on the depth of Rayleigh wave amplitudes in the half space can be described as follows:

$$A_Z(\omega, Z) = \frac{k^2 - 2}{k^2} \left(e^{-r_\alpha Z} + \frac{2}{k^2 - 2} e^{-r_\beta Z} \right), \qquad (2)$$

where

$$r_\alpha = \frac{2\pi}{l} \sqrt{1 - \gamma^2 k^2}; \quad k = V_R / V_S$$

$$r_\beta = \frac{2\pi}{l} \sqrt{1 - k^2}; \quad \gamma = V_S / V_P \quad l - \text{Rayleigh wave length}.$$

Making flexible assumptions about retaining the energy when the Rayleigh wave is crossing the vertical boundary, we calculated a relation of amplitudes in waves for the surface in both half spaces based on equation (2). We varied parameters V_S, V_P and V_R for both half spaces trying to approach the calculated result to the experimental amplitude's relations for the anomalous zone and surrounding space of different frequencies. At that we kept the relations between V_S, V_P and V_R within the definite limits existing for the real velocities. We found during the modelling that the amplitude distribution is most sensible to the V_S parameter.

In our calculations we proceeded from values in surrounding space $V_{S1} = 100$ m/s and $V_{S1} = 300$ m/s and estimated the values of V_{S2} and V_{R2} for the anomalous zone. Then we calculated a length of the Rayleigh wave in the anomalous zone and estimated its penetration to the depth using the relation: $H - 1/2\, l$. Results of our estimations are presented in Figure 6b. This figure indicates the distribution of velocities in the intrusive body when approaching the surface. The fragments of the dashed lines illustrate the range of possible velocities for the given depth in one case and the range of depths within which the given velocity can be met in another case. This simple modelling for V_{S2} (shear-waves velocity in the anomalous zone) produced a resulting value typical for basalt (see Fig. 6a.). This is in good agreement with the conclusion of CAMACHO et al., (2001) regarding the existence, position and substance of the intrusive body.

The range of velocities in the surrounding space we chose ($V_{S1} = 100$ m/s and $V_{S1} = 300$ m/s) was not accidental, and was obtained as a result of two

◄

Figure 6
(a) Distribution of gravity stations and the map of the Bouguer gravimetric anomaly. The value used for density of terrain correction is 2,480 kg/m^3 (adopted from CAMACHO et al., 2001). (b) Results of simulation of the distribution of V_{S2} velocity versus depth for the different V_{S1} in the revealed anomalous zone. The V_{S1} magnitudes (given in m/s) accepted for calculations is shown near each curve.

experiments. The first one was described earlier and consisted of the estimation of V_{S1} velocity beyond the anomalous zone from the Rayleigh amplitudes comparison between the surface site and the deepened site in the lava tunnel. It produced the result $V_{S1} = 100$ m/s. The second experiment consisted of direct measurements of velocities of microseisms (beyond the anomalous zone as well). This was done with the help of a small aperture seismic array and resulted in $V_{S1} = 300$ m/s.

7. Conclusions

We performed a microseismic survey at the territory of the Lanzarote Island, Canary Archipelago. For this we used a three-component seismometer which we moved from one site to another. The initial target consisted of searching for the appropriate place to organize the long-term seismic station. We used a velocimeter–type channels with the band 0.3–12.5 Hz for our measurements. Later on a data analysis showed that the microseism's parameters stably depend on peculiarities of the subsurface geological structure.

To interpret correctly the survey data we studied the properties of the microseisms signal. First, we ascertained that microseisms on the island are mainly presented with the surface waves of the Rayleigh-type. It was also illustrated experimentally that frequency-dependent amplitudes decrease with the depth. A microseisms particle motion analysis showed that (for our case at least) the microseisms source has no definite location and looks like a distributed source. Second, we studied a stationarity of microseisms analyzing a standard deviation in increasing the time window and found that after approximately 800 seconds the signal is receiving stationary properties.

During the survey we measured power spectra with 64 times stacking and analyzed a distribution of individual frequency components thereafter for the surface sites. We found that spatial distribution of the spectral components ranging 0.3–1.2 Hz amazingly coincides with the detailed gravity survey data obtained in the independent study. The gravimetric data point to a deep intrusive body in the central part of the island. The microseismic survey results can be explained similarly .

We tried to estimate shear velocities in the surrounding rocks and in the intrusive body as well as the density estimation from gravimetric data. We derived values typical for basalt rocks, which is in agreement with the conclusions of the gravimetric study.

Acknowledgements

We thank Academician V.N. Strakhov, Director of Institute of Physics of the Earth, Russian Academy of Science for organizational support of this research. We are also grateful to the staff of Casa de los Volcanes (Lanzarote) for their assistance. The Contract between the Russian Academy of Sciences and the Spanish Council for Scientific Research has supported this research.

REFERENCES

AKI, K. (1957), *Space and Time Spectra of Stochastic Waves with Special Reference to Microtremors*, Bull. Earthquake Res. Inst. Tokyo Univ. *35*, 415–457.

ARNOSO, J., FERNANDEZ, J., and VIEIRA, R. (2001), *Interpretation of Tidal Gravity Anomalies in Lanzarote, Canary Islands*, J. Geody. *31*(4), 341–354.

ASTEN, M. (1978), *Geological Control on the Three-component Spectra of Rayleigh-wave Microseisms*, Bull. Seismol. Soc. Am. *68*(5), 1623–1636.

BANDA, E., DANOBEITIA, J. J., SURINACH, E., and ANSORGE, J. (1981), *Features of Crustal Structure under Canary Islands*, Earth Planet. Sci. Lett., *55*, 11–24.

BARD, P. (1999), *Microtremor Measurements: A Tool for the Effect Estimation?* The Effects of Surface Geology on Seismic Motion, (Balkeman, Rotterdam), 1999 ISBN 90 5809 030 2.

BENDAT, J. S., and PIERSOL, A. G. (1966), *Random Data: Analysis and Measurement Procedures*, (Wiley-Interscience), New York, 463 pp.

CAMACHO, A. G., MONTESINOS, F. G., VIEIRA, R., and ARNOSO, J. (2001), *Modeling of Crustal Anomalies of Lanzarote (Canary Islands) in Light of Gravity Data*, Geophys. J. Int. *147*, 1–22.

CHAVEZ-GARCIA, F. J., CUENCA, J., and LERMO, J. (1995), *Seismic microzonation in Mexico. The examples of Mexico City, Oaxaca and Puebla*, Proc. 5th Int. Conf. on Seism. Zonation, Nice, France,. vol. I, 699–706.

FIELD, E. H., HOUGH, S. E., and JACOB, K. H. (1990), *Using Microtremors to Assess Potential Earthquake Site Response: A Case Study in Flushing Meadows, New York City*, Bull. Seismol. Soc. Am. *80*, 1456–1480.

GORBATIKOV, A. V. and BARABANOV, V. L. (1994), *Experience in Use of Microseisms to Evaluate the State of the Upper Crust*, Phys. Solid Earth, Eng. Transl. *29*(7), 640–644.

GUTENBERG, B.,. *Grundlagen der Erdbebenkunde*, Velag von Gebruder Borntraeger, Berlin (1927).

HORIKE, M. (1985), *Inversion of Phase Velocity of Long–period Microtremors to the S-wave Velocity Structure down to the Basement in Urbanized Areas*, J. Phys. Earth *33*, 59–96.

HOUGH, S. E., SEEBER, L., ROVELLI, A., MALAGNINI, L., DECESARE, A., SEVEGGI, G., and LERNER-LAM, A. (1992), *Ambient Noise and Weak-motion Excitation of Sediment Resonances: Results from the Tiber Valley, Italy*, Bull. Seism. Soc. Am. *82*, 1186–1205.

IRIKURA, K. and KAWANAKA, T. (1980), *Characteristics of Microtremors on Ground with Discontinuous Underground Structure*, Bull. Disas. Prev. Inst. Kyoto Univ., *30-3*, 81–96.

KAGAMI, H., OKADA, S., SHIONO, K., ONER, M., DRAVINSKI, M., MAL, A. K. (1986), *Observation of 1 to 5 Second Microtremors and their Application to Earthquake Engineering. Part III. A Two-dimensional Study of Site Effects in S. Fernando Valley*, Bull. Seismol. Soc. Am. *76*, 1801–1812.

KANAI, K. and TANAKA, T. (1954), *Measurement of the Microtremor*, Bull. Earthq. Res. Inst. Tokyo Univ. *32*, 199–209.

KATZ, L. J. and BELLON, R. S. (1978), *Microtremor Site Analysis Study at Beatty, Nevada*, Bull. Seismol. Soc. Am. *68*, 757–765.

LACOSS R. T., KELLY, E. J. and TOKSOZ, M. N. (1969), *Estimation of Seismic Noise Structure Using Arrays*, Geophys. *1*(34), 21–38.

LERMO, J. and CHAVEZ-GARCIA, F. J. (1994), *Are Microtremors Useful in Site Response Evaluation?* Bull. Seismol. Soc. Am., *84*(5), 1350–1364.

LEVSHIN, A., RATNIKOVA, L. and BERGER, J. (1992), *Peculiarities of Surface-wave Propagation across Central Eurasia*, Bull. Seismol. Soc. Am. *82*, 2464–2493.

MACFARLANE, D. J. and RIDLEY, W. I. (1969), *An interpretation of Gravity Data for Lanzarote, Canary Islands*, Earth Planet. Sci. Lett. *6*, 431–436.

MATSUSHIMA, T. and OKADA, H. (1990), *Determination of Deep Geological Structures under Urban Areas Using Long-Period Microtremors*, Butsuri-Tansa, *43, 1*, 21–33.

NAKAMURA, Y. (1989), *A method for Dynamic Characteristics Estimation of Surface Using Microtremor on the Ground Surface*, QR of RTRI *30*(1), February, 25–33.

OHTA, Y., KAGAMI, H., GOTO, N. and KUDO, K. (1978), *Obsevation of 1 to 5 Second Microtremors and their Application to Earthquake Engineering. Part I: Comparison with Long- period Accelerations at the Tokachi-Oki Earthquake of 1968*, Bull. Seismol. Soc. Am. *68*, 767–779.

OMORI, F. (1908), *On Microtremors,* Res. Imp. Earthquake Inv. Comm. *2*, 1–6.

OMOTE, S. and NAKAJIMA, N. (1973), *Some Considerations for the Relation between Microtremors and Underground Structure*, Bull. Int. Inst. Seism. Earthq Eng. *11*, 9–19.

SAKAJIRI, N. (1982), *Experimental Study on Fundamental Characteristics of Long-period Microtremors*, Bull. H.I.T. *2*, 112–154.

SATO, T., KAWASE, H., MATSUI, M. and KATAOKI, S. (1991), *Array measurements of high frequency microtremors for underground structure estimation. In Proc. 4th. Conf. n Seismic Zonation,.* Stanford, California, vol. *II*, 409–415.

SEO, K., HAILE, M., KURITA, K., YAMAZAKI, K. and NAKAMURA, A., *Study of Site Effect in Kobe Area Using Microtremors,* X-th World Conf. on *Earthquake Engineering*, Acapulco, #1656, (Elselvier Science Ltd).(1996)

UDWADIA, F. E. and TRIFUNAC, M. D. (1973), *Comparison of Earthquake and Microtremor Ground Motions in El Centro, California*, Bull. Seismol. Soc. Am. *63*, 1227–1253.

VIEIRA, R., VAN RUYMBEKE, M., ARNOSO, J. and TORO, C. (1991), *The Lanzarote Underground Laboratory*, Cahiers du Centre Europeen Geodynamique de Seismologie *4*, 71–86.

(Received February 15, 2002, revised December 23, 2002, accepted January 29, 2003)

 To access this journal online:
http://www.birkhauser.ch

Pure appl. geophys. 161 (2004) 1579–1596
0033–4553/04/071579–18
DOI 10.1007/s00024-004-2522-5

© Birkhäuser Verlag, Basel, 2004

❙ Pure and Applied Geophysics

Microtremor Analyses at Teide Volcano (Canary Islands, Spain): Assessment of Natural Frequencies of Vibration Using Time-dependent Horizontal-to-vertical Spectral Ratios

JAVIER ALMENDROS[1], FRANCISCO LUZÓN[1,2],
and ANTONIO POSADAS[1,2]

Abstract—We use time-dependent horizontal-to-vertical spectral ratios (HVSR) of microtremors to determine the dominant frequencies of vibration of the geological structures beneath several recording sites in the vicinity of Teide volcano (Canary Islands, Spain). In the microtremors, the time-dependent HVSRs (ratiograms) are a useful tool to discriminate between the presence of real dominant frequencies linked to resonances of the subsurface structure and the spurious appearance of peaks due to local transients. We verified that the results are repeatable, in the sense that microtremors recorded at the same site but at different times yield a very similar HVSR function. Two types of results are found: (1) sites where there is no resonance of the propagating microtremors, and therefore no value of a dominant frequency can be assessed; and (2) sites where a stationary peak in the HVSR is found and a dominant frequency related to resonance of the shallow structure can be estimated. These resonant frequencies show substantial spatial variations even for nearby sites, which reflects the complexity of the shallow velocity structure in the Las Cañadas area. Large dominant frequencies occur near the caldera walls and also at a few locations that coincide with the intersections of the inferred rims of the three calderas forming Las Cañadas. Small dominant frequencies also occur near the caldera rim, and may be due to discontinuities in the caldera wall and/or to local velocity anomalies. Intermediate frequencies are mostly found in the eastern part of the caldera, where a tentative profile of the basement depth has been obtained. Intermediate frequencies have also been measured south of Ucanca and south of Montaña Blanca. In view of the present results, we conclude that the use of ratiograms constitutes an improvement of the HVSR method and provides an appropriate tool to investigate the shallow velocity structure of a volcanic region.

Key words: Surface geology, microtremors, resonant period, ratiogram, Teide volcano.

1. Introduction

The Canary Islands consist of seven islands and several islets located at about 28°N, 16°W in the Atlantic Ocean (Fig. 1). They lie in an intraplate setting on the northwestern African continental shelf. Several theories have been proposed to

[1] Instituto Andaluz de Geofísica, Universidad de Granada, Campus Universitario de Cartuja s/n 18071 Granada, Spain.
[2] Departamento de Física Aplicada, Universidad de Almería, Cañada de San Urbano s/n, 04120 Almería, Spain. E-mails: alm@iag.ugr.es; fluzon@ual.es; aposadas@ual.es

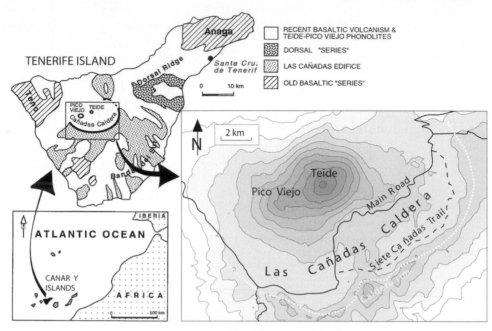

Figure 1

Composite map showing the region under study. (lower left) Position of the Canary Islands near the
northwestern African coast. (upper left) Simplified geological map of Tenerife Island, showing the position
of Teide volcano and Las Cañadas caldera. (lower right) Las Cañadas caldera. Microtremor records were
obtained at several sites along the main road (solid black line) and at Siete Cañadas Trail (dotted black
line). The dotted white line shows the approximate position of the caldera rim.

explain the origin of the Canarian volcanism (ANGUITA and HERNÁN, 2000),
including a mantle hotspot (MORGAN, 1983; CARRACEDO *et al.*, 1998; DAÑOBEITIA
and CANALES, 2000) and the uplift of tectonic blocks (ARAÑA and ORTIZ, 1991),
among others. In Tenerife, subaereal volcanism started more than 12 Ma ago
(ANCOCHEA *et al.*, 1990). The activity was concentrated mainly along the NE-SW and
NW-SE faults that cross the center of the island, where a great volcanic edifice was
built some 3 Ma ago (MARTÍ *et al.*, 1994). The collapse of this edifice, most likely by
multiple collapses rather than by a single giant event (MARTÍ and GUDMUNDSSON,
2000) produced the Las Cañadas caldera, which constitutes one of the best exposed
calderas in the world. It has an approximately elliptical shape, with dimensions of 16
and 9 km in east-west and north- south directions, respectively. The caldera floor has
an elevation of about 2 km, and it is surrounded by walls clearly visible for 27 km. A
gravitational failure of part of the north wall left the caldera open to the north.
Subsequent episodes of volcanism on the northwest of caldera gave birth to the Teide-
Pico Viejo stratovolcano (ABLAY and MARTÍ, 2000) and contributed to the filling of
the caldera. There is also evidence of the importance of landslides in the volcanic
evolution of Tenerife (ANCOCHEA *et al.*, 1990; ABLAY and HÜRLIMANN, 2000).

Teide is considered as potentially hazardous due to the proximity of populated areas such as the Orotava Valley. This may be one of the reasons for its selection as the Decade Volcano and the European Laboratory Volcano by the IAVCEI and the European Science Foundation, respectively. Several eruptions have occurred in Tenerife within historical times, for example in Siete Fuentes (1704), Fasnia and Montaña Arenas (1705), Montaña Negra (1706), Chahorra (1798), and Chinyero (1909) on the northwestern flank of the volcano. All these eruptions were preceded by premonitory seismicity (ARAÑA and CARRACEDO, 1978), which emphasizes the importance of a systematic geophysical monitoring.

The present study is intended to improve our understanding of the relationship between the shallow geological structure and microtremors in a volcanic environment. The experiments consisted of microtremor measurements at different points in the vicinity of the Teide volcano. These data are used to estimate the dominant frequencies of vibration of the structures beneath the recording sites, which may be used to characterize the local geological setting.

2. Instruments and Data

A single seismic station was used for the microtremor measurements. It was composed of a short-period, three-component seismometer with natural period of one second, a 24-bit A/D converter with GPS time, sampling each channel at 100 sps, and a laptop computer to control the system and store the data. We recorded data during three different surveys. Due to the rough topography and the difficult access to most parts of the Las Cañadas caldera, sites were generally selected on the basis of logistic ease. On July 2000, data were recorded at 56 locations along the main road and the Siete Cañadas trail crossing the Las Cañadas caldera (Fig. 1); record duration was set to three minutes. Additional data were obtained by recording 3-minute microtremor samples every hour during 24 hours near the Parador Nacional. The second survey was performed during July 2001, and consisted of a repetition of the 2000 survey to obtain new 3-minute microtremor records at the same 56 positions. Finally, in October 2001, 5-minute data samples were recorded at 28 sites in a denser grid on the eastern part of the Las Cañadas caldera. Measurements were repeated three times for one of the sites to control the stability of the results. Distances between nearby recording sites ranged between 0.5 and 1.0 km.

The microtremor spectrum contains energy basically below 10 Hz, with a main peak observed at 1 Hz. The main source of microtremors is most likely the oceanic microseismic noise, which usually peaks at periods of a few seconds. This kind of noise has been found to peak between 3 and 8 s at the oceanic islands of Hawaii (DAWSON et al., 1998). At Tenerife, ALMENDROS et al. (2000) documented the presence of oceanic noise at frequencies below 1 Hz. The peak observed in our data probably corresponds to the shoulder of the microseismic noise filtered by our short-period instruments.

3. Method

3.1. The HVSR Technique

The horizontal-to-vertical spectral ratio (HVSR) method considers that the amplification produced by a surface layer can be obtained by evaluating the ratio between the horizontal and vertical spectral amplitudes of microtremors recorded at the site. This method is known as the Nakamura's technique since the publication of his paper (NAKAMURA, 1989). Nevertheless, the idea was introduced by NOGOSHI and IGARASHI (1971) who showed the coincidence between the lowest-frequency maximum of the HVSR of Rayleigh waves and the fundamental resonance frequency of the site. The method assumes that microtremors are composed of Rayleigh waves which propagate in a surface layer over a half-space (LERMO and CHÁVEZ-GARCÍA, 1994; DRAVINSKI *et al.*, 1996). The motion at the interface between the layer and the half-space is not affected by the source effect; moreover, the horizontal and vertical motions at the interface are approximately equal due to the ellipticity of the Rayleigh waves. Therefore the site effect can be computed by the spectral ratio of horizontal versus vertical components of motion at the surface, that is,

$$HVSR(f) = \frac{\sqrt{A_{\text{east}}(f)^2 + A_{\text{north}}(f)^2}}{A_{\text{vertical}}(f)} \tag{1}$$

where A_C represents the spectral amplitude of the C component.

KONNO and OHMACHI (1998) extended the problem considering a multi-layered system; they reinforced the technique, which up to that moment had some theoretical gaps. Other studies have shown that the HVSR can reveal the fundamental resonant frequency of the structure beneath the site (LERMO and CHÁVEZ-GARCÍA, 1994; FIELD and JACOB, 1995; LACHET *et al.*, 1996; SEEKINS *et al.*, 1996; COUTEL and MORA, 1998). The method works reasonably well when the structure of the site under study can be approximated by a 1-D model. Recent works have investigated the applicability of this technique by the simulation of seismic waves in specific structures such as sedimentary basins (AL YUNCHA and LUZÓN, 2000; DRAVINSKI *et al.*, 1996). LUZÓN *et al.* (2001) performed a numerical study of the propagation of seismic waves in 2-D flat sedimentary basins, and concluded that the HVSR can reasonably predict the natural frequency when there is a high-impedance contrast between the sedimentary inclusion and the bedrock.

Despite its restrictions, Nakamura's method has become a widely used tool, with interest in seismological and engineering applications. For example, it has been used to map the thickness of sediments in the Segura Valley, Spain (DELGADO *et al.*, 2000), and the Lower Rhine Embayment, Germany (IBS-VON SEHT and WOHLENBERG, 1999); to characterize seismic hazards in small scale for seismic microzonation (see BARD, 1999 and references therein); or to estimate the topographical effects of ground shaking at mountains (ZASLAVSKY and SHAPIRA, 2000). The main reason for such a success is its simplicity both in field operations and data analysis. A few minutes of microtremor data is enough. Microtremors are ubiquitous in time and space, and it is not necessary to wait for the occurrence of earthquakes. A single three-component station is the only instrument required. Then, routine spectral techniques can be easily applied to obtain estimates of the dominant frequency of vibration of the underlying structure. These frequencies of vibration are strongly related to the physical properties of the site under study, i.e., layer thicknesses, densities, or wave velocities. Estimates of these frequencies could be thus useful to characterize the physical properties of a geological structure.

3.2. Application of the Method

The procedure generally used to calculate the HVSR consists in the application of eq. (1) to the average amplitude spectra of the three components of motion. However, we have tested that the routine application of this technique may lead to errors. Local perturbations of the wavefield may happen during the recording periods and be recorded together with the microtremor data. Usually, these transients are easily identified in the seismograms and/or spectra, and the analysis can be performed only on those data free of perturbations in order to obtain reliable results. But variations in the microtremor wavefields might be subtler in such a way that neither visual nor spectral analysis are able to reveal them. In this case, spurious peaks would appear in the average spectra. These peaks could be large enough to affect the spectral ratio and produce unrealistic results.

To avoid this problem, we used a different approach aimed to the estimate of a time-dependent HVSR. Our method consists in the application of eq. (1) to successive data windows along the traces. This procedure yields several HVSR functions that can be represented as a two-dimensional contour plot versus frequency and time. This plot, that we name ratiogram, represents the evolution of the HVSR in the same way that a spectrogram represents the evolution of the spectrum versus frequency and time.

Figure 2 shows an example of application of the HVSR method using the two approaches described above. Following the standard technique, average spectra are calculated (Fig. 2b, top) and the HVSR is obtained as the ratio between the horizontal and vertical spectral amplitudes (Fig. 2b, bottom); we can see the presence of a dominant peak at about 5 Hz and we could conclude that the site produces amplification for this frequency. However, the time-dependent HVSR with the second approach demonstrates that this peak is due to a single transient at about 60 s

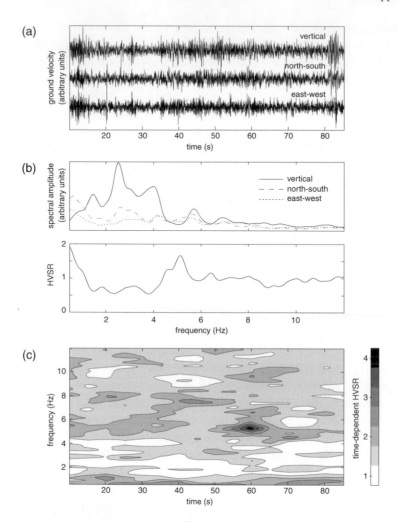

Figure 2

Example of the application of the HVSR method. (a) Three-component microtremor data. (b) Amplitude spectra of the three components (top) and HVSR (bottom) calculated using the standard procedure. (c) Ratiogram representing the HVSR as a function of frequency and time.

(Fig. 2c). This plot shows that the HVSR is not stationary at this site, at least at the time of our measurement, and therefore it makes no sense to estimate a value for the dominant frequency.

When the time-dependent HVSR is stationary, at least during particular time periods, an average HVSR can be obtained by stacking the corresponding HVSRs. This average usually coincides with the one calculated by the standard technique, except for problematic records as shown in Figure 2. The difference between the two techniques consists in the averaging procedure. In the standard method, we average the spectra corresponding to several windows along the traces, and then apply eq. (1)

just once to obtain the HVSR. In our approach, we apply eq. (1) several times to calculate the HVSR for each window, and then average them to obtain the site response. In our opinion, the second approach should be preferred since more stability is expected in the HVSR (that in theory depends only on the site structure) than in the microtremor amplitude spectra which can show temporal variations. Moreover, we take into account not only which peak of the average HVSR is highest, but also whether or not this peak appears stable in time in the ratiogram. A high peak that is not stationary is not considered in the assessment of the dominant frequency, something that is impossible to take into account in the standard approach.

We calculated HVSRs with the second approach described above. We select a window of 20.48 s and slide it at intervals of 5 s along the traces. For each window we calculated the amplitude spectra of the three components using an FFT algorithm, and smoothed using a 0.7-Hz Hanning window. Then we applied eq. (1) to obtain the HVSR. We repeated this procedure for successive time windows over the duration of the microtremor records to obtain a ratiogram for the site. In a ratiogram, the presence of a dominant frequency is marked by a horizontal, ridge-like structure. All noise-free HVSRs representing stable dominant frequencies were selected and averaged to yield the average HVSR, whose peak determines the dominant frequency for the site. Taking into account the response of our instrument and the frequency content of microtremors recorded in the Las Cañadas area, we restricted *a priori* the possible dominant frequencies to the range 1–10 Hz. This condition is not too restrictive, since this is the range of frequencies usually obtained in this kind of studies.

4. Results

Figure 3 shows examples of stationary ratiograms obtained at two different sites. In the first case (Fig. 3a), a clear dominant frequency of 3.6 Hz appears throughout the duration of the records. In the second case (Fig. 3b), the average HVSR does not present a resonant frequency but a flat response with an amplification level approximately equal to one. Ratiograms like these have been calculated for the entire data set using the method described in the last section.

The shape of the HVSR is usually repeatable; microtremors obtained at the same site at different times show similar HVSR functions. For example, Figure 4a shows average HVSRs of 24 data samples recorded hourly at the same site (July 2000 survey). A very stable peak at about 5 Hz is found. Two other peaks are also evidenced between 6 and 9 Hz, but their positions and values change slightly with time. Since the velocity structure under the recording site does not vary, these differences between HVSRs corresponding to different hours may suggest that the source effect has not been completely removed in the spectral ratio. Figure 4b shows

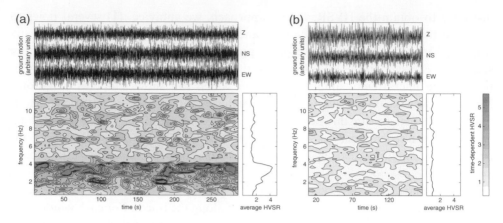

Figure 3

Two examples of ratiograms and average HVSRs obtained from microtremors recorded at sample locations. In each case, the top pannel shows the three-components of ground motion; the bottom pannel corresponds to the ratiogram calculated using the procedure described in the text; and the right pannel shows the average HVSR. The scale on the right represents the values of the time-dependent HVSR in both ratiograms. (a) Example with a stationary peak at a frequency of about 3.6 Hz. (b) Example with no resonant peaks.

another example of HVSRs calculated for data recorded at the same site during the October 2001 survey. Although the microtremors were recorded after lapses of several hours, a dominant frequency at about 3 Hz is clearly seen in the three cases. Finally, Figure 4c shows two examples of HVSRs for data recorded at the same sites with a temporal lapse of one year. The upper panel corresponds to a site where a dominant frequency could be determined at about 7 Hz. The bottom panel shows the average HVSR for a site where there is no vertical resonance of the propagating waves and therefore no dominant frequency could be established. In both cases, we find the same patterns for data recorded during the July 2000 and July 2001 surveys. This repeatability is the behavior observed at about 85% of the recording sites, for which both HVSR functions obtained during the July 2000 and July 2001 surveys at each site lead either to a flat amplification spectrum, or to the determination of the same values of dominant frequency within a limit of about 20%. There are a few cases, however, where we have found a flat HVSR for the July 2000 data and a more or less dominant peak for the July 2001 data, or *vice versa*. These inconsistencies could be attributed to near-site variations of the wavefield that have not been detected and have slipped in during the analysis.

The results corresponding to the dominant frequencies obtained for the three seismic surveys are summarized in Figure 5. The unfortunately large percentage of sites where no information could be obtained was mostly due to failure of a channel during the recording process and/or the presence of non-stationary noise in the vicinity of the recording site. Specifically, this happened at 12 of the 57 recording sites of the July 2000 survey; 21 of the 56 sites of the July 2001 survey; and 1 of the 28 sites

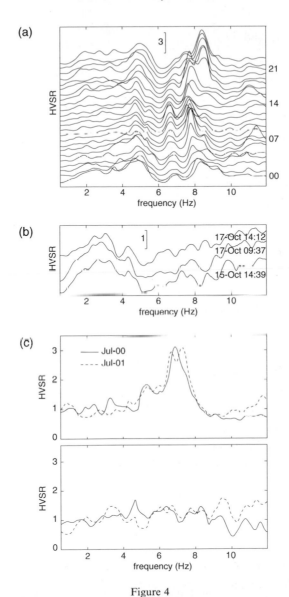

Figure 4

Average HVSRs obtained at the same site at different times: (a) from the analysis of microtremors recorded hourly for 24 hours during the July 2000 survey, (b) at three different times during the October 2001 survey, and (c) at the same sites during the July 2000 and July 2001 surveys.

of the October 2001 survey. Flat HVSRs, such as the example in Figure 3b, have been found at several recording sites during the July 2000, July 2001 and October 2001 surveys (for 16, 13 and 1 sites, respectively). The lack of stationary peaks can be interpreted as the absence of resonance of the structures under the recording sites, at least in the frequency range investigated. For the rest of the sites, the shallow

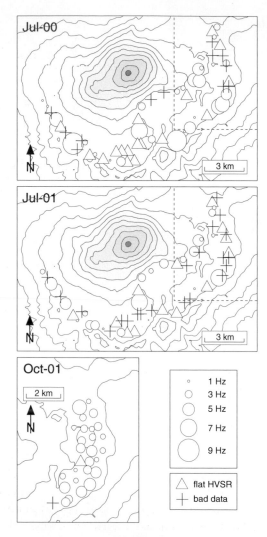

Figure 5

Maps of Las Cañadas caldera showing the results of the HVSR method for microtremor data obtained during the three surveys. Crosses indicate sites where no information could be obtained due to instrument failures or presence of non-stationary noise; triangles mark sites where flat response spectra were found. Finally, circles show those sites where dominant frequencies were determined. The region shown in the bottom left panel is marked by dashed lines on the top and middle panels.

structure beneath the station efficiently induces the resonance and stationary peaks are found. These HVSRs allowed us to assess values of a dominant frequency ranging from 1 to 8 Hz.

Figure 6 combines in a single plot the average values of the dominant frequency obtained at the different sites during the three surveys. The most striking characteristic of the distribution of dominant frequencies obtained is their high

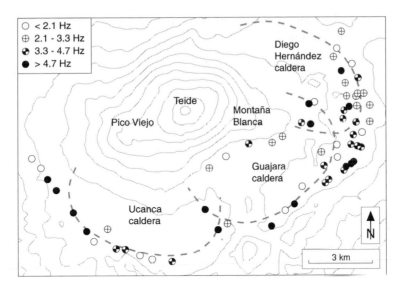

Figure 6

Map showing the average dominant frequencies measured during the three surveys (see Fig. 5). The dashed grey lines correspond to the inferred rims of the Ucanca, Guajara, and Diego Hernández calderas.

spatial variability. For example, in some cases there are very different frequencies at nearby sites located just 300 m apart.

5. Discussion

Several studies show the relationship between the velocity structure beneath the recording site and the dominant frequency obtained from HVSR analysis (see for example NAVARRO *et al.*, 2001). As a first-order approximation, we can apply a simple formula that relates the fundamental frequency to the thickness and seismic velocity of a medium composed of a single layer over a half-space:

$$f_0 = \frac{\beta}{4H} , \tag{2}$$

where β is the average S-wave velocity and H is the thickness of the layer. When the sites are located over the same structure we can assume that the S-wave velocity is approximately constant from site to site. This is usually the case in most sedimentary basins. The former relation thus yields a map of sediment thickness from estimates of the resonant frequency (e.g., IBS-VON SEHT and WOHLENBERG, 1999). Inversely, if we assume that in a certain region the surface layers have approximately a constant thickness, the dominant frequency at each point would be then related to an S-wave velocity and a map of surface velocities could be obtained.

However, in a volcanic environment like Las Cañadas caldera we cannot assume any of the former assumptions. Exhaustive studies of the surface geology in Las Cañadas (ANCOCHEA *et al.*, 1990; MARTÍ *et al.*, 1994; ANCOCHEA *et al.*, 1999; ABLAY and MARTÍ, 2000) have produced detailed information about the composition, location, and extent of each exposed rock type and its role in the volcanic evolution of the region. Borehole data provide additional constraints on the subsurface structure of Las Cañadas. Current interpretations (e.g., ABLAY and MARTÍ, 2000) assume that the caldera fill is made by superposition of a variety of eruptive products originated at different vents related to the Teide-Pico Viejo edifice. These deposits are thick (> 500 m) in the central and eastern parts of the caldera, and thin (< 100 m) in the western part. Although, in principle, the different layers are assumed to be horizontal, no detailed information about thickness or horizontal extent is available. The diversity of volcanic materials which form the caldera prevents us from using the constant *S*-wave simplification; it would imply the assumption of a homogeneous velocity for the entire caldera, something that is not justified at all in view of the heterogeneities found by geological studies. Likewise, the absence of correlation between the observed dominant frequencies and the mapped geological units suggests that the constant depth simplification is not useful either. Even when we assume a smaller thickness of the volcanic deposits in the western caldera (ABLAY and MARTÍ, 2000) we find that some sites apparently located on the same geological structure show very different dominant frequencies.

In reality, it is likely that both *S*-wave velocity and layer thickness in Las Cañadas change considerably from site to site, over distance scales shorter than our sampling interval. This idea is consistent with the results shown in Figure 6. Dominant frequencies in the caldera vary sharply, even for nearby locations just a few hundred meters apart. Short-scale heterogeneity in the spatial distribution of dominant frequencies has been found in other volcanic areas as well. We have a clear example in the work by MORA *et al.* (2001), who applied the standard HVSR method to investigate the presence of site effects at the Arenal volcano, Costa Rica; they used a dense linear array (with a spacing of 150 m) and found that dominant frequencies may vary drastically in distances of just a few hundred meters, in correspondence with changes in the velocity structure.

The difficulty in the interpretation of this kind of results arises because we have to consider the effects of layer thickness and wave velocity simultaneously. To be able to interpret the measured values of dominant frequencies, further constraints on the shallow structure of the Las Cañadas caldera are required. In this sense, several geophysical studies have been carried out to investigate the structure and evolution of the Las Cañadas edifice. For example, gravimetric measurements have revealed the presence of density anomalies (VIEIRA *et al.*, 1986; CAMACHO *et al.*, 1991, 1996; ARAÑA *et al.*, 2000; ABLAY and KEAREY, 2000); a shallow, low-density region is located under the Teide-Pico Viejo edifice in the northern part of the caldera. This anomaly could be due to the presence of a low-density caldera fill and the abundance of light materials erupted during recent volcanism. High-density anomalies are observed near the caldera walls. Gravity modeling suggests that these anomalies could extend several kilometers in depth (ABLAY and KEAREY, 2000) and be caused by the influence of dense materials related to the old basaltic shield (CAMACHO et al., 1996). A dense fill could be also responsible for the positive gravity anomalies in the southern part (ABLAY and KEAREY, 2000). Magnetic surveys have revealed the existence of anomalies (BLANCO, 1997; GARCÍA et al., 1997). A negative anomaly coinciding with the Teide summit has been interpreted as an effect of hydrothermal alteration due to fluids at high temperatures (ARAÑA et al., 2000). The presence of volcanic fluids has been evidenced by measurements of gas emission rates (HERNÁNDEZ et al., 2000). Another negative anomaly, coinciding with the Diego Hernández caldera, has been attributed to an accumulation of phonolitic materials infilling the caldera, which show very low magnetization compared with basalts (ARAÑA et al., 2000). Seismic studies have been also performed to investigate the structure of Tenerife. Recent works have established the presence of a high-velocity zone southwest of Las Cañadas (WATTS et al., 1997; CANALES et al., 2000). This zone is coincident with a high-density body inferred from gravimetric studies and interpreted as the core of an old, large mafic volcano (CANALES et al., 2000; ABLAY and KEAREY, 2000). The eastern part of Las Cañadas was the site of a seismic experiment by DEL PEZZO et al. (1997), who analyzed the coda of local earthquakes using two seismic arrays. They showed that the area is characterized by a highly heterogeneous shallow structure that produces strong scattering of the seismic waves. However, the locations of the scatterers could not be estimated.

All these studies constitute important contributions to our understanding of the general structure and volcanic evolution of Las Cañadas, although their resolution is too low for a direct comparison with the distribution of dominant frequencies. Nevertheless, they may help us understand the few patterns that we have been able to find in our results. For example, the largest dominant frequencies usually occur near the caldera rim (see the distribution of black dots in Fig. 6). Large frequencies are generally associated to thin layers (eq. (2)); therefore our observation is reasonable since we expect that the volcanic deposits become thinner when we move outward the eruptive centers and toward the caldera walls. There are two exceptions, one located south of Ucanca and another along the northeast caldera wall (Fig. 6). They may be attributed to a decrease in the seismic velocity that somewhat compensates the decrease in layer thickness, yielding moderate dominant frequencies. In support of this interpretation, both areas display loose pyroclastic deposits that would be characterized by low S-wave velocities. A few large dominant frequencies are measured within the caldera, far from the caldera walls (Fig. 6). They seem to correlate with the intersections of the ancient caldera rims (Ucanca-Guajara and Guajara-Diego Hernández) corresponding to the three inferred caldera collapses that formed Las Cañadas (MARTÍ et al., 1994; MARTÍ and GUDMUNDSSON, 2000). If this is the case, our method would be sensitive to the presence of buried, compact remains of the Las Cañadas edifice. This might help us answer the question of whether the Cañadas caldera originated by a northward-directed landslide or by multiple vertical collapses (ANCOCHEA *et al.*, 1990; MARTÍ *et al.*, 1994; ABLAY and HÜRLIMANN, 2000; MARTÍ and GUDMUNDSSON, 2000). However, our sampling is too coarse to ensure a direct relationship. Further experiments with denser spatial sampling would be required. Very small dominant frequencies are likewise measured at several sites near the caldera rim (Fig. 6, white dots). They might correspond to discontinuities in the structure of the caldera wall, or local variations of the *S*-wave velocity of the caldera fill. Small-to-intermediate dominant frequencies are found in the central caldera, immediately south of Montaña Blanca (Fig. 6). This result is consistent with greater thickness of the volcanic deposits near the eruptive vents, as evidenced for example by the interpretation of borehole data (ABLAY and MARTÍ, 2000).

The easternmost region of Las Cañadas shows dominant frequency measurements that are more self-consistent and show less spatial variability than the rest of the caldera (Fig. 6). This may be due in part to the shorter sampling interval, which produces a clearer image of the distribution of dominant frequencies, and also to better conditions of the sites for application of the Nakamura method. For example, DEL PEZZO *et al.* (1997) speculated about the possibility of a strong velocity contrast between the uppermost and the underlying layers in the Diego Hernández area. A negative magnetic anomaly found in this part of the caldera (BLANCO, 1997; GARCÍA *et al.*, 1997) has been interpreted as a relatively thick deposit of phonolitic materials (ARAÑA *et al.*, 2000). Both results favor the generation of resonance and therefore the measurement of a dominant frequency.

Figure 7a shows a closer view of the results in the eastern caldera, where the spatial sampling is densest. Dotted lines surround regions where the measured dominant frequencies are similar. When we follow the AA' profile, we find a smooth variation of dominant frequencies, ranging in average between 2.7 Hz in zone I and 6.0 in zone III. If we assume a constant seismic velocity, our interpretation of these variations in dominant frequency is that the thickness of the volcanic deposits changes along the profile. Justification for this hypothesis comes from the fact that surface materials are far more homogeneous in this small region than in the caldera as a whole. They are mostly composed of phonolitic lavas and scorias (ABLAY and MARTÍ, 2000). Figure 7b shows the basement depth profile obtained from eq. (2) for an *S*-wave velocity of 1 km/s, that is the average velocity obtained from array measurements by DEL PEZZO *et al.* (1997). We obtain realistic basement depths between 40 and 95 m, approximately. The smallest layer thickness, corresponding to zone III, could correlate with the region where the ancient rims of the Guajara and Diego Hernández calderas would have intersected (Fig. 6). Alternatively, we could assume a constant thickness approximation to consider that the variations in dominant frequency are related to changes in the seismic velocity of the medium. This hypothesis is supported by the flatness of the caldera floor before the formation of the Teide-Pico Viejo edifice and subsequent filling of the caldera with new volcanic

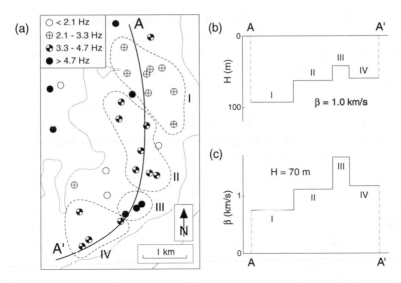

Figure 7

(a) Dominant frequencies obtained at the eastern part of the Las Cañadas caldera. Dashed lines identify regions where the measured dominant frequencies are similar. The solid line AA' represents a profile discussed in Figures 7b and 7c. (b,c) Interpretation of the dominant frequency data corresponding to the AA' profile for the simplified cases of constant S-wave velocity of $\beta = 1.0$ km/s (b), and constant layer thickness of $H = 70$ m (c).

materials (ABLAY and MARTÍ, 2000). In this case, eq. (2) yields the seismic velocity profile shown in Figure 7c, for a constant layer thickness of 70 m. However, there is no evident correspondence between these velocities and the surface geological units. Although the subsurface geology may be somewhat different from the exposed layer, the relative homogeneity of the materials in this region makes us think that the effect of layer thickness on the dominant frequencies might dominate over the effect of wave velocity. Therefore, we believe that the profile shown in Figure 7b may be closer to reality.

6. Conclusions

We have proposed a modification of Nakamura's method to microtremor data recorded near Teide volcano. This new approach consists of the calculation of time-dependent HVSR functions, or ratiograms, to investigate the temporal evolution of the spectral ratios and the effect of local perturbations. The interpretation of these ratiograms allows us to determine the dominant frequencies, excluding those cases where the time-dependent HVSRs are not stationary. In this way, we are able to discriminate between real resonant frequencies and spurious peaks due to local transients in the data. We performed various tests of the stability of our estimates of the average HVSR functions, checking the similarity between HVSRs for data recorded at the same site but at different times. We demonstrate that at a site, the HVSRs show similar shapes even when the measurements are separated by a time interval of a year. This fact is independent of the presence of resonant peaks, and suggests that the HVSRs depend basically of the geological structure at each recording location.

The application of the HVSR method provides new clues about the shallow velocity structure of the caldera. The distribution of dominant frequencies, although heterogeneous, may be approximately explained in terms of thickness and average seismic velocity of the volcanic deposits. However, some questions remain open, mostly due to lack of coverage. A finer sampling interval would be desirable for further applications of the method. From the results obtained in this study we conclude that ratiograms constitute an improvement of the HVSR method, due to their capability to help us understand the temporal evolution of the spectral ratio. This technique can be easily implemented, and may be applied not only to the problems or structures considered up to now by other authors, but to the characterization of the subsurface geology in volcanic ambients as well.

Acknowledgements

This work was partially supported by CICYT, Spain, under Grants AMB99-1015-CO2-02 and REN2001-2418-CO4-02/RIES. Additional support was provided by projects REN2001-3833/ANT, REN2001-2814-C04-04/RIES and REN2002-4198-C02-02/RIES, and by the research teams RNM-194 and RNM-104 of Junta de Andalucía, Spain. We thank the Teide Natural Park authorities and all institutions and individuals that collaborated in the microtremor experiments: Juan, Benito, and María José Blanco from Centro Geofísico de Canarias; Olga Muñoz; Benito Martín from Instituto Andaluz de Geofísica; and Serafín Limonchi and César F. Aguilera from Universidad de Almería. Ramón Ortiz provided useful hints for the interpretation of seismic data. We also acknowledge José Morales, Gerardo Alguacil, and Jesús Ibáñez for useful comments. We thank the Martín-Barry family for the review of the manuscript. The reviews of Joan Martí and an anonymous reviewer helped to improve this paper.

REFERENCES

ABLAY, G. and HÜRLIMANN, M. (2000), *Evolution of the North Flank of Tenerife by Recurrent Giant Landslides*, J. Volcan. Geotherm. Res. *103*, 135–159.

ABLAY, G. and KEAREY, P. (2000), *Gravity Constraints on the Structure and Volcanic Evolution of Tenerife, Canary Islands*, J. Geophys. Res. *105*, 5783–5796.

ABLAY, G. J. and MARTÍ, J. (2000), *Stratigraphy, Structure, and Volcanic Evolution of the Pico Teide-Pico Viejo Formation, Tenerife, Canary Islands*, J. Volcan. Geotherm. Res. *103*, 175–208.

ALMENDROS, J., IBÁÑEZ, J., ALGUACIL, G., MORALES, J., DEL PEZZO, E., LA ROCCA, M., ORTIZ, R., ARAÑA, V., and BLANCO, M. J. (2000), *A Double Seismic Antenna Experiment at Teide Volcano: Existence of Local Seismicity and Lack of Evidences of Volcanic Tremor*, J. Volcan. Geotherm. Res. *103*, 439–462.

AL YUNCHA, Z. and LUZÓN, F. (2000), *On the Horizontal-to-Vertical Spectral Ratio in Sedimentary Basins*, Bull. Seismol. Soc. Am. *90*, 1101–1106.

ANCOCHEA, E., HUERTAS, M. J., CANTAGREL, J. M., COELLO, J., FÚSTER, J. M., ARNAUD, N., and IBARROLA, E. (1999), *Evolution of the Cañadas Edifice and its Implications for the Origin of the Cañadas Caldera (Tenerife, Canary Islands)*, J. Volcan. Geotherm. Res. *88*, 177–199.

ANCOCHEA, E., FÚSTER, J., IBARROLA, E., CENDRERO, A., COELLO, J., HERNÁN, F., CANTAGREL, J., and JAMOND, C. (1990), *Volcanic Evolution of the Island of Tenerife (Canary Islands) in the Light of new K-Ar Data*, J. Volcan. Geotherm. Res. *44*, 231–249.

ANGUITA, F. and HERNÁN, F. (2000), *The Canary Islands: A unifying model*, J. Volcan. Geotherm. Res. *103*, 1–26.

ARAÑA, V. and CARRACEDO, J. C., *Tenerife*, In *Los volcanes de las Islas Canarias* (ed. Rueda, Madrid 1978).

ARAÑA, V. and ORTIZ, R. (1991), *The Canary Islands: Tectonics, magmatism and geodynamic framework*. In *Magmatism in Extensional Structural Settings: The Phanerozoic African Plate* (eds. Kampunzu, A. and Lubala, R.) (Springer-Verlag 1991) pp. 209–249.

ARAÑA, V., CAMACHO, A. G., GARCÍA, A., MONTESINOS, F. G., BLANCO, I., VIEIRA, R., and FELPETO, I. (2000), *Internal Structure of Tenerife (Canary Islands) Based on Gravity, Aeromagnetic, and Volcanological Data*, J. Volcan. Geotherm. Res. *103*, 43–64.

BARD, P.-Y., *Microtremor measurements: A tool for site effect estimation?* In *The Effects of Surface Geology on Seismic Motion* (eds. Irikura, Kudo, Okada and Sasatani) (Balkema, Rotterdam 1999) pp. 1251–1279.

BLANCO, I., *Análisis e interpretación de las anomalías magnéticas de tres calderas volcánicas: Decepción (Shetland del Sur, Antártida), Furnas (San Miguel, Azores) y Las Cañadas del Teide (Tenerife, Canarias)* (Ph.D. Thesis, Universidad Complutense de Madrid, 1997).

CAMACHO, A. G., MONTESINOS, F. G., and VIEIRA, R. (1991), *Microgravimetric Model of the Las Cañadas Caldera (Tenerife)*, J. Volcan. Geotherm. Res. *47*, 75–88.

CAMACHO, A. G., MONTESINOS, F. G., and VIEIRA, R., *Gravimetric structure of the Teide volcano environment*. In *Proceed. of the Second Workshop on European Lab. Volcanoes* (Santorini, Greece 1996) pp. 605–613.

CANALES, J. P., DAÑOBEITIA, J. J., and WATTS, A. B. (2000), *Wide-angle Seismic Constraints on the Internal Structure of Tenerife, Canary Islands*, J. Volcan. Geotherm. Res. *103*, 65–81.

CARRACEDO, J. C., DAY, S., GUILLOU, H., RODRÍGUEZ, E., CANAS, J. A., and PÉREZ, F. J. (1998), *Hotspot Volcanism Close to a Passive Continental Margin*, Geol. Mag. *135*, 591–604.

COUTEL, F. and MORA, P. (1998), *Simulation Based Comparison of Four Site-response Estimation Techniques*, Bull. Seismol. Soc. Am. *88*, 30–42.

DAÑOBEITIA, J. J., and CANALES, J. P. (2000), *Magmatic Underplating in the Canary Archipelago*, J. Volcan. Geotherm. Res. *103*, 27–42.

DAWSON, P., DIETEL, C., CHOUET, B., HONMA, K., OHMINATO, T., and OKUBO (1998), *A Digitally Telemetered Broadband Seismic Network at Kilauea Volcano, Hawaii*, US Geol. Surv. Open-file Report, *122*, pp. 98–108

DELGADO, J., LÓPEZ-CASADO, C., ESTÉVEZ, A. C., GINER, J., CUENCA, A., and MOLINA, S. (2000), *Mapping Soft Soils in the Segura River Valley (SE Spain): A Case Study of Microtremors as an Exploration Tool*, J. Appl. Geophys. *45*, 19–32.

DEL PEZZO, E., LA ROCCA, M., and IBÁÑEZ, J. (1997), *Observations of High-frequency Scattered Waves Using dense Arrays at Teide Volcano*, Bull. Seismol. Soc. Am. *87*, 1637–1647.

DRAVINSKI, M., DING, G., and WEN, K. L. (1996), *Analysis of Spectral Ratios for Estimating Ground Motion in Deep Basins*, Bull. Seismol. Soc. Am. *86*, 646–654.

FIELD, E. H. and JACOB, K. H. (1995), *A Comparison and Test of Various Site-response Estimation Techniques, Including Three that are not Reference-site Dependent*, Bull. Seismol. Soc. Am. *85*, 1127–1143.

GARCÍA, A., BLANCO, I., TORTA, M., and SOCÍAS, I. (1997), *High-resolution Aeromagnetic Survey of the Teide Volcano (Canary Islands): A Preliminary Analysis*, Annali di Geofisica *40*, 329–340.

HERNÁNDEZ, P., PÉREZ, N., SALAZAR, J., SATO, M., NOTSU, K., and WAKITA, H. (2000), *Soil Gas CO₂, CH₄, and H₂ distribution in and around Las Cañadas Caldera, Tenerife, Canary Islands, Spain*, J. Volcan. Geotherm. Res. *103*, 425–438.

IBS-VON SEHT, M. and WOHLENBERG, J. (1999), *Microtremor Measurements Used to Map Thickness of Soft Sediments*, Bull. Seismol. Soc. Am. *89*, 250–259.

KONNO, K. and OHMACHI, T. (1998), *Ground-motion Characteristics Estimated from Spectral Ratio between Horizontal and Vertical Components of Microtremor*, Bull. Seismol. Soc. Am. *88*, 228–241.

LACHET, C., HAZFELD, D., BARD, P. Y., THEODULIDIS, N., PAPAIOANNOU, C., and SAVVAIDIS, A. (1996), *Site Effects and Microzonation in the City of Thessaloniki (Greece) Comparison of Different Approaches*, Bull. Seismol. Soc. Am. *86*, 1692–1703.

LERMO, J. and CHÁVEZ-GARCÍA, F. J. (1994), *Are Microtremors Useful in Site Response Evaluation?* Bull. Seismol. Soc. Am. *84*, 1350–1364.

LUZÓN, F., AL YUNCHA, Z., SÁNCHEZ-SESMA F. J., and ORTIZ-ALEMÁN, C. (2001), *A Numerical Experiment on the Horizontal-to-Vertical Spectral Ratio in Flat Sedimentary Basins*, Pure Appl. Geophys. *158*, 2451–2461.

MARTÍ, J., MITJAVILA, J. and ARAÑA, V. (1994), *Stratigraphy, Structure and Geochronology of the Cañadas Caldera*, Geological Magazine *131*, 715–727.

MARTÍ, J. and GUDMUNDSSON, A. (2000), *The Las Cañadas Caldera (Tenerife, Canary Islands): An Overlapping Collapse Caldera Generated by Magma-chamber Migration*, J. Volcan. Geotherm. Res. *103*, 161–173.

MORA, M. M., LESAGE, P., DOREL, J., BARD, P. Y., MÉTAXIAN, J. P., ALVARADO, G. E., and LEANDRO, C. (2001), *Study of Seismic Site Effects Using H/V Spectral Ratios at Arenal Volcano, Costa Rica*, Geophys. Res. Lett. *28*, 2991–2994.

MORGAN, W. J. (1983), *Hotspot Tracks and the Early Rifting of the Atlantic*, Tectonophysics *94*, 123–139.

NAKAMURA, Y. (1989), *A Method for Dynamic Characteristics Estimation of Subsurface Using Microtremor on the Ground Surface*, Q. Rept. Railway Tech. Res. Inst. *30*, 25–33.

NAVARRO, M., ENOMOTO, T., SÁNCHEZ, F., MATSUDA, I., IWATATE, T., POSADAS, A., LUZÓN, F., VIDAL, F., and SEO, K. (2001), *Surface Soil Effects Study Using Short-period Microtremor in Almeria City, Southern Spain*, Pure Appl. Geophys. *158*, 2481–2497.

NOGOSHI, M. and IGARASHI T. (1971), *On the Amplitude Characteristics of Microtremor (Part 2)*, J. Seismol. Soc. Japan, *24*, 26–40, (in Japanese with English abstract).

SEEKINS, L. C., WENNERBERG, L., MARGHERITI, L., and LIU, H. P. (1996), *Site Amplification at Five Locations in S. Francisco: A Comparison of S Waves, Coda, and Microtremor*, Bull. Seismol. Soc. Am. *86*, 627–635.

VIEIRA, R., TORO, C., and ARAÑA, V. (1986), *Microgravimetric Survey in the Caldera of Teide, Tenerife, Canary Islands*, Tectonophysics *130*, 249–257.

WATTS, A. B., PEIRCE, C., COLLIER, J., DALWOOD, R., CANALES, J. P., and HENSTOCK, T. J. (1997), *A Seismic Study of Lithospheric Flexure in the Vicinity of Tenerife, Canary Islands*, Earth Planet. Sci. Lett. *146*, 431–447.

ZASLAVSKY, Y. and SHAPIRA, A. (2000), *Experimental Study of Topographic Amplification Using the Israel Seismic Network*, J. Earthq. Engin. *4*, 43–65.

(Received April 2, 2002, revised December 23, 2002, accepted January 31, 2003)

 To access this journal online:
http://www.birkhauser.ch

Pure appl. geophys. 161 (2004) 1597–1611
0033–4553/04/071597–15
DOI 10.1007/s00024-004-2523-4

© Birkhäuser Verlag, Basel, 2004

❙ Pure and Applied Geophysics

Tilt Observations in the Normal Mode Frequency Band at the Geodynamic Observatory Cueva de los Verdes, Lanzarote

A. V. Kalinina[1], V. A. Volkov[1], A. V. Gorbatikov[1], J. Arnoso[2],
R. Vieira[2], and M. Benavent[2]

Abstract—We have conducted observations with the aid of a seismo-tiltmeter station, which is based on the Ostrovsky pendulum and installed at the Geodynamic Observatory Cueva de los Verdes at Lanzarote Island since 1995. In this station the signal is separated into two frequency bands – tidal tilts (from 0 to 5 mHz) and ground oscillations in the frequency range of free Earth's normal modes (from 0.2 to 5 mHz). The later band, called accelerometer channel, has additional amplification. We analyzed the background records in the frequency range of Earth's free oscillations from August 2000 to September 2001, as well as, Earth's normal modes after strong earthquakes. We found several distinctive persistent peaks in the spectra of background oscillations. Both amplitudes of distinguished peaks and noises have seasonal variations. We found that spectra of background oscillations are different in the frequency interval between 1.4 and 2.5 mHz for North- South and East-West components.

Key words: Background Earth's oscillations, Earth's normal modes, seismo-tiltmeter, tidal tilts, Lanzarote, Canary.

1. Introduction

We installed an automatic seismo-tiltmeter station (ASTS) at the Geodynamic Observatory Cueva de Los Verdes at Lanzarote Island in 1995. These observations constitute part of multidisciplinary studies carried out on the island by the Institute of Astronomy and Geodesy (VIEIRA *et al.*, 1991; ARNOSO *et al.*, 2001a, b, c). In ASTS, the signal is separated into two frequency bands: band of tidal tilts (from 0 to 5 mHz) and band of free Earth's oscillations (from 0.2 to 5 mHz). The latter band, called accelerometer channel, has additional amplification.

The general scope of our investigations at Lanzarote was as follows: (1) to study connections between signal parameters observed with ASTS and local volcanic and seismic activity, and if possible also to find correlations with global Earth's processes;

[1] Joint Institute of Physics of the Earth, Russian Academy of Sciences, B. Gruzinskaya, 10, 123995 Moscow, Russia. E-mails: kalinina@uipe-ras.scgis.ru; volkov@uipe-ras.scgis.ru; avgor70@mail.ru
[2] Instituto de Astronomía y Geodesia (CSIC-UCM), Universidad Complutense de Madrid, Facultad de Matemáticas. 28040-Madrid, Spain. E-mails: vieira@iagmat1.mat.ucm.es; arnoso@iagmat1.mat.ucm.es; mbena@iagmat1.mat.ucm.es

(2) to study the horizontal components of the ground surface oscillations ranging from 0.2 to 5 mHz, mindful of both Earth's normal modes after strong earthquakes and Earth's background oscillations; (3) to study the tidal and long-period deformations.

The preliminary results of observations in Cueva de Los Verdes with the ASTS is presented in a paper by KALININA *et al.* (2000), in which the authors mentioned the presence of background oscillations with dominant periods about 7 and 10 min in quiet seismic periods. The more detailed study of background oscillations is the subject of the present paper.

Many researchers have conducted investigations of the Earth's background oscillations in the range of normal modes and longer periods. Based on their records of background oscillations with superconducting gravimeters, NAWA *et al.* (1998) reported that spectral analysis of the data at the Sayowa station in Antarctica yielded evidence for incessantly excited normal modes of the Earth's free oscillations. The same group in a companion paper (SUDA and NAKAJIMA 1998) reported that analysis of ten-year gravity data from the IDA network yielded spectral peaks from 2 to 7 mHz corresponding to the fundamental spheroidal modes which are not of earthquake origin. Similar results have been reported in other papers (e.g., KOBAYASHI and NISHIDA, 1998; KOBAYASHI *et al.*, 2001; TANIMOTO *et al.*, 1998). The atmosphere was suggested as a source of excitation, taking into account that a known set of earthquakes solely was not sufficient to explain the observed spectra. IMANISHI (1998) compared data from superconducting gravimeters at the Matsushiro and Sayowa stations and found that peaks in the Sayowa spectra below 3 mHz could be the signals of some local geophysical phenomena.

DOLGIKH *et al.* (1983) conducted the observations with a laser strain-meter at Stenin Island (the Far East) and reported that they observed free Earth's oscillations normal modes in the background and could identify individual peaks in the spectra. Weak earthquakes were proposed as a source of generation of these background free oscillations. NESTEROV *et al.* (1990) drew the same conclusion based on their observations in Crimea.

Investigations of background oscillations in free normal modes range and longer periods also demonstrate that local geophysical phenomena could be reflected in the signals. DAVYDOV and DOLGIKH (1995) registered the long-period background oscillations (longer than 1 hour) with a 52.5 m length laser strainmeter in the Far East region and concluded that oscillations were caused by the loading effect of Japanese Sea free oscillations. NESTEROV *et al.* (1990) explained the background oscillations registered in Crimea ranging over longer periods by the influence of seiche oscillations of bays. PETROVA (2000) considered that background oscillations ranging 0.005–0.5 mHz, observed at the Eurasian continent, reflect the dynamic of the continent oscillations. An attempt was even made to find a correlation between the increase of long-period background oscillations intensity and strong earthquakes preparation process (PETROVA and VOLKOV, 1996).

2. Observation Site and Instruments

Lanzarote Island belongs to the Canary Archipelago. It is located at the edge of the West African Continental Margin. The Canary Islands are an old volcanic structure sited on top of the Jurassic oceanic crust. Their origin is still under discussion and therefore several geophysical research projects have been conducted in the archipelago (e.g., BOSSHARD and MACFARLANE, 1970; BANDA et al., 1981; CANALES and DANOBEITIA, 1998; DANOBEITIA et al., 1994; RANERO et al., 1997). A number of genetic models has been proposed for this region, for instance, the hotspot theory (WILSON, 1973) or the connection to the Alpine orogeny, which reached its maximum activity in this zone during the Miocene (ANGUITA and HERNAN 1975). MAZAROVICH (2000) proposed the model where mantle plume rises to the lithosphere bottom and then spreads out under the entire area generating magma chambers. The molten material from the chambers rises to the crust and forms the basis of the Islands.

A general geological study of Lanzarote can be found in FUSTER et al. (1968) and MARINONI (1991). For a better understanding of the inner structure and evolution of Lanzarote, several geophysical and geodynamical studies were carried out on the island by the Institute of Astronomy and Geodesy. Two underground geodynamic stations monitoring several geophysical parameters were set up (VIEIRA et al., 1991; ARNOSO et al., 2001a). One of the stations was located above the geothermal anomaly area in Timanfaya National Park.

The other station named Cueva de los Verdes was located inside a lava tunnel of the La Corona Volcano northeast of the island. The tunnel is about 15 m wide and about 5–6 m high at the place of installation. The station was installed about 1.5 km apart from the tunnel entrance 18 m deep from the ground surface. The distance of the station from the nearest shoreline is about 1.6 km. Average air temperature inside the tunnel is 18 °C and the daily variations of the temperature reach 0.1 °C due to air penetrating the open upper end. The average relative air humidity is 80%.

The observatory was equipped with a LaCoste & Romberg-G, number 434, gravimeter, two-component water-tube tiltmeter, quartz-bar strainmeter, vertical pendulum, horizontal pendulums, seismometers, rock and air thermometers and atmospheric barograph. In May 1995 we additionally installed a seismo-tiltmeter station (Fig. 1). ASTS is a modernized version of the well-known Ostrovsky pendulum (OSTROVSKY, 1961). Here the photoelectric pendulum displacement converter was changed to the capacity converter. The electrodynamic type feedback in displacements was added to decrease the magnitude of nonlinearity in the pendulum system. The automatic control system of the pendulum position was added to extend the effective dynamic range of measurements up to ± 20 arc sec without manual adjusting. Besides two additional channels (accelerometer channels) were added to record in the frequency range of the Earth's free oscillations. These channels were designed as second-order band-pass filters 0.2–5 mHz followed by amplifiers.

Figure 1
Seismo–tiltmeter station ASTS installed in the Geodynamic Observatory Cueva de los Verdes.

Main technical parameters of ASTS are as follows:

1. The dynamic range of tilts recorded is $\pm 10^{-5}$ radians without automatic compensation and $\pm 10^{-4}$ radians with automatic compensation; frequency range (with AFR deviations ± 3 dB) is 0–5 mHz; the transducer coefficient from input tilt to output voltage is 10^{6} volt/radian.

2. The dynamic range of accelerometer channels is $\pm 10^{-5}$ m/s^2 and the frequency range (with AFR deviations ± 3 dB) is 0.2–5 mHz; the transducer coefficient from input accelerations to output voltage is 10^{6} volt $\times$ s^2/m.

3. Maximum output voltage with 10 kΩ loading is ± 10 volt.

4. Pendulum-free period is 5 s $\pm 10\%$.

5. The station tilt channels sensitivity was determined with the help of a controlled tilt platform at the factory. Accelerometer channels sensitivity was calculated on the basis of tilt sensitivity at the factory as well. For periodical verification of these parameters' stability ASTS has pulse calibrators producing current pulse of 720 second duration and synchronized with hour time marks.

The data acquisitions system contains a personal computer and a 16-channel, 16-bit analog–digital converter (ADC). The output-sampling rate is 1/10 s. Each sample is calculated as average from 70 samples with an initial sampling rate of 10 Hz.

The station pendulums were installed on a $1 \times 1 \times 0.3$ m^3 concrete basement and the components were aligned along NS and EW directions using a geodetic compass within 0.2 degree. The pendulums were protected from convection air noise by putting a foam plastic case above them, and a pyramidal protecting hood of polyethylene film was stretched on wood bars. These simple precautions reduced the noise by one order of magnitude.

3. The Results of Observations

First we want to emphasize that we have achieved a stable qualitative functioning of ASTS. Figure 2a demonstrates an example of tidal tilt record. This fragment contains the strong earthquake which occurred near the coast of Peru on 23.06.2001 with a magnitude M = 8.4. Even at first sight one can recognize oscillations after the earthquake during half day. The record of the event with an accelerometer channel is exhibited on Figure 2b.

The accelerometer channel data were discrete Fourier transformed after multiplication by a Hanning window. The analyzed time interval contained 34,9 hours (12,590 samples with an 10 s sampling rate). Figures 2c and d show the amplitude spectra of both components (in nm/s^2). Thin lines represent the spectra of Peru earthquake coda with an analyzed interval beginning 2.8 hours later than the arrival time. Thick lines show amplitude spectra of the background for comparison. The interval for background analysis was chosen just before the earthquake. The theoretical eigenfrequencies are indicated by vertical dashed lines and taken from ALSOP et al. (1961).

One can clearly observe the generation of known modes of the Earth's free oscillations after the strong earthquake. Amplitudes of these modes reach a magnitude of 2–3 nm/s^2 at the observational site and the level of the background varies between 0.1–1 nm/s^2 in the frequency range higher than 1.0 mHz. As we can see from the figures, the noise level at the station is high enough due to its proximity to the ocean, but the identifiable normal modes are observed after the strong earthquake with acceptable signal-to-noise ratio. However, we had more interest in revealing the connections between observed signal parameters and local geophysical processes, and thus further in the paper we would concentrate on the statistical properties in the analysis of the background oscillations.

It is easy to see (e.g., Fig. 2c and d) that even a single-time spectrum of the background reveals increasing amplitudes for certain frequencies. To test if this increase is statistically stable we calculated a dynamic spectrum for the time interval from August 2000 to September 2001 (see Fig. 3). Both components were considered. For the analysis, the sequent row of 3000 samples window (equal to 8.33 hours) was multiplied by Hanning window and Fourier transformed. Each next window was taken from the initial data row by shifting it forward on half of its length. On Figure 3 the X-axis depicts time in months during the observational period and the Y-axis shows frequencies of the dynamic spectrum. Saturation reflects the logarithmic amplitudes of the spectrum. White vertical stripes means

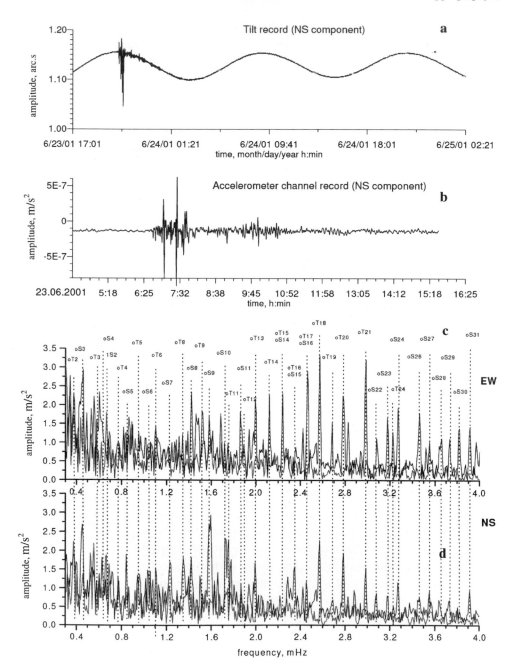

Figure 2

(a) Example of tilt record containing the earthquake from Peru offshore on 23.06.2001, M = 8.4, H = 33 km. (b) Example of accelerometer record containing the earthquake from Peru offshore on 23.06.2001, M = 8.4, H = 33 km. (c, d) Spectra before (thick line) and after (thin line) the earthquake on 23.06.2001 (Peru offshore), M = 8.4, H = 33 km; (c) NS component and (d) EW component.

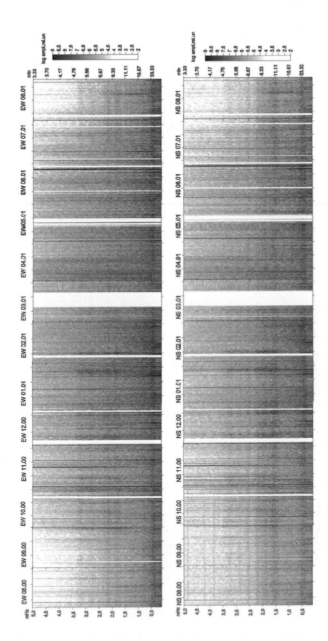

Figure 3

Dynamic spectra of the background oscillations from August 2000 to September 2001 for NS and EW components.

gaps in the data. One can see a system of well-distinguished horizontal stripes which form a pretty periodical pattern beginning from the certain frequency of about 2 mHz. A bright stripe with frequency 0.28 mHz lies slightly apart from others. The stripes have blinking intensity and sink in the noises during winter. Thus, we can conclude that horizontal components of the background oscillations have some permanently present periodicities in the frequency range 0.2–5 mHz at our observational site.

Figure 4a clearly demonstrates the behavior of these periodicities in amplitudes during the observational period. Here the logarithmic amplitudes of spectra in relative units are in along the vertical axis. Frequencies in mHz are along the X-axis in descending order. Y-axis indicates time in months beginning August 2000 and ending September 2001. The upper part of the figure contains spectra for NS component and the lower part for EW component. Each spectrum in Figure 4a was calculated as an average for one of the spectral series prepared earlier for Figure 3 within a month. We see that the background noises virtually in the entire spectrum have visible variation with its maximum coming during winter. The frequency peaks corresponding to the stripes on Figure 3 have the variation as well, and we can see that frequency peaks for the NS and EW components differ from each other. To view this difference with a better resolution we calculated the Fourier transform for the total raw data with preliminary Hanning windowing. The duration of the observational period constituted 404 days and corresponding frequency resolution in the resulting spectrum was 2.865×10^{-8} Hz. Transformed spectra smoothed with 400-sampled running window are presented in Figures 4b and c. A vertical segment on the figures shows the $\pm 2\sigma$ confidence interval. These two figures again demonstrate the difference between NS and EW spectra. For the NS component the visible peaks come to the frequencies approximately: 0.2, 1.7, 2.3 and 3.0 mHz and for the EW component – to the frequencies 0.2, 2.0, 2.3 and 3.0 mHz. These peaks are not narrow and their width in average constitutes 0.2 mHz. Figure 4d exhibits a ratio between NS and EW components. The vertical segment with arrows shows the $\pm 2\sigma$ confidence interval. The ratio has a slight general trend beginning from the magnitude approximately 0.7 and finishing with the magnitude about 1.0 for higher frequencies (not shown on the figure). Though the 0.2 mHz peaks for both components have large amplitudes and their ratio remains smooth, nonetheless for the frequency interval between 1.4 and 2.5 mHz the peak amplitudes are considerably smaller and their ratio has two sharp splashes.

The result is rather unexpected. Seeking the explanation of the difference between NS and EW peaks on frequencies 1.7 and 2.3 mHz, we must exclude the instrumental distortions since we have at our disposal the results of a technical experiment. Assuming an unbelievable fact that each accelerometer channel has its own uncontrolled resonance on these frequencies, the pendulums being directed on a parallel path should produce different spectra. In the experiment we turned the EW

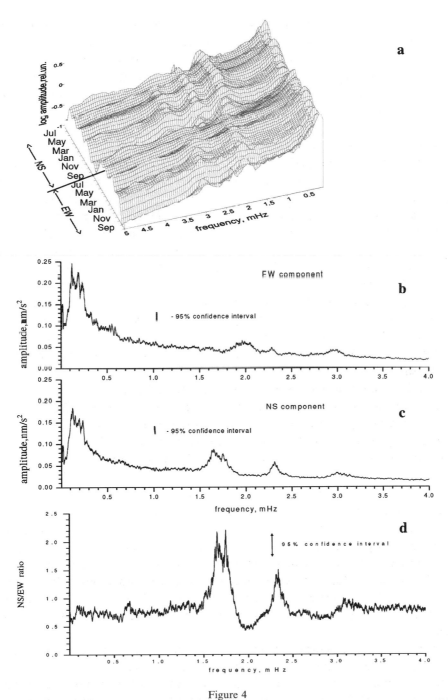

Figure 4
(a) Monthly averaged spectra for NS and EW components and the observational period from August 2000 to September 2001. (b, c) Amplitude spectra for the total data row from August 2000 to September 2001 for NS (b) and EW (c) components. (d) NS/EW spectral ratio.

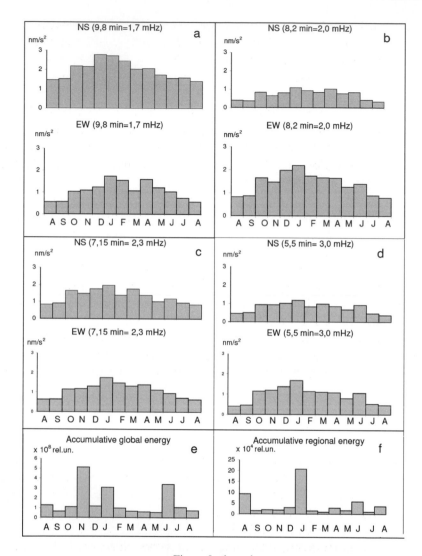

Figure 5a, b, c, d

A course of monthly averaged amplitude values for chosen spectral components for NS and EW components. X-axis is marked up with months of 2000–2001.

component parallel to NS one and made a one-week test record. The result of the record processing showed, however, that amplitude spectra for NS and EW components coincided to within 2%.

Let us consider again the behavior of the monthly averaged spectra. Figure 5 contains temporal cross sections for the chosen frequencies of the surfaces shown in Figure 4a. Figure 5a shows a course of the NS and EW averaged amplitudes for the frequency 1.7 mHz (in nm/s^2). Figure 5b – the same for the frequency 2.0 mHz,

Figure 5c – for 2.3 mHz and Figure 5d – for 3.0 mHz. We can see that amplitudes at all the chosen frequencies and both components have identical courses. Maximum values are during the period of December 2000 to February 2001 and minimum values in the summer months. The amplitude values themselves can be different for different components on the same frequency (see Figs. 5a and b).

To explain the presence of stable peaks and the difference between NS and EW components in the frequency range of 1.4–2.6 mHz we analyzed different data at our disposal. We compared the amplitude courses with cumulative released seismic energy for the entire globe and for the region surrounding the Canary Archipelago from 15 to 45 degrees north and from 0 to 30 degrees west (taken from the USGS NEIC catalogue) (Figs. 5e and f). From Figures 5a–f we see no clear correlation between the amplitudes course and cumulative seismic energy. For instance, the correlation coefficient between NS amplitudes on frequency 1.7 mHz (Fig. 5a) and global seismic energy (Fig. 5e) is equal 0.20; the same with regional seismic energy is 0.29, whereas the correlation coefficient between the amplitudes of NS and EW components for the same frequency (on the basis of the whole observational period (!)) is 0.79. We can conclude from a comparison of emitted seismic energy and the temporal course of spectral components that seismic activity can hardly play a part as a constitutive mechanism generating the observed background oscillations.

Spectral analysis of data from atmospheric pressure and air temperature conducting observations at the same site revealed no peak in the mentioned frequency band.

4. Discussion

We can consider that distinguishably normal modes of the Earth's free oscillations after strong earthquakes can to some extent play a part as a quality test for our "station-site" combination functioning. Estimations of the background oscillations amplitudes show that the latter exceed the conventional magnitudes for the background free Earth's oscillations, whose character magnitudes would be about 1 nGal (e.g., IMANISHI, 1998; NAWA et al., 1998; SUDA et al., 1998). Even the spectrum estimation on the temporal base of 404 days with subsequent smoothing with the 400-samples window (see Figs. 4b and c) only brings us near this magnitude. Thus the background we register hardly deals with background Earth's free oscillations such as those mentioned by (e.g., NAWA et al., 1998; SUDA et al., 1998; KOBAYASHI and NISHIDA, 1998) and possessing the global character. However, we see the global feature in our records which is the seasonal variation of the background amplitudes in the entire spectrum. We will probably not lack much if we assume the ocean proximity to the observational site stipulates this fact.

The most surprising fact in our observations is the difference between NS and EW components in the frequency fragment 1.4–2.6 mHz. We think that the presence of stable peaks and the observed difference between components could be explained by some local reasons solely independent from the concrete generating mechanism. Here we see three variants of the generating mechanism at least. The first one is connected with the lava tunnel itself and its mechanical reactions (torsions, etc.). The second one is connected with oceanic currents and the third one with atmospheric flows. Spectral analysis of atmospheric pressure and air temperature data revealed no peak in the mentioned frequency band. This suggests that we exclude the mechanisms of the first and third origins, since the absence of micro-variations in the fields of temperature and air pressure inside the tunnel implies the oscillations of the island on the mentioned frequencies as a whole. The second hypothesis, based on the ocean currents, incidentally conforms to the seasonal behavior of observed background amplitudes.

Here we should note that two ways exist for the ocean currents to produce oscillations on that frequencies. The first way is connected with the generation of large ocean curls, whose character frequencies are transmitted to the island. The second way probably could be related with the mechanical reaction to ocean currents of the island edifice and some surrounding area. We cannot distinguish between these two mechanisms due to the absence of corresponding data. Nonetheless we suggest that the mechanism of mechanical reaction of asymmetric structure to ocean currents is more preferable because it could explain the observed difference in the records of components NS and EW. This idea is supported by the fact that the Canary archipelago has a distinguished lateral elongation. We can estimate the character size of the reacting area, taking into account the revealed frequencies and the character velocities of the seismic waves as, for example $V_S \sim 2000$ m/s in this area (see WATTS *et al.*, 1997). This simple estimation gives approximately 1000 km as a maximum size (which corresponds incidentally to the character size of the archipelago). However, the detailing of these estimations is a question for future studies.

The NS and EW components differ not only in peaks within 1.4–2.6 mHz fragment, but also in general components ratio (Fig. 4d). We cannot now offer a model explaining the presence of the trend in the components ratio, however we want to point out that on low frequencies this ratio tends to a magnitude 0.7, which is equal to the ratio NS/EW for observed tidal amplitudes for M_2 wave. On higher frequencies this ratio tends to 1.0, which is conventional for the horizontal components in microseisms. We cannot explain the trend in the ratio with the difference in the frequency responses of NS and EW components, because the technical experiment mentioned above showed identical spectra in all the frequency working range with co-axial registration.

5. Conclusions

We installed seismo-tiltmeter station (ASTS) in the Geodynamic Observatory at Lanzarote Island. In this station the signal is separated into two frequency bands — tidal tilts (from 0 to 5 mHz) and ground oscillations in the frequency range of free Earth's normal modes (from 0.2 to 5 mHz). The latter band, called the accelerometer channel, has additional amplification. We analyzed the records of background in the frequency range of Earth's free oscillations from August 2000 to September 2001.

Sufficiently satisfactory signal to noise ratio of normal modes of the Earth's free oscillations after strong earthquakes allows us to conclude that our "station-site" combination is functioning well.

We conclude on the basis of the data analysis that spectra of horizontal oscillations in the frequency range of the Earth free oscillations have numerous spectral peaks, with stable frequencies in the absence of strong earthquakes. The amplitudes of these peaks vary, sometimes sinking in the background noise.

Both peak and noise amplitudes have a seasonal dependence. Maximum amplitudes occur during winter. We compared amplitudes of the peaks and amplitudes of the noise with released seismic energy and the absence of an obvious correlation between these processes is found. We also found no peaks with in the spectra of atmospheric pressure and air temperature in the frequency band of 1.4–2.6 mHz.

Oppositely, difference exists between NS and EW components in the frequency range of 1.4–2.6 mHz. Probably it is the presence of stable peaks and the difference could be explained by local reasons solely independent from a concrete generating mechanism. We suggest that the ocean currents could be one possible real generating mechanism.

We surmise that such a difference between NS and EW components is very important for future utilization of this fact. This is probably connected with particulars of the regional structures and their asymmetry, nevertheless it should be thoroughly verified in future studies. Clearly observations with other identical instruments on other islands of the archipelago could be helpful.

Acknowledgements

We thank Academician V.N. Strakhov, Director of Institute of Physics of the Earth, Russian Academy of Science for organizational support of this research. We are also grateful to the staff of Casa de los Volcanes (Lanzarote) for the help given and for the many years of technical maintenance of the observational instruments. Projects AMB99-0824 and REN2001-2271/RIES of the Spanish CICYT and the Contract between the Russian Academy of Sciences and the Spanish Council for Scientific Research have supported this research.

REFERENCES

ALSOP, L., SUTTON, G., and EWING, M. (1961), *Free Oscillations of the Earth Observed on Strain and Pendulum Seismographs*, J. Geophys. Res. *66*, 631.

ANGUITA, F. and HERNAN, F. (1975), *A Propagating Fracture Model versus a Hot Spot Origin for the Canary Islands*, Earth Planet. Sci. Lett. *27*, 11–19.

ARNOSO, J., FERNANDEZ, J., and VIEIRA, R. (2001a), *Interpretation of Tidal Gravity Anomalies in Lanzarote, Canary Islands*, J. Geodynamics *31*(4), 341–354.

ARNOSO, J., VIEIRA, R., VELEZ, E., VAN RUYMBEKE, M., and VENEDIKOV, A. (2001b), *Studies of Tides and Instrumental Performance of Three Gravimeters at Cueva De Los Verdes (Lanzarote, Spain)*, J. Geodetic Soc. Japan *47*(1), 70–75.

ARNOSO, J., VIEIRA, R., VELEZ, E., WEIXIN, C., SHILING, T., JUN, J., and VENEDIKOV, A. (2001c), *Monitoring Tidal and Non-tidal Tilt Variations in Lanzarote Island (Spain)*, J. Geodetic Soc. Japan *47*(1), 456–462.

BANDA, E., DANOBEITIA, J. J., SURINACH, E., and ANSORGE, J. (1981), *Features of Crustal Structure under Canary Islands*, Earth Planet. Sci. Lett. *55*, 11–24.

BOSSHARD, E. and MACFARLANE, D. J. (1970), *Crustal Structure of the W Canary Islands from Seismic Refraction and Gravity Data*, J. Geophys. Res. *75*, 4901–4918.

CANALES, J. P. and DANOBEITIA, J. J. (1998), *The Canary Islands Swell: A Coherence Analysis of Bathymetry and Gravity*, Geophys. J. Inst. *132*, 479–488.

DANOBEITIA, J. J., CANALES, J. P., and DEHGHANI, G. A. (1994), *An Estimation of the Elastic Thickness of the Lithosphere in the Canary Archipelago Using Admittance Functions*, Geophys. Res. Lett. *21*, 2649–2652.

DAVYDOV, A. and DOLGIKH, G. (1995), *Recording of Ultra Low-frequency Oscillations by a 52.5 m Laser Strainmeter*, Izv. Acad. Sc. USSR, Physics of the Solid Earth *31*(3), 248–251.

DOLGIKH, G., KOPVILLEM, U., and PAVLOV, A. (1983), *Observation of Periods of Free Oscillations of the Earth with a Laser Deformometer*, Izv. Acad. Sc. USSR, Physics of the Solid Earth *19*, 2 79–83.

FUSTER, J. M., FERNANDEZ SANTIN, S, and SAGREDO, J. (1968), *Geology and Volcanology of the Canary Islands, Lanzarote*, Instituto Lucas Mallada, CSIS, Madrid.

IMANISHI, Y. (1998), *Comment on "Incessant Excitation of the Earth's Free Oscillations" by Nawa et al.*, Earth Planet. Space *50*, 883–885.

KALININA, A., VOLKOV, V., VIEIRA, R., ARNOSO, J., BULOSHNIKOV, A., GORBATIKOV, A., and NIKOLAEV, A. (2000), *Preliminary Results from Tiltmeter Records of Earthquakes Made at the Cueva De Los Verdes Geodynamical Observatory on Lanzarote Island, Spain*, Volc. Seis. *22*, 73–86.

KOBAYASHI, N. and NISHIDA, K. (1998), *Continuous Excitation of Planetary Free Oscillations by Atmospheric Disturbances*, Nature *395*, 357–360.

KOBAYASHI, N., NISHIDA, K., and FUKAO, Y. (2001), *Continuous Excitation of Earth's free oscillations*. In: Proc. Int. Conf. *Long-term Observations in the Oceans*, Fuji, Yamanashi, Japan, pp. 30–32.

MARINONI, L. B. (1991), *Evoluzione geologica e sttructurale dell'isola di Lanzarote (Islas Canarias)*, Ph.D. Thesis, University of Milan, Milan.

MAZAROVICH, A. O., *Geology of the Central Atlantic: Fractures, Volcanic Edifices and Oceanic Bottom Deformations* (Nauchnyi Mir, Moscow 2000).

NAWA, K., SUDA, N., FUKAO, Y., SATO, T., AOYAMA, Y., and SHIBUYA, K. (1998), *Incessant Excitation of the Earth's Free Oscillations*, Earth Planet. Space *50*, 3–8.

NESTEROV, V., GOLOVIN, S., and NASOKIN, V. (1990), *Measurements of Long-period Oscillations of the Earth Using Laser Interferometer – Strainmeters*, Izv. Acad.Sc. USSR, Physics of the Solid Earth *26*, 326–330.

OSTROVSKY, A. E. (1961), *Tiltmeter with Photo-electric Registration*, BIM *25*, 500 and BIM *26*, 540.

PETROVA, L. and VOLKOV, V. (1996), *Dynamic Features of Seismo-gravitational Oscillations of the Earth*, Report Acad. Sc. *41*(12), 683–686.

RANERO, C. R., BANDA, E., and BUHL, P. (1997), *The Crustal Structure of the Canary Basin: Accretion Process at Slow Spreading Centers*, J. Geophys. Res. *102*, 10,185–10,201.

SUDA, N., NAWA, K., and FUKAO, Y. (1998), *Earth's Background Free Oscillations*, Science *279*, 2085–2091.

TANIMOTO, T., UM, J., NISHIDA, K., and KOBAYASHI, N. (1998), *Earth's Continuous Oscillations Observed on Seismically Quiet Days*, Geophys. Res. Lett., *25*(10), 1553–1556.

VIEIRA, R., VAN RUYMBEKE, M., FERNANDEZ, J., ARNOSO, J., and TORO, C. (1991), *The Lanzarote Underground Laboratory*, Cahiers du Centre Europeen de Geodynamique et de Seismologie *4*, 71–86.

WATTS, A. B., PEIRCE, C., COLLIER, J., DALWOOD, R., CANALES, J. P., and HENSTOCK, T. J. (1997), *A Seismic Study of Lithospheric Flexure in the Vicinity of Tenerife, Canary Islands*, Earth Planet. Sci. Lett. *146*, 431–447.

WILSON, J. T. (1973), *Mantle Plumes and Plate Motions*, Tectonophysics *19*, 149–164.

(Received February 15, 2002, revised February 5, 2003, accepted February 10, 2003)

 To access this journal online:
http://www.birkhauser.ch